Argument-Driven Inquiry in PHYSICAL SCIENCE

LAB INVESTIGATIONS for GRADES 6–8

Argument-Driven Inquiry in PHYSICAL SCIENCE

LAB INVESTIGATIONS for GRADES 6–8

Jonathon Grooms, Patrick J. Enderle, Todd Hutner, Ashley Murphy, and Victor Sampson

NSTApress
National Science Teachers Association
Arlington, Virginia

Claire Reinburg, Director
Wendy Rubin, Managing Editor
Rachel Ledbetter, Associate Editor
Donna Yudkin, Book Acquisitions Coordinator

ART AND DESIGN
Will Thomas Jr., Director

PRINTING AND PRODUCTION
Catherine Lorrain, Director

NATIONAL SCIENCE TEACHERS ASSOCIATION
David L. Evans, Executive Director
David Beacom, Publisher

1840 Wilson Blvd., Arlington, VA 22201
www.nsta.org/store
For customer service inquiries, please call 800-277-5300.

19 18 17 16 4 3 2 1

Library of Congress Cataloging-in-Publication Data

Names: Grooms, Jonathon, 1981- author. | Enderle, Patrick, author. | Hutner, Todd, 1981- author. | Murphy, Ashley, 1988- author. | Sampson, Victor, 1974- author. | National Science Teachers Association.
Title: Argument-driven inquiry in physical science : lab investigations for grades 6-8 / Jonathon Grooms, Patrick J. Enderle, Todd Hutner, Ashley Murphy, and Victor Sampson.
Description: Arlington, VA : National Science Teachers Association, [2016] | Includes bibliographical references and index.
Identifiers: LCCN 2016027981 (print) | LCCN 2016030475 (ebook) | ISBN 9781938946233 (print) | ISBN 1938946235 (print) | ISBN 9781681403724 (e-book) | ISBN 1681403722 (e-book)
Subjects: LCSH: Physical sciences--Methodology--Study and teaching (Middle school) | Physical sciences--Experiments. | Inquiry-based learning.
Classification: LCC Q182.3 .G76 2016 (print) | LCC Q182.3 (ebook) | DDC 500.2071/2--dc23
LC record available at https://lccn.loc.gov/2016027981

CONTENTS

SECTION 3—Physical Science Core Idea 2
Motion and Stability: Forces and Interactions

INTRODUCTION LABS

SECTION 6—Appendixes

PREFACE

There is a push to change the way science is taught in the United States, arising from a different idea of what it means to know, understand, and be able to do in science. As described in *A Framework for K–12 Science Education* (National Research Council [NRC] 2012) and the *Next Generation Science Standards* (NGSS Lead States 2013), science education should be structured to emphasize ideas *and* practices to

> ensure that by the end of 12th grade, *all* students have some appreciation of the beauty and wonder of science; possess sufficient knowledge of science and engineering to engage in public discussions on related issues; are careful consumers of scientific and technological information related to their everyday lives; are able to continue to learn about science outside school; and have the skills to enter careers of their choice, including (but not limited to) careers in science, engineering, and technology. (NRC 2012, p. 1)

Instead of teaching with the goal of helping students learn facts and concepts, science teachers are now charged with helping their students become *proficient* in science by time they graduate from high school. To be considered proficient in science, the NRC (2012) suggests that students need to understand four core ideas in the physical sciences,[1] be aware of seven crosscutting concepts that span the various disciplines of science, and learn how to participate in eight fundamental scientific practices. These important practices, crosscutting concepts, and core ideas are summarized in Figure 1 (p. xii).

As described by the NRC (2012), new instructional approaches will be needed to assist students in developing these proficiencies. In answer to this call, this book provides 22 laboratory investigations designed using an innovative approach to lab instruction called argument-driven inquiry (ADI). This approach and the labs based on it are aligned with the content, crosscutting concepts, and scientific practices outlined in Figure 1. Because the ADI model calls for students to give presentations to their peers, respond to questions, and then write, evaluate, and revise reports as part of each lab, the lab activities described in this book will also enable students to develop the disciplinary-based literacy skills outlined in the *Common Core State Standards* for English language arts (National Governors Association Center for Best Practices and Council of Chief State School Officers 2010). Use of these labs, as a result, can help teachers align their teaching with current recommendations for making physical science more meaningful for students and instruction more effective for teachers.

---

[1] Throughout this book, we use the term *physical sciences* when referring to the core ideas of the *Framework* (in this context the term refers to a broad collection of scientific fields), but we use the term *physical science* when referring to courses at the middle school level (as in the title of the book).

FIGURE 1

The three dimensions of the framework for the *Next Generation Science Standards*

Scientific Practices	Crosscutting Concepts
1. Asking questions and defining problems	1. Patterns
2. Developing and using models	2. Cause and effect: Mechanism and explanation
3. Planning and carrying out investigations	3. Scale, proportion, and quantity
4. Analyzing and interpreting data	4. Systems and system models
5. Using mathematics and computational thinking	5. Energy and matter: Flows, cycles, and conservation
6. Constructing explanations and designing solutions	6. Structure and function
7. Engaging in argument from evidence	7. Stability and change
8. Obtaining, evaluating, and communicating information	

Physical Sciences Core Ideas

- PS1: Matter and its interactions
- PS2: Motion and stability: Forces and interactions
- PS3: Energy
- PS4: Waves and their applications in technologies for information transfer

Source: Adapted from NRC 2012, p. 3.

References

National Governors Association Center for Best Practices and Council of Chief State School Officers (NGAC and CCSSO). 2010. *Common core state standards.* Washington, DC: NGAC and CCSSO.

National Research Council (NRC). 2012. *A framework for K–12 science education: Practices, crosscutting concepts, and core ideas.* Washington, DC: National Academies Press.

NGSS Lead States. 2013. *Next Generation Science Standards: For states, by states.* Washington, DC: National Academies Press. *www.nextgenscience.org/next-generation-science-standards.*

ACKNOWLEDGMENTS

The development of this book was supported by the Institute of Education Sciences, U.S. Department of Education, through grant R305A100909 to Florida State University. The opinions expressed are those of the authors and do not represent the views of the institute or the U.S. Department of Education.

ABOUT THE AUTHORS

Jonathon Grooms is an assistant professor of curriculum and pedagogy in the Graduate School of Education and Human Development at The George Washington University. He received a BS in secondary science and mathematics teaching with a focus in chemistry and physics from Florida State University (FSU). Upon graduation, Jonathon joined FSU's Office of Science Teaching, where he directed the physical science outreach program "Science on the Move." He also earned a PhD in science education from FSU. His research interests include student engagement in scientific argumentation and students' application of argumentation strategies in socioscientific contexts. To learn more about his work in science education, go to *www.jgrooms.com.*

Patrick J. Enderle is an assistant professor of science education in the Department of Middle and Secondary Education at Georgia State University. He received his BS and MS in molecular biology from East Carolina University. Patrick spent some time as a high school biology teacher and several years as a visiting professor in the Department of Biology at East Carolina University. He then attended FSU, from which he graduated with a PhD in science education. His research interests include argumentation in the science classroom, science teacher professional development, and enhancing undergraduate science education. To learn more about his work in science education, go to *http://patrickenderle.weebly.com.*

Todd Hutner is a research associate in the Center for STEM Education (see *http://stemcenter.utexas.edu*) at The University of Texas at Austin (UT-Austin). He received a BS and an MS in science education from FSU and a PhD in curriculum and instruction from UT-Austin. Todd also taught high school physics and chemistry for four years. He specializes in teacher learning, teacher practice, and educational policy in science education.

Ashley Murphy attended FSU and earned a BS with dual majors in biology and secondary science education. Ashley spent some time as a middle school biology and science teacher before entering graduate school at UT-Austin, where she is currently working toward a PhD in STEM (science, technology, engineering, and mathematics) education. Her research interests include argumentation in middle and elementary classrooms. As an educator, she frequently employed argumentation as a means to enhance student understanding of concepts and science literacy.

Victor Sampson is an associate professor of STEM education and the director of the Center for STEM Education at UT-Austin. He received a BA in zoology from the University of Washington, an MIT from Seattle University, and a PhD in curriculum and instruction with a specialization in science education from Arizona State University. Victor also taught high school biology and chemistry for nine years. He specializes in argumentation in science education, teacher learning, and assessment. To learn more about his work in science education, go to *www.vicsampson.com.*

INTRODUCTION

The Importance of Helping Students Become Proficient in Science

The new aim of science education in the United States is for all students to become proficient in science by the time they finish high school. It is essential to recognize that science proficiency involves more than an understanding of important concepts, it also involves being able to *do* science. *Science proficiency,* as defined by Duschl, Schweingruber, and Shouse (2007), consists of four interrelated aspects. First, it requires an individual to know important scientific explanations about the natural world, be able to use these explanations to solve problems, and understand new explanations when they are introduced to the individual. Second, it requires an individual to be able to generate and evaluate scientific explanations and scientific arguments. Third, it requires an individual to understand the nature of scientific knowledge and how scientific knowledge develops over time. Finally, and perhaps most important, an individual who is proficient in science should be able to participate in scientific practices (such as designing and carrying out investigations and arguing from evidence) and communicate in a manner that is consistent with the norms of the scientific community.

In the past decade, however, the importance of learning how to participate in scientific practices has not been acknowledged in the standards of many states. Many states have also attempted to make their science standards more rigorous by adding more content to them or lowering the grade level at which content is introduced rather than emphasizing depth of understanding of core ideas and crosscutting concepts, as described by the National Research Council (NRC) in *A Framework for K–12 Science Education* (NRC 2012). The result of the increased number of content standards and the pressure to cover them to prepare students for high-stakes tests that target facts and definitions is that teachers have "alter[ed] their methods of instruction to conform to the assessment" (Owens, 2009, p. 50). Teachers, as a result, tend to move through the science curriculum quickly to ensure that they have introduced all the content found in the standards before the administration of the tests, which leads them to cover many topics in a shallow fashion rather than to delve into a smaller number of core ideas in a way that promotes a coherent and deep understanding. The unintended consequence of this approach has been a focus on content (learning facts) rather than on developing scientific habits of mind or learning how to use core ideas and the practices of science to explain natural phenomena.

Despite this focus on more content and high-stakes accountability for science learning, students do not seem to be gaining proficiency in science. According to *The Nation's Report Card: Science 2009* (National Center for Education Statistics 2011), only 21% of all 12th-grade students who took the National Assessment of Educational Progress in science scored at the proficient level. The performance of U.S. students on international assessments is even bleaker, as indicated by their scores on the science portion of the Programme for International Student Assessment (PISA). The Organisation for Economic Co-operation and Development (OECD) began administering the PISA in 1997 to assess and compare education systems. Since 1997, students in more than 70 countries have taken the PISA. The test is designed to assess reading, math, and science achievement and is given every three years. The mean score for students in the United States on the science portion of the PISA in 2012 was below the international mean (500), and there has been no significant change in the U.S. mean score since 2000; in fact, the U.S. mean score in 2012 was slightly less than it was in 2000 (OECD 2012; see Table 1). Students in many different countries, including China, Korea, Japan, and Finland, consistently score higher than students in the United States. These results suggest that U.S. students are not learning what they need to be considered proficient in science, even though teachers are covering a great deal of material and being held accountable for it.

TABLE 1

PISA scientific literacy performance for U.S. students

Year	U.S. mean score*	U.S. rank/ Number of countries assessed	Top-performing countries (score)		
			1	2	3
2000	499	14/27	Korea (552)	Japan (550)	Finland (538)
2003	491	22/41	Finland (548)	Japan (548)	Hong Kong–China (539)
2006	489	29/57	Finland (563)	Hong Kong–China (542)	Canada (534)
2009	499	15/43	Japan (552)	Korea (550)	Hong Kong–China (541)
2012	497	36/65	Shanghai-China (580)	Hong Kong–China (555)	Singapore (551)

*The mean score of the PISA is 500 across all years.

Source: OECD 2012.

Additional evidence of the consequences of emphasizing breadth over depth comes from empirical research in science education that supports the notion that broad, shallow coverage neglects the practices of science and hinders the development of science proficiency (Duschl, Schweingruber, and Shouse 2007; NRC 2005, 2008). As noted in the *Framework* (NRC 2012),

> K–12 science education in the United States fails to [promote the development of science proficiency], in part because it is not organized systematically across multiple years of school, emphasizes discrete facts with a focus on breadth over depth, and does not provide students with engaging opportunities to experience how science is actually done. (p. 1)

Based on their review of the available literature, the NRC recommends that science teachers delve more deeply into core ideas to help their students develop improved understanding and retention of science content. The NRC also calls for students to be given more experience participating in the practices of science, with the goal of enabling students to better engage in public discussions about scientific issues related to their everyday lives, be consumers of scientific information, and have the skills and abilities needed to enter science or science-related careers. We think the school science laboratory is the perfect place to focus on core ideas and engage students in the practices of science and, as a result, help them develop the knowledge and abilities needed to be proficient in science.

How School Science Laboratories Can Help Foster the Development of Science Proficiency

Investigators have shown that lab activities[1] have a standard format in U.S. secondary-school classrooms (Hofstein and Lunetta 2004; NRC 2005). This format begins with the teacher introducing students to a concept through direct instruction, usually a lecture and/or reading. Next, students complete a confirmatory laboratory activity, usually following a "cookbook recipe" in which the teacher provides a step-by-step procedure to follow and a data table to fill out. Finally, students are asked to answer a set of focused analysis questions to ensure that the lab has illustrated, confirmed, or otherwise verified the targeted concept(s). This type of approach does little to promote science proficiency because it often fails to help students think critically about the concepts, engage in important scientific practices (such as designing an investigation, constructing explanations, or arguing from evidence), or develop scientific habits of mind (Duschl, Schweingruber, and Shouse 2007; NRC 2005). Further, this approach does not do much to improve science-specific literacy skills.

Changing the focus of lab instruction can help address these challenges. To implement such a change, teachers will have to emphasize "how we know" in the physical sciences (i.e., how new knowledge is generated and validated) equally with "what we know" about behavior of matter on Earth (i.e., the theories, laws, and unifying

[1] We use the NRC's definition of a school science lab activity, which is "an opportunity for students to interact directly with the material world using the tools, data collection techniques, models, and theories of science" (NRC 2005, p. 3).

concepts). Because it is an essential practice of science, the NRC calls for *argumentation* (which we define as a process of proposing, supporting, evaluating, and refining claims on the basis of reason) to play a more central role in the teaching and learning of science. The NRC (2012) provides a good description of the role argumentation plays in science:

> Scientists and engineers use evidence-based argumentation to make the case for their ideas, whether involving new theories or designs, novel ways of collecting data, or interpretations of evidence. They and their peers then attempt to identify weaknesses and limitations in the argument, with the ultimate goal of refining and improving the explanation or design. (p. 46)

This means that the focus of teaching will have to shift more to scientific abilities and habits of mind so that students can learn to construct and support scientific knowledge claims through argument (NRC 2012). Students will also have to learn to evaluate the claims and arguments made by others.

A part of this change in instructional focus will need to be a change in the nature of lab activities (NRC 2012). Students will need to have more experiences engaging in scientific practices so that lab activities can become more authentic. This is a major shift away from labs driven by prescribed worksheets and data tables to be completed. These activities will have to be thoughtfully constructed so as to be educative and help students develop the required knowledge, skills, abilities, and habits of mind. This type of instruction will require that students receive feedback and learn from their mistakes; hence, teachers will need to develop more strategies to help students learn from their mistakes.

The argument-driven inquiry (ADI) instructional model (Sampson and Gleim 2009; Sampson, Grooms, and Walker 2009, 2011) was designed as a way to make lab activities more authentic and educative for students and thus help teachers promote and support the development of science proficiency. This instructional model reflects research about how people learn science (NRC 1999) and is also based on what is known about how to engage students in argumentation and other important scientific practices (Berland and Reiser 2009; Erduran and Jimenez-Aleixandre 2008; McNeill and Krajcik 2008; Osborne, Erduran, and Simon 2004; Sampson and Clark 2008).

Organization of This Book

The remainder of this book is divided into six sections. Section 1 includes two chapters: the first describes the ADI instructional model, and the second describes the development and components of the ADI lab investigations. Sections 2–5 contain the lab investigations, including notes for the teacher, student handouts, and checkout questions for students. Four appendixes contain standards alignment matrixes, timeline and proposal options for the investigations, and a peer-review guide and instructor rubric for assessing the investigation reports.

Safety Practices in the Science Laboratory

It is important for science teachers to make hands-on and inquiry-based lab activities safer for students and teachers. Teachers therefore need to have proper safety equipment in the classroom/laboratory in the form of engineering controls such as ventilation, fume hoods, fire extinguishers, eye wash, and showers. They also need to ensure that students use appropriate personal protective equipment (PPE; e.g., sanitized indirectly vented chemical-splash goggles meeting ANSI/ISEA Z87.1 standard, chemical-resistant aprons and nonlatex gloves) during all components of laboratory activities (i.e., setup, hands-on investigation, and takedown). Teachers also need to adopt legal safety standards and better professional practices and enforce them inside the classroom and/or laboratory. Finally, teachers must review and comply with all safety policies and procedures, including but not limited to appropriate chemical management, that have been established by their school district or school.

Throughout this book, safety precautions are provided for each investigation. Teachers should follow these safety precautions to provide a safer learning experience for students. The safety precautions associated with each activity are based, in part, on the use of the recommended materials and instructions, legal safety standards, and better professional safety practices. We also recommend that students review the National Science Teacher Association's document *Safety in the Science Classroom, Laboratory, or Field Sites* under the direction of the teacher before working in the laboratory for the first time. This document is available online at *www.nsta.org/docs/SafetyInTheScienceClassroomLabAndField.pdf.* The students and their parents or guardians should then sign the document to acknowledge that they understand the safety procedures that must be followed during a lab activity.

As a final note, remember that the lab activity is composed of three sections: the setup, the hands-on investigation, and takedown. PPE and safety procedures apply to all three sections!

Disclaimer: The safety precautions for each activity are based in part on use of the recommended materials and instructions, legal safety standards, and better professional practices. Selection of alternative materials or procedures for these activities may jeopardize the level of safety and therefore is at the user's own risk.

References

Berland, L., and B. Reiser. 2009. Making sense of argumentation and explanation. *Science Education* 93 (1): 26–55.

Duschl, R. A., H. A. Schweingruber, and A. W. Shouse, eds. 2007. *Taking science to school: Learning and teaching science in grades K–8*. Washington, DC: National Academies Press.

Erduran, S., and M. Jimenez-Aleixandre, eds. 2008. *Argumentation in science education: Perspectives from classroom-based research*. Dordrecht, The Netherlands: Springer.

Hofstein, A., and V. Lunetta. 2004. The laboratory in science education: Foundations for the twenty-first century. *Science Education* 88: 28–54.

McNeill, K., and J. Krajcik. 2008. Assessing middle school students' content knowledge and reasoning through written scientific explanations. In *Assessing science learning: Perspectives from research and practice,* eds. J. Coffey, R. Douglas, and C. Stearns, 101–116. Arlington, VA: NSTA Press.

National Center for Education Statistics. 2011. *The nation's report card: Science 2009.* Washington, DC: U.S. Department of Education.

National Research Council (NRC). 1999. *How people learn: Brain, mind, experience, and school.* Washington, DC: National Academies Press.

National Research Council (NRC). 2005. *America's lab report: Investigations in high school science*. Washington, DC: National Academies Press.

National Research Council (NRC). 2008. *Ready, set, science: Putting research to work in K–8 science classrooms*. Washington, DC: National Academies Press.

National Research Council (NRC). 2012. *A framework for K–12 science education: Practices, crosscutting concepts, and core ideas*. Washington, DC: National Academies Press.

Organisation for Economic Co-operation and Development (OECD). 2012. OECD Programme for International Student Assessment. *www.oecd.org/pisa.*

Osborne, J., S. Erduran, and S. Simon. 2004. Enhancing the quality of argumentation in science classrooms. *Journal of Research in Science Teaching* 41 (10): 994–1020.

Owens, T. 2009. Improving science achievement through changes in education policy. *Science Educator* 18 (2): 49–55.

Sampson, V., and D. Clark. 2008. Assessment of the ways students generate arguments in science education: Current perspectives and recommendations for future directions. *Science Education* 92 (3): 447–472.

Sampson, V., and L. Gleim. 2009. Argument-driven inquiry to promote the understanding of important concepts and practices in biology. *American Biology Teacher* 71 (8): 471–477.

Sampson, V., J. Grooms, and J. Walker. 2009. Argument-driven inquiry: A way to promote learning during laboratory activities. *The Science Teacher* 76 (7): 42–47.

Sampson, V., J. Grooms, and J. Walker. 2011. Argument-driven inquiry as a way to help students learn how to participate in scientific argumentation and craft written arguments: An exploratory study. *Science Education* 95 (2): 217–257.

SECTION 1
Using Argument-Driven Inquiry

CHAPTER 1

Argument-Driven Inquiry

Stages of Argument-Driven Inquiry

Each of the eight stages in the argument-driven inquiry (ADI) instructional model is designed to ensure that the experience is authentic (students have an opportunity to engage in the practices of science) *and* educative (students receive the feedback and explicit guidance that they need to improve on each aspect of science proficiency). Figure 2 summarizes the eight stages.

FIGURE 2

Stages of the argument-driven inquiry instructional model

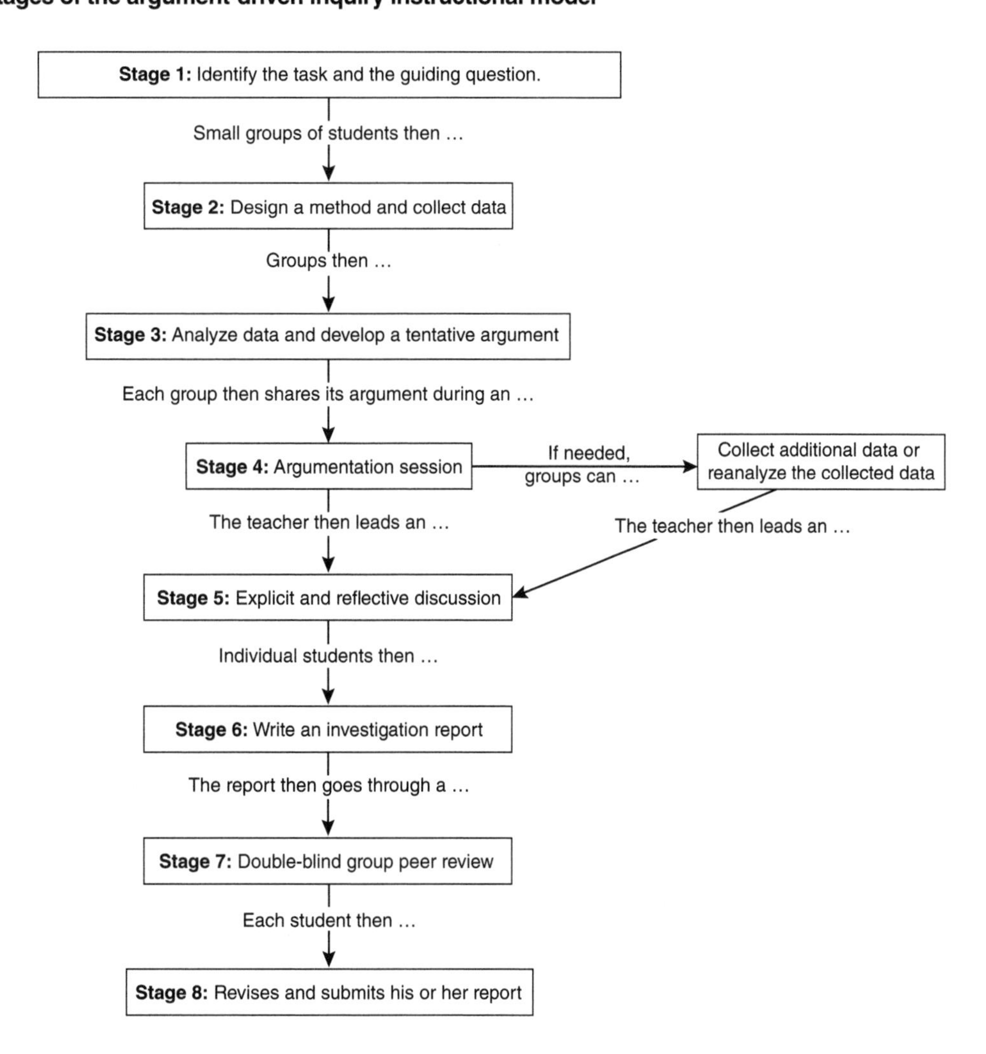

Stage 1: Identification of the Task and the Guiding Question; Tool Talk

In the ADI instructional model each lab activity begins with the teacher identifying a phenomenon to investigate and offering a guiding question for the students to answer. The goal of the teacher at this stage of the model is to capture the students' interest and provide them with a reason to complete the investigation. To aid in this, teachers should provide each student with a copy of the Lab Handout. This handout includes a brief introduction that provides a puzzling phenomenon or a problem to solve and a guiding question to answer. This handout also includes information about the nature of the artifact they will need to produce (i.e., an argument on a large whiteboard, a piece of paper, presentation slides, or other medium), some helpful tips on how to get started, and criteria that will be used to judge argument quality (e.g., the sufficiency of the explanation and the quality of the evidence).

One engaging way to begin each lab includes selecting a different student to read each section of the handout out loud. After each section is read, the teacher should pause to clarify expectations, answer questions, and provide additional information as needed.

It is also important for the teacher to hold a "tool talk" during this stage, taking a few minutes to explain how to use specific lab equipment, how to use a computer simulation, or even how to use software to analyze data. Teachers need to hold a tool talk because students are often unfamiliar with lab equipment; even if they are familiar with the equipment, they will often use it incorrectly or in an unsafe manner. A tool talk can also be productive during this stage because students often find it difficult to design a method to collect the data needed to answer the guiding question (the task of stage 2) when they do not understand how to use the available materials. The teacher should also review specific safety protocols and precautions as part of the tool talk.

Once all the students understand the goal of the activity and how to use the available materials, the teacher should divide the students into small groups (we recommend three students per group) and move on to the second stage of the model.

Stage 2: Design a Method and Collect Data

In stage 2, small groups of students develop a method to gather the data they need to answer the guiding question and carry out that method. How students complete this stage depends on the nature of the investigation. Some investigations call for groups to answer the guiding question by designing a controlled experiment, whereas others require students to analyze an existing data set (e.g., a database or information sheets). If students need assistance in designing their method, teachers can have students complete an investigation proposal. These proposals guide students through the process of developing a method by encouraging them to think about what type of data they will need to collect, how to collect it, and how to analyze it. We have included three different investigation proposals in Appendix 3 (p. 415) of this book that students can use to design their investigations. Teachers can direct students to use Investigation Proposal A or Investigation Proposal B when they need to test alternative explanations or claims as part of their investigation and Investigation Proposal C when they need to collect systematic observations for a descriptive investigation.

The overall intent of this stage is to provide students with an opportunity to interact directly with the natural world (or, in some cases, with data drawn from the natural world) using appropriate tools and data collection techniques and to learn how to deal with the ambiguities of empirical work. This stage of the model also gives students a chance to learn why some approaches to data collection or analysis work better than others and how the method used during a scientific investigation is based on the nature of the question and the phenomenon under investigation. At the end of this stage, students should have collected all the data they need to answer the guiding question.

Stage 3: Data Analysis and Development of a Tentative Argument

The next stage of the instructional model calls for students to develop a tentative argument in response to the guiding question. To do this, each group needs to be encouraged to first make sense of the measurements (e.g., temperature and mass) and/or observations (e.g., appearance and location) they collected during stage 2 of the model. Once the groups have analyzed and interpreted their data, they can create a tentative argument (the "initial argument" in the Lab Handout). The argument consists of a claim, the evidence they are using to support their claim, and a justification of their evidence. The *claim* is their answer to the guiding question. The *evidence* consists of the data (measurements or observations) they collected, an analysis of the data, and an interpretation of the analysis. The *justification of the evidence* is a statement that defends their choice of evidence by explaining why it is important and relevant and makes the concepts or assumptions underlying the analysis and interpretation explicit. The components of a scientific argument are illustrated in Figure 3 (p. 6).

To illustrate each of the three structural components of a scientific argument, consider the following example. This argument was made in response to the guiding question, "What type of metal are objects A, B, and C?"

> *Claim*: Objects A and B are tin. Object C is lead.
>
> *Evidence*: The density of object A is 7.44 g/cm^3, and the density of object B is 7.34 g/cm^3. These objects have almost the same density as the known density of tin, which is 7.36 g/cm^3. The density of object C is 11.12 g/cm^3. This object has almost the same density as the known density of lead, which is 11.34 g/cm^3.
>
> *Justification of the evidence*: Density is a physical property of matter and remains constant, regardless of the amount of the object present. Therefore, density can be used to identify the substance that makes up an unknown object. The difference in the calculated densities and the known densities is likely due to measurement error.

FIGURE 3

Framework for the components of a scientific argument and criteria for evaluating the merits of the argument

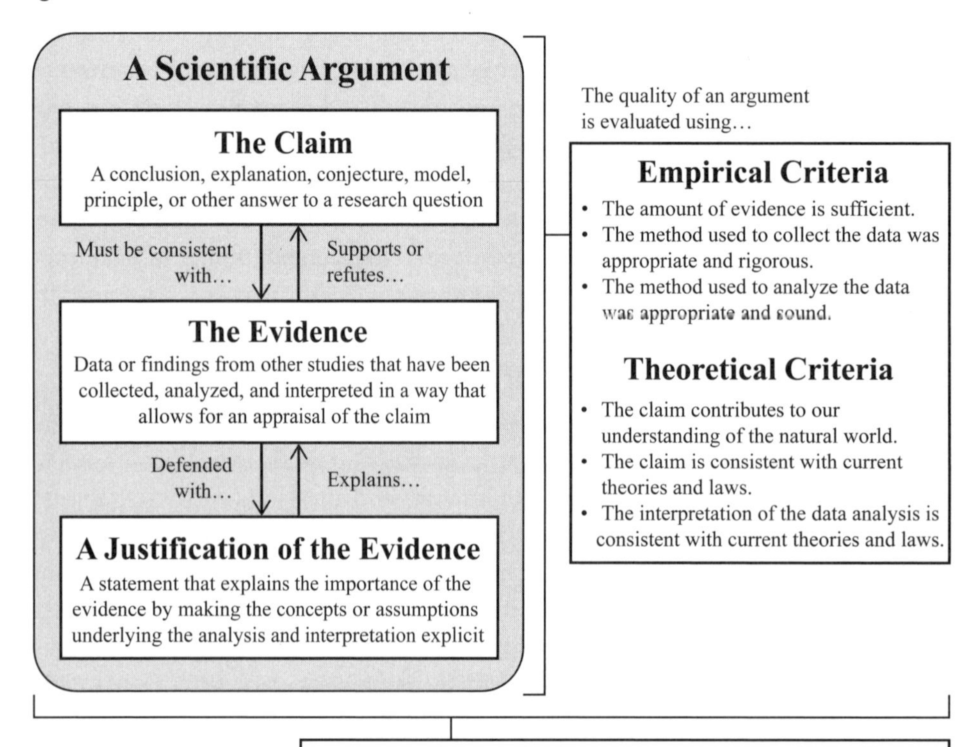

The claim in this argument provides an answer to the guiding question. The author then uses genuine evidence to support the claim by providing an analysis of the data collected (density of each substance) and an interpretation of the analysis (an inference based on known and unknown density values). Finally, the author provides a justification of the evidence in the argument by making explicit the underlying concept and assumptions (density as an inherent physical property and the likelihood that the difference between the calculated and known values is due to measurement error) guiding the analysis of the data and the interpretation of the analysis.

It is important for students to understand that, in science, some arguments are better than others. An important aspect of science and scientific argumentation involves the evaluation of the various components of the arguments put forward by others. Therefore, the framework provided in Figure 3 also highlights two types of criteria that students can and should use to evaluate an argument in science: empirical criteria and theoretical criteria. *Empirical criteria* include

- how well the claim fits with all available evidence,
- the sufficiency of the evidence,
- the quality of the evidence,
- the appropriateness of the method used to collect the data, and
- the appropriateness of the method used to analyze the data.

Theoretical criteria, on the other hand, refer to standards that are important in science but are not empirical in nature; examples of these criteria are

- the sufficiency of the claim (i.e., Does it include everything needed?);
- the usefulness of the claim (i.e., Does it allow us to engage in new inquiries or understand a phenomenon?);
- the consistency of the claim and the reasoning in terms of other accepted theories, laws, or models; and
- the manner in which the data analysis was conducted.

What counts as quality within these different components, however, varies from discipline to discipline (e.g., physics, chemistry, geology, biology) and within the specific fields of each discipline (e.g., astrophysics, biophysics, nuclear physics, optics, thermodynamics). This variation is due to differences in the types of phenomena investigated, what counts as an accepted mode of inquiry (e.g., descriptive studies, experimentation, computer modeling), and the theory-laden nature of scientific inquiry. It is important to keep in mind that what counts as a quality argument in science is discipline- and field-dependent.

To allow for the critique and refinement of the tentative argument during the next stage of ADI, each group of students should create their tentative argument in a medium that can easily be viewed by the other groups. We recommend using a 2' x 3' whiteboard or large piece of butcher paper. Students should lay out each component of the argument on the board or paper. Figure 4 shows the general layout for a presentation of an argument, and Figure 5 (p. 8) provides an example of argument crafted by students. Students can also create their tentative arguments using presentation software such as Microsoft's PowerPoint or Apple's

FIGURE 4

The components of an argument that should be included on a whiteboard (outline)

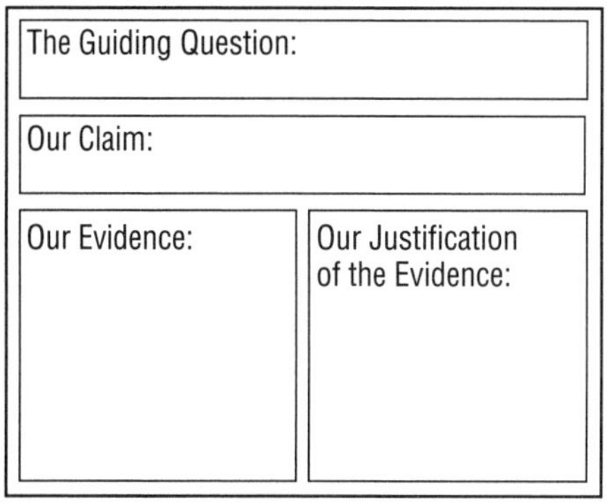

The Guiding Question:	
Our Claim:	
Our Evidence:	Our Justification of the Evidence:

FIGURE 5

An example of a student-generated argument on a whiteboard

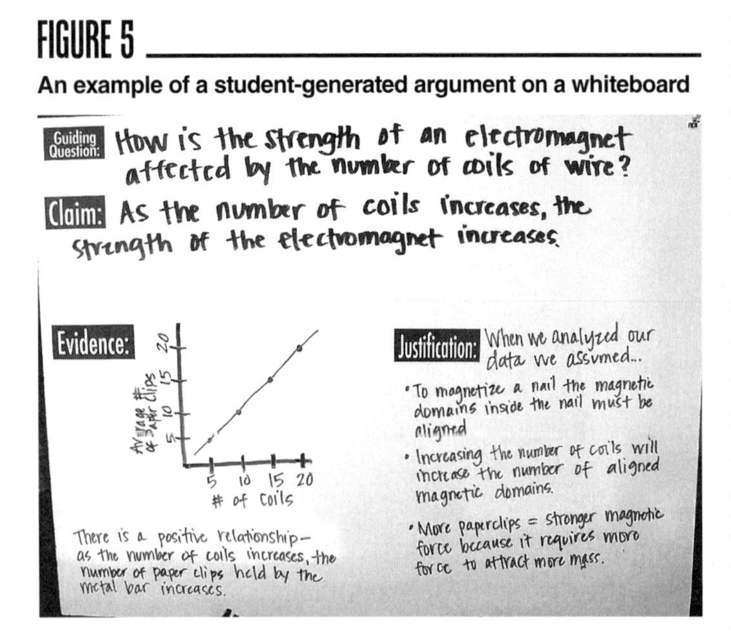

Keynote and devote one slide to each component of an argument. The choice of medium is not important, as long as students are able to easily modify the content of their argument as they work and it enables others to easily view their argument.

The intention of this stage of the model is to provide the student groups with an opportunity to make sense of what they are seeing or doing during the investigation. As students work together to create a tentative argument, they must talk with each other and determine how to analyze the data and how to best interpret the trends, differences, or relationships that they identify. They must also decide if the evidence (data that have been analyzed and interpreted) that they chose to include in their argument is relevant, sufficient, and convincing enough to support their claim. This process, in turn, enables the groups of students to evaluate competing ideas and weed out any claim that is inaccurate, does not fit with all the available data, or contains contradictions.

This stage of the model is challenging for students because they are rarely asked to make sense of a phenomenon based on raw data, so it is important for teachers to actively work to support their sense-making. In this stage, teachers should circulate from group to group to act as a resource person for the students, asking questions that urge them to think about what they are doing and why. To help students remember the goal of the activity, ask "What are you trying to figure out?"; to encourage them to think about whether or not the data are relevant, ask "Why is that information important?"; to help them remember to use rigorous criteria to evaluate the merits of a tentative idea, ask "Does that fit with all the data or what we know about a particular phenomenon?"

It is important to remember that at the beginning of the school year, students will struggle to develop arguments and will often rely on inappropriate criteria, such as plausibility (e.g., "That sounds good to me") or fit with personal experience (e.g., "But that is what I saw on TV once"), as they attempt to make sense of their data. However, with enough practice, *students will improve their skills.* This is an important principle underlying the ADI instructional model.

Stage 4: Argumentation Session

The fourth stage of ADI is the argumentation session. In this stage, each group is given an opportunity to share, evaluate, and revise their tentative arguments with the other groups (see Figure 6). This stage is included in the model for three reasons:

FIGURE 6

A student presents her group's argument to students from other groups during the argumentation session.

1. Scientific argumentation (i.e., arguing from evidence) is an important practice in science because critique and revision leads to better outcomes.
2. Research indicates that students learn more about the content and develop better critical-thinking skills when they are exposed to the alternative ideas, respond to the questions and challenges of other students, and evaluate the merits of competing ideas (Duschl et al. 2007; National Research Council 2012).
3. Students learn how to distinguish between ideas using rigorous scientific criteria and are able to develop scientific habits of mind (such as treating ideas with initial skepticism, insisting that the reasoning and assumptions be made explicit, and insisting that claims be supported by valid evidence) during the argumentation sessions.

This stage, as a result, provides the students with an opportunity to learn from and about scientific argumentation. It is important to note, however, that supporting and promoting productive interactions between students inside the classroom can be difficult because the practice of arguing from evidence is foreign to most students when they first begin participating in ADI. To aid these interactions, students are required to generate their arguments in a medium that can be seen by others. By looking at whiteboards, paper, or slides, students tend to focus their attention on evaluating evidence and the core ideas that were used to justify the evidence, rather than attacking the source of the ideas. As a result, this strategy often makes the discussion more productive and makes it easier for student to identify and weed out faulty ideas. It is also important for the students to view the argumentation session as an opportunity to learn. The teacher, therefore, should describe the argumentation session as an opportunity for students to collaborate with their peers and a chance to give each other feedback so that the quality of all the arguments can be

FIGURE 7

A modified gallery walk format is used during the argumentation session to allow multiple groups to share their arguments at the same time.

improved, rather than as an opportunity determine who is right or wrong.

To ensure that all students remain engaged during the argumentation session, we recommend that teachers use a modified "gallery walk" format, rather than a whole-class presentation format. In the modified gallery walk format, one or two members of the group stay at their workstation to share their groups' ideas, while the other group members go one at a time to different groups to listen to and critique the arguments developed by their classmates (see Figure 7). This type of format ensures that all ideas are heard and more students are actively involved in the process. We recommend that the students who are responsible for critiquing arguments visit at least three different groups during the argumentation session. We also recommend that the presenters keep a record of the critiques made by their classmates and any suggestions for improvement. The students who are responsible for critiquing the arguments should also be encouraged to keep a record of good ideas or potential ways to improve their own arguments as they travel from group to group.

Just as is the case in earlier stages of ADI, it is important for the classroom teacher to be involved in (without leading) the discussions during the argumentation session. Once again, the teacher should move from group to group to keep students on task and model good scientific argumentation. The teacher can ask the presenter(s) questions such as "How did you analyze the available data?" or "Were there any data that did not fit with your claim?" to encourage students to use empirical criteria to evaluate the quality of the arguments. The teacher can also ask the presenter(s) to explain how the claim they are presenting fits with the theories, laws, or models of science or why the evidence they used is important. In addition, the teacher can also ask the students who are listening to the presentation questions such as "Do you think their analysis is accurate?" or "Do you think their interpretation is sound?" or even "Do you think their claim fits with what we know about X?" These questions can serve to remind students to use empirical and theoretical criteria to evaluate an argument during the discussions. Overall, it is the goal of the teacher at this stage of the lesson to encourage students to think about how they know what they know and why some claims are more valid or acceptable in science. This stage of the model, however, is not the time to tell the students that they are right or wrong.

At the end of the argumentation session, it is important to give the students time to meet with their original group so they can discuss what they learned by interacting with individuals from the other groups and revise their tentative arguments. This process can begin with the presenter(s) sharing the critiques and the suggestions for improvement that they heard during the argumentation session. The students who visited the other groups during the argumentation can then share their ideas for making the arguments better, based on what they observed and discussed at other stations. Students often realize at this point in the process that the way they collected or analyzed data during stage 2 was flawed in some way. The teacher should therefore encourage students to collect new data or reanalyze the data they collected as needed. At the end of this stage, each group should have a final argument that is much better than their initial one.

Stage 5: Explicit and Reflective Discussion

The teacher should lead a whole-class explicit and reflective discussion during stage 5 of ADI. The intent of this discussion is to give students an opportunity to think about and share what they know and how they know it. This stage enables the classroom teacher to ensure that all students understand the core ideas at the heart of the investigation and to help students think about ways to improve their participation in scientific practices such as planning and carrying out investigations, analyzing and interpreting data, and arguing from evidence. At this point in the instructional model, the teacher should also discuss one or two relevant crosscutting concepts and one or two aspects of the nature of science or scientific inquiry.

It is important to stress that an explicit and reflective discussion is not a lecture; it is an opportunity for students to think about important ideas and practices and to share what they have learned or do not understand through teacher-to-student and student-to-student talk. The more students talk during this stage of the model, the more meaningful the experience will be for them.

Teachers should begin the explicit and reflective discussion by asking students to explain what they know about the core idea at the heart of the lab (the core idea can be found in the "Introduction" section of the Lab Handout). The teacher can then encourage the students to think about how this core idea helped them explain the phenomenon under investigation and how they can use this idea to provide a justification of the evidence in their arguments. The teacher can also encourage the students to explain what they learned about the phenomenon under investigation. The teacher should not tell the students what results they should have obtained or what information should be included in each argument. Instead, the teachers should focus on ensuring that everyone in the class understands the content by providing a context for students to share their ideas and explain their thinking. Remember, this stage of ADI is a *discussion,* not a lecture.

Students do not learn how to participate in the practices of planning and carrying investigations, analyzing and interpreting data, and arguing from evidence by following directions that are given to them on a handout or by answering analysis questions. This is one reason that cookbook laboratory activities do little to help students learn how to participate in scientific practices. In ADI, students are expected to design their own investigations, decide how to analyze and interpret data, and support their claims with evidence. It is important to keep in mind that these practices are complex and students cannot be expected to master them without giving students opportunities to try, fail, and then learn from their mistakes.

To encourage students to learn from their mistakes, students must have an opportunity to reflect on what went well and what went wrong during their investigation. Therefore, as part of the explicit and reflective discussion, the teacher should encourage the students to reflect how they designed their investigation, analyzed and interpreted data, and argued from evidence and what they could do to improve in the future. To accomplish this task, we recommend asking students the following questions:

1. What were some of the strengths of your investigation? What made it scientific?
2. What were some of the weaknesses of your investigation? What made it less scientific?
3. If you were to do this investigation again, what would you do to address the weaknesses in your investigation? What could you do to make it more scientific?

Two additional questions are asked in the design challenge labs: "Did you meet the goal of the design challenge?" and "Did you ensure that your solution is consistent with the design parameters?" The teacher can use the students' responses to all of these questions to highlight what does and does not count as quality or rigor in science and to offer advice about ways to improve. This feedback during the explicit and reflective discussion helps students gradually improve their abilities to participate in the practices of designing and carrying out investigations, analyzing and interpreting data, and arguing from evidence.

Next, the teacher should discuss one or two crosscutting concepts, using what the students did during the investigation to illustrate how these concepts can be used to explain natural phenomena. For example, a teacher might decide to talk about the relationship between structure and function or the importance of tracking the flow of energy or matter in a system. The teacher needs to highlight these ideas in an explicit manner and encourage students to reflect on them in a wide range of contexts before students will begin use them as a way to understand the world around them. Without this type of sustained focus, students will not learn the value of these crosscutting concepts as way to explain natural phenomena. They also need to discuss these important ideas over and over again before they will begin to use them in future investigations.

The teacher should end this stage of the ADI instructional model with an explicit discussion of one or two aspects of the nature and practices of science, building on the students' own investigations to illustrate these important concepts (NGSS Lead States 2013). Examples of what a teacher might include are

- the diversity of methods that scientists can use to collect data,
- what does and does not count as an experiment in science,
- how scientists must be open to changing their minds in the face of empirical evidence, and
- the role that creativity and imagination play during an investigation.

Teachers might compare and contrast observations and inferences, data and evidence, qualitative data and quantitative data, or theories and laws. This stage provides a golden opportunity for explicit instruction about the nature of scientific knowledge and how this knowledge develops over time, in a context that is meaningful to students. Current research suggests that students only develop an appropriate understanding of the nature and practices of science when teachers discuss these concepts in an *explicit* fashion (Abd-El-Khalick and Lederman 2000; Lederman and Lederman 2004; Schwartz, Lederman, and Crawford 2004).

Stage 6: Writing the Investigation Report

Stage 6 is included in the ADI model because writing is an important part of doing science. Scientists must be able to read and understand the writing of others, as well as evaluate its worth. They also must be able to share the results of their own research through writing. In addition, writing helps students learn how to articulate their thinking in a clear and concise manner, encourages metacognition, and improves student understanding of the content (Wallace, Hand, and Prain 2004). Finally, and perhaps most important, writing makes each students' thinking visible to the teacher (which facilitates assessment) and enables the teacher to provide students with the educative feedback they need to improve.

In stage 6, each student is required to write an individual investigation report using his or her group's argument. The report should be centered on three fundamental questions:

- What question were you trying to answer and why?
- What did you do to answer your question and why?
- What is your argument?

Teachers should encourage students to use tables or graphs to help organize their evidence and require them to reference this information in the body of the report. Stage 6 is important because it allows them to learn how to construct an explanation, argue from evidence, and communicate information. It also enables students to master the disciplinary-based writing skills outlined in the *Common Core State Standards* in English Language Arts (*CCSS ELA;* National Governors Association Center for Best Practices and

Council of Chief State School Officers 2010). The report can be written during class or can be assigned as homework.

The format of the report is designed to emphasize the persuasive nature of science writing and to help students learn how to communicate in multiple modes (words, figures, tables, and equations). The three-question format is well aligned with the components of a traditional laboratory report (i.e., introduction, procedure, results and discussion) but allows students to see the important role argument plays in science. We strongly recommend that teachers *limit the length of the investigation report* to two double-spaced pages or one single-spaced page. This limitation encourages students to write in a clear and concise manner, because there is little room for extraneous information. This limitation is less intimidating than a more lengthy report requirement, and it lessens the work required in the subsequent stages.

Stage 7: Double-Blind Group Peer Review

During stage 7, each student is required to submit to the teacher two to three typed copies of their investigation report. Students do not place their names on the report; instead they use an identification number to maintain anonymity—to ensure that reviews are based on the ideas presented and not the person presenting the ideas.

We recommend that students be placed into groups of three (these groups can be different from the groups that students worked in during stages 1–4). Each group then is given a set of reports (i.e., the two or three report copies submitted by a single student) and a peer-review guide (see Appendix 4, p. 419). They are then asked to review the set of reports as a team, using the peer-review guide (see Figure 8). The peer-review guide contains specific criteria that are to be used by the group as they cooperatively evaluate the quality of each section of an investigation report, as well as the mechanics of the writing. There is also space for the reviewers to provide the author with feedback about how to improve the report. Once a group finishes reviewing a report, they are given another set to review. Each group is responsible for reviewing three different sets of reports during this stage.

When students are grouped together in threes, they only need to review three sets of reports in a period. Be sure to give students only 15 minutes to review each set of reports (we recommend setting a timer to help manage time). When students are grouped into three and given 15 minutes to complete each review, the entire peer-review process can be completed in one 50-minute class period (3 sets of reports × 15 minutes = 45 minutes).

Reviewing each report as a group is an important component of the peer-review process because it provides students with a forum to discuss what counts as high quality or acceptable and, in so doing, forces them to reach a consensus during the process. This method also helps prevent students from checking off "good" for each criterion on the peer-review guide without thorough consideration of the merits of the paper. It is also important for students to provide constructive and specific feedback to the author when areas of the paper are found to not meet the standards established by the peer-review guide. The peer-review process provides students with an opportunity to read good and bad examples of the reports. This helps the students learn new ways to organize and present information, which in turn will help them write better on subsequent reports.

FIGURE 8

A group of students reviewing a report written by a classmate, using the peer-review guide

This stage of the model also gives students more opportunities to develop reading skills that are needed to be successful in science. Students must be able to determine the central ideas or conclusions of a text and determine the meaning of symbols, key terms, and other domain-specific words. In addition, students must be able to assess the reasoning and evidence that an author includes in a text to support his or her claim and, when they read a scientific text, compare or contrast findings presented in a text with those from other sources. Students can develop all these skills, as well as the other discipline-based reading standards found in the *CCSS ELA*, when they are required to read and critically review reports written by their classmates.

Stage 8: Revision and Submission of the Investigation Report

The final stage in the ADI instructional model is to revise the report based on the suggestions given during the peer review. If the report met all the criteria, the student may simply submit the paper to the teacher with the original peer-reviewed "rough draft" and peer-review sheet attached, ensuring that his or her name replaces the identification number. Students whose reports are found by the peer-review group to be acceptable can maintain the option to revise it if they so desire after reviewing the work of other students.

If the report was found unacceptable by the group during peer review, the author is required to rewrite his or her report using the reviewers' comments and suggestions as a guideline. Once the report is revised, it is turned in to the teacher for evaluation with the original rough draft and the peer-review sheet attached. The author is required to explain what he or she did to improve each section of the report in response to the reviewers' suggestions (or explain why he or she decided to ignore the reviewers' suggestions) in the author response section of the peer-review sheet. The teacher can then provide a score on the peer-review sheet "Instructor Score" column and use these ratings to assign an overall grade.

This approach provides students with a chance to improve their writing mechanics and develop their reasoning and understanding of the content. This process also offers students the added benefit of reducing academic pressure by providing support in obtaining the highest possible grade for their final product.

The Role of the Teacher During Argument-Driven Inquiry

If the ADI instructional model is to be successful and student learning is to be optimized, the role of the teacher during a lab activity designed using this model must be different than the teacher's role during a more traditional lab. The teacher *must* act as a resource for the students, rather than as a director, as students work through each stage of the activity; the teacher must encourage students to think about *what they are doing* and *why they made that decision* throughout the process. This encouragement should take the form of probing questions that teachers ask as they walk around the classroom, such as "Why do you want to set up your equipment that way?" or "What type of data will you need to collect to be able to answer that question?"

Teachers must restrain from telling or showing students how to "properly" conduct the investigation. However, teachers must emphasize the need to maintain high standards for a scientific investigation by requiring students to use rigorous standards for what counts as a good method or a strong argument in the context of science.

Finally, and perhaps most important for the success of an ADI activity, teachers must be willing to let students try and fail, and then help them learn from their mistakes. Teachers should not try to make the lab investigations included in this book "student-proof" by providing additional directions to ensure that students do everything right the first time. We have found that students often learn more from an ADI lab activity when they design a flawed method to collect data during stage 2 or analyze their results in an inappropriate manner during stage 3, because their classmates quickly point out these mistakes during the argumentation session (stage 4), and it leads to more teachable moments.

Because the teacher's role in an ADI lab is different from what typically happens in laboratories, we've provided a chart describing teacher behaviors that are consistent and inconsistent with each stage of the instructional model (see Table 2). This table is organized by stage, because what the students and the teacher need to accomplish during each stage is different. It might be helpful to keep this table handy as a guide when you are first attempting to implement the lab activities found in the book.

TABLE 2

Teacher behaviors during the stages of the ADI instructional model

Stage	What the teacher does that is ... Consistent with ADI model	Inconsistent with ADI model
1: Identification of the task and the guiding question; "tool talk"	• Sparks students' curiosity • "Creates a need" for students to design and carry out an investigation • Organizes students into collaborative groups • Supplies students with the materials they will need • Holds a "tool talk" to show students how to use equipment or to illustrate proper technique • Reviews relevant safety precautions and protocols • Provides students with hints	• Provides students with possible answers to the research question • Tells students that there is one correct answer • Provides a list of vocabulary terms or explicitly describes the content addressed in the lab
2: Designing a method and collecting data	• Encourages students to ask questions as they design their investigations • Asks groups questions about their method (e.g., "Why did you do it this way?") and the type of data they expect from that design • Reminds students of the importance of specificity when completing their investigation proposals	• Gives students a procedure to follow • Does not question students about the method they design or the type of data they expect to collect • Approves vague or incomplete investigation proposals
3: Data analysis and development of a tentative argument	• Reminds students of the research question and what counts as appropriate evidence in science • Requires students to generate an argument that provides and supports a claim with genuine evidence • Asks students what opposing ideas or rebuttals they might anticipate • Provides related theories and reference materials as tools	• Requires only one student to be prepared to discuss the argument • Moves to groups to check on progress without asking students questions about why they are doing what they are doing • Does not interact with students (uses the time to catch up on other responsibilities) • Tells students the right answer
4: Argumentation session	• Reminds students of appropriate behaviors in the learning community • Encourages students to ask questions of peers • Keeps the discussion focused on the elements of the argument • Encourages students to use appropriate criteria for determining what does and does not count	• Allows students to negatively respond to others • Asks questions about students' claims before other students can ask • Allows students to discuss ideas that are not supported by evidence • Allows students to use inappropriate criteria for determining what does and does not count
5: Explicit and reflective discussion	• Encourages students to discuss what they learned about the content and how they know what they know • Encourages students to discuss what they learned about the nature of science • Encourages students to think of ways to be more productive next time	• Provides a lecture on the content • Skips over the discussion about the nature of science and the nature of scientific inquiry to save time • Tells students "what they should have learned" or "this is what you all should have figured out"

Table 2 ***(continued)***

Stage	What the teacher does that is ...	
	Consistent with ADI model	**Inconsistent with ADI model**
6: Writing the investigation report	• Reminds students about the audience, topic, and purpose of the report • Provides the peer-review guide in advance • Provides an example of a good report and an example of a bad report	• Has students write only a portion of the report • Moves on to the next activity/topic without providing feedback
7: Double-blind peer group review	• Reminds students of appropriate behaviors for the review process • Ensures that all groups are giving a quality and fair peer review to the best of their ability • Encourages students to remember that while grammar and punctuation are important, the main goal is an acceptable scientific claim with supporting evidence and justification • Holds the reviewers accountable	• Allows students to make critical comments about the author (e.g., "This person is stupid") rather than their work (e.g., "This claim needs to be supported by evidence") • Allows students to just check off "Yes" on each item without providing a critical evaluation of the report
8: Revision and submission of the investigation report	• Requires students to edit their reports based on the reviewers' comments • Requires students to respond to the reviewers' ratings and comments • Has students complete the checkout questions after they have turned in their report	• Allow students to turn in a report without a completed peer-review guide • Allow students to turn in a report without revising it first

References

Abd-El-Khalick, F., and N. G. Lederman. 2000. Improving science teachers' conceptions of nature of science: A critical review of the literature. *International Journal of Science Education* 22: 665–701.

Duschl, R. A., H. A. Schweingruber, and A. W. Shouse, eds. 2007. *Taking science to school: Learning and teaching science in grades K-8*. Washington, DC: National Academies Press.

Lederman, N. G., and J. S. Lederman. 2004. Revising instruction to teach the nature of science. *The Science Teacher* 71 (9): 36–39.

National Governors Association Center for Best Practices and Council of Chief State School Officers (NGAC and CCSSO). 2010. *Common core state standards*. Washington, DC: NGAC and CCSSO.

National Research Council (NRC). 2012. *A framework for K–12 science education: Practices, crosscutting concepts, and core ideas*. Washington, DC: National Academies Press.

NGSS Lead States. 2013. *Next Generation Science Standards: For states, by states*. Washington, DC: National Academies Press. *www.nextgenscience.org/next-generation-science-standards*.

Schwartz, R. S., N. Lederman, and B. Crawford. 2004. Developing views of nature of science in an authentic context: An explicit approach to bridging the gap between nature of science and scientific inquiry. *Science Education* 88: 610–645.

Wallace, C., B. Hand, and V. Prain, eds. 2004. *Writing and learning in the science classroom*. Boston: Kluwer Academic Publishers.

CHAPTER 2

Lab Investigations

This book includes 22 physical science lab investigations designed around the argument-driven inquiry (ADI) instructional model. Please note that these investigations are not designed to replace an existing curriculum, but to transform the laboratory component of a science course. A teacher can use these investigations as a way to introduce students to new content ("introduction labs") or as a way to give students an opportunity to apply a theory, law, or unifying concept introduced in class in a novel situation ("application labs"). To facilitate curriculum and lesson planning, the lab investigations have been aligned with *A Framework for K–12 Science Education* (National Research Council [NRC] 2012) and with the *Common Core State Standards* for English language arts and mathematics (*CCSS ELA and CCSS Mathematics;* National Governors Association Center for Best Practices and Council of Chief State School Officers 2010); see Standards Matrixes A and B in Appendix 1 (p. 405).

Many of the ideas for the investigations in this book came from existing resources; however, we modified these existing activities to fit with the ADI instructional model. Once the ADI lab investigations were created, several practicing physicists reviewed them for content accuracy. Two science teachers then piloted the labs in several sections of a middle school physical science course (including general and honors sections). After the pilot year, each lab investigation and all related instructional materials (such as the investigation proposals and peer-review guides) were revised based on the feedback of these two teachers and information from student assessments to increase their effectiveness. The modified labs were then piloted and modified for a second time by other middle school science teachers at other locations and refined again. The final iteration of each lab is included in this book.

The lab investigations were developed as part of a three-year research project funded by the Institute of Education Sciences (grant R305A100909). The goals of this project were to develop a set of ADI lab activities for a variety of science disciplines (including physical science), refine the ADI instructional model, and examine what students learn when they complete eight or more ADI labs of the course over a school year. Our research indicates that after participating in at least eight ADI investigations, students have much better inquiry and writing skills and make substantial gains in their understanding of important physical science content and the nature of science and scientific inquiry (Grooms, Enderle, and Sampson 2015; Sampson et al. 2013). To learn more about the research associated with the ADI instructional model, visit *www.argumentdriveninquiry.com.*

Teacher Notes

Each middle school science teacher must decide when and how to use a laboratory experience to best support student learning. To help with this decision making, we have included Teacher Notes for each investigation. These notes include information about the purpose of the lab, the time needed to implement each stage of the model for that lab, the materials needed, and hints for implementation. We have also included a "Topic Connections" section that shows how each ADI lab activity is aligned with the NRC *Framework* and the *CCSS ELA* and *CCSS Mathematics.* In the following subsections, we will describe the information provided in each section of the Teacher Notes.

Purpose

This section discusses the main idea of the lab and indicates whether the activity is an introduction lab or an application lab. Please note that because of the ADI structure, in both cases very *little* emphasis needs to be placed on making sure the students "get the vocabulary" or "know their stuff" before the lab investigation begins. Instead, with the combination of the information provided in the Lab Handout, students' evolving understanding of the actual practice of science, and the various resources available to your students (i.e., the book, the internet, and you), your students will develop an understanding of the content *as they work through the activity.* The "Purpose" section also highlights the nature of science (NOS) or nature of scientific inquiry (NOSI) concepts that should be emphasized during the explicit and reflection discussion stage of the activity.

The Content

This section of the Teacher Notes provides an overview of the concept that is being introduced to the students during the investigation or the concept that students will need to apply during the investigation.

Timeline

Unlike most traditional laboratories, which can be completed in a single class period, ADI labs typically take four to five days to complete. The amount of time it will take to complete each lab will vary, depending on how long it takes to collect data and whether or not the students write in class or at home. The length of time associated with each ADI lab investigation may be longer for the first few labs your students conduct but will grow shorter as your students become familiar with the practices used in the model (e.g., argumentation, designing investigations, writing reports). We therefore provide suggestions about which stages of ADI you should be able to complete in a 50-minute class period.

Please note that lab days do *not* have to be completed on consecutive school days. Some of the lab stages do not take an entire class period to complete, especially once students are acclimated to the ADI instructional model. In other cases, teachers can allow students to have more than one night to complete their work (such as writing the investigation report), especially in early laboratories.

Materials and Preparation

This section of the Teacher Notes describes the lab supplies (i.e., consumables and equipment) and instructional materials (e.g., Lab Handout, investigation proposal, and peer-review guide) needed to implement the lab activity. The lab supplies listed in this section are designed for one group, but multiple groups can share if resources are scarce. We have also included specific suggestions for some lab supplies, because these were the supplies that worked best during the field tests. However, if needed, substitutions can be made. As always, be sure to test all new lab supplies before conducting the lab with the students, because using new materials often has unexpected consequences.

This section also describes the *setup* of the investigation, by which we mean the work that teachers must do *before* the students become involved in the investigation. Please note that some of the laboratories require preparation up to 24 hours in advance. Make sure to read this section at least two days before conducting your investigation in the classroom, to be sure you are prepared.

Safety Precautions and Laboratory Waste Disposal

This section of the Teacher Notes provides an overview of potential safety hazards, as well as safety protocols that should be followed to make the laboratory safer for students. These protocols are based on legal safety standards and better professional safety practices. Teachers should also review and follow all local policies and protocols used within their school district and/or school (e.g., the district chemical hygiene plan and the Board of Education safety policies).

Topics for the Explicit and Reflective Discussion

This section of the Teacher Notes is the conceptual heart of the laboratory. It provides an overview of important content, relevant crosscutting concepts, and facets of NOS and NOSI to discuss during the investigation. Equally important, this section provides advice for teachers about how to encourage students to reflect on ways to improve the design of their investigation in the future.

Hints for Implementing the Lab

These laboratories have been tested by many teachers many times. As a result, we have collected hints from the teachers for each stage of the ADI process. These hints are designed to help you avoid some of the pitfalls earlier teachers have experienced and make the investigation run smoothly. In some labs, the hints include tips for making the investigation safer (in addition to the safety precautions listed earlier).

Topic Connections

This section is designed to inform curriculum and lesson planning by highlighting the scientific practices, crosscutting concepts, and core ideas from the *Framework* addressed through each lab activity. This table also outlines the supporting ideas and the *Common Core State Standards* (*CCSS ELA* and *CCSS Mathematics*) that are addressed by the lab.

Instructional Materials

The instructional materials included in this book are reproducible copy masters that are designed to support students as they participate in an ADI lab activity. The materials needed for each lab include a Lab Handout, the peer-review guide, and the Checkout Questions. An investigation proposal is recommended for each lab. Some labs also include images that should be viewed in color for complete understanding; these images can be found at *www.nsta.org/adi-physicalscience.*

Lab Handout

At the beginning of each lab activity, each student should be given a copy of the Lab Handout. This handout provides information about the phenomenon that they will investigate and a guiding question for the students to answer. The handout also provides hints for students to help them design their investigation in the "Getting Started" section, information about what to include in their initial argument, and the requirements for the investigation report.

Investigation Proposal

To help students design better investigations, we have developed and included three different types of investigation proposals in this book (see Appendix 3, p. 415). These investigation proposals are optional, but we have found that students design and carry out much better investigations when they are required to fill out a proposal and then get teacher feedback about their method before they begin. We provide recommendations about which investigation proposal (A, B, or C) to use for a particular lab as part of the Teacher Notes. If a teacher decides to use an investigation proposal as part of a lab activity, we recommend providing one copy for each group. The Lab Handout for students also has a heading "Investigation Proposal Required?" that is followed by "Yes" and "No" check boxes. The teacher should be sure to have students check the appropriate box on the Lab Handout when introducing the lab activity.

Peer-Review Guide

The peer-review guide is designed to make the criteria that are used to judge the quality of an investigation report explicit. We recommend that teachers make one copy for each student and provide it to the students before they begin writing their investigation report. This will ensure that students understand how they will be evaluated. Then during the double-blind group peer-review stage of the model, each group should fill out the peer-review guide as they review the reports of their classmates (each group will need to review the reports of three different students). The reviewers should rate the report on each criterion and then provide advice to the author about ways to improve. Once the review is complete, the author needs to revise his or her report and respond to the reviewers' rating and comments in the appropriate sections in the peer-review guide. The peer-review guide should be submitted to the instructor along with the rough draft and the final version of the report for a final evaluation. To score the report, the teacher can simply fill out the "Instructor Score" column of the peer-review guide and then total the scores.

Checkout Questions

To facilitate classroom assessment, we have included a set of Checkout Questions for each lab activity. The questions target the key ideas and the NOS and NOSI concepts that are addressed in the lab. Students should complete the Checkout Questions on the same day that they turn in their final report. One handout is needed for each student. The students should complete these questions on their own. The teacher can then use the students' responses, along with the report, to determine if the students learned what they needed to during the lab, and then reteach as needed.

References

Grooms, J., P. Enderle, and V. Sampson. 2015. Coordinating scientific argumentation and the Next Generation Science Standards through argument driven inquiry. *Science Educator* 24 (1): 45–50.

National Governors Association Center for Best Practices and Council of Chief State School Officers (NGAC and CCSSO). 2010. *Common core state standards.* Washington, DC: NGAC and CCSSO.

National Research Council (NRC). 2012. *A framework for K–12 science education: Practices, crosscutting concepts, and core ideas.* Washington, DC: National Academies Press.

Sampson, V., P. Enderle, J. Grooms, and S. Witte. 2013. Writing to learn and learning to write during the school science laboratory: Helping middle and high school students develop argumentative writing skills as they learn core ideas. *Science Education* 97 (5): 643–670.

SECTION 2

Physical Sciences Core Idea 1

Matter and Its Interactions

Introduction Labs

LAB 1

Teacher Notes

Lab 1. Thermal Energy and Matter

What Happens at the Molecular Level When Thermal Energy Is Added to a Substance?

Purpose

The purpose of this lab is to *introduce* students to the relationship between kinetic energy, thermal energy, and the states of matter. This lab gives students an opportunity to track the movement of energy into and out of a system. Students will also learn about the difference between laws and theories in science and the role imagination and creativity play in the development of scientific theories.

The Content

All matter is made of atoms, and all atoms are constantly in motion. This is true for gases, liquids, and solids. The energy associated with motion is kinetic energy. The average kinetic energy for a collection of atoms or molecules has an impact on two very important properties of matter: temperature and state.

The temperature of a substance is the average kinetic energy of all the atoms or molecules in that substance. The greater the temperature of a substance, the greater the average kinetic energy of the atoms or molecules in the substance. This does not mean, however, that all the atoms or molecules have the same kinetic energy or velocity. Some are moving faster than the average, others are moving slower than the average, and some may even be moving at the average velocity.

The second property that is affected by the kinetic energy of the atoms or molecules is the *state* of matter (i.e., solid, liquid, or gas). The state of matter is determined by *intermolecular bonds,* or the attractive forces between molecules (which is different from *intramolecular bonds*—ionic and covalent bonds—that hold compounds together). In a solid, there are many intermolecular bonds between atoms. When a substance is a solid, these intermolecular bonds keep molecules close together. Thus, instead of moving freely, molecules in a solid vibrate around a fixed point (this vibration happens at the atomic scale, so we do not see solids vibrating). See Figure 1.1 for an illustration of behavior of molecules in gases, liquids, and solids.

FIGURE 1.1

Illustration of behavior of molecules in gases, liquids, and solids

Gas Liquid Solid

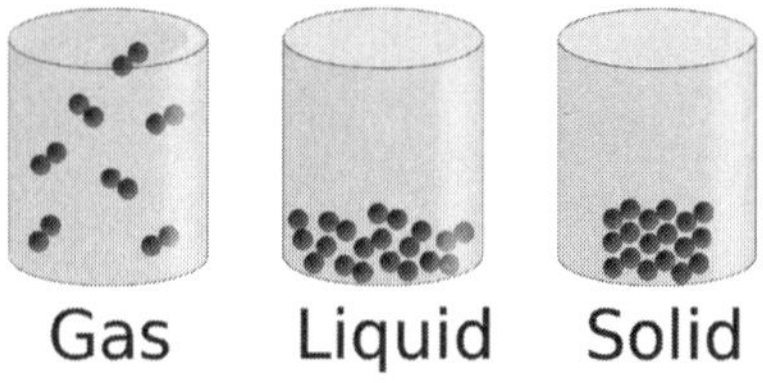

When heat energy is added to a solid, the molecules gain kinetic energy and start to vibrate more quickly. Eventually, the kinetic energy of the vibrating molecules becomes great enough to break some of the intermolecular bonds. When this happens, the substance changes from a solid to a liquid. A liquid then has groups of molecules held together by intermolecular bonds moving freely. Finally, if more heat energy is added to the substance, the molecules of the liquid will continue to gain kinetic energy and the remaining intermolecular bonds will break. When this happens, the substance will change from a liquid to a gas. See Figure 1.2 for an illustration of the transitions that take place between the various states of matter as the *enthalpy* (or total heat energy) of the system increases.

FIGURE 1.2

Transitions between the states of matter as the enthalpy of the system increases

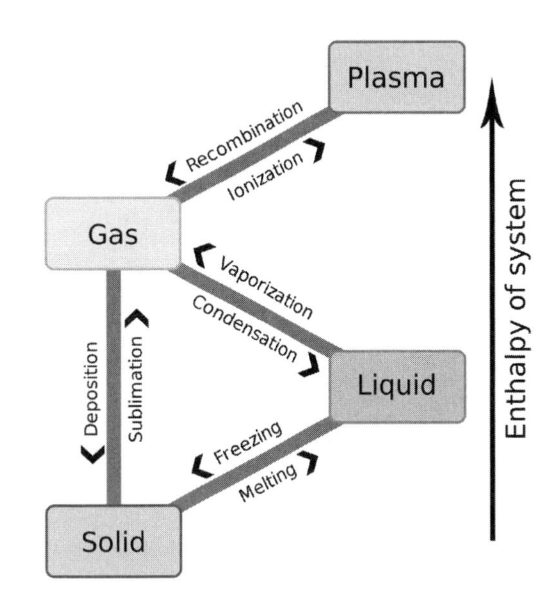

The energy required to break the intermolecular bonds of a certain substance is a constant. For water, the energy required to break the intermolecular bonds when it transitions from solid to liquid is 18.6 J/g or 334.77 J/mol (J = joule; g = gram; and mol = mole). However, the greater the mass of a substance, the greater the amount of heat energy that must be added to complete the phase change. As an example, assume that there are two samples of ice at 0°C, with one sample having a mass of 10 g and the other sample having a mass of 50 g. For the 10 g sample, the amount of heat energy required to melt the sample is 10 g × 18.6 J/g, or 186 J. For the 50 g sample, the amount of heat energy required to melt the sample is 50 g × 18.6 J/g, or 930 J. If heat energy is added to each sample at the same rate and the samples start at the same temperature, then the amount of time it will take for the ice to melt is dependent on the mass of the sample of ice.

So, what happens to a sample of ice that starts at a temperature of –10°C when heat is added to it? First, the temperature of the sample will increase from –10°C to 0°C. When the sample reaches 0°C, added heat will break the intermolecular bonds, resulting in the solid ice melting to liquid water. *During the melting process, the temperature of the water does not change.* Once all of the ice melts, adding heat to the water will once again increase the temperature of the water, this time from 0°C to 100°C. When the water reaches 100°C, adding more heat will result in the remaining intermolecular bonds breaking and the water changing phase from liquid to gas (i.e., the water will boil). *During the boiling process, the temperature of the water does not change.* Once all of the water boils, adding more heat energy will cause the temperature of the gaseous water to increase again. See Figure 1.3 (p. 30) for a graphical representation of this process.

FIGURE 1.3

Graph of heat added versus temperature of water

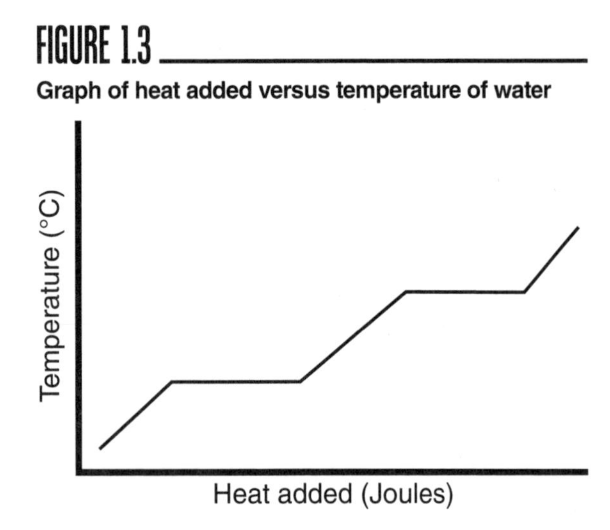

Timeline

The instructional time needed to complete this lab investigation is 230–280 minutes. Appendix 2 (p. 411) provides options for implementing this lab investigation over several class periods. Option A (280 minutes) should be used if students are unfamiliar with scientific writing, because this option provides extra instructional time for scaffolding the writing process. You can scaffold the writing process by modeling, providing examples, and providing hints as students write each section of the report. Option B (230 minutes) should be used if students are familiar with scientific writing and have developed the skills needed to write an investigation report on their own. In option B, students complete stage 6 (writing the investigation report) and stage 8 (revising the investigation report) as homework.

Materials and Preparation

The materials needed to implement this investigation are listed in Table 1.1. The equipment can be purchased from a science supply company such as Carolina, Flinn Scientific, or Ward's Science. We recommend that you use a set routine for distributing and collecting the materials during the lab investigation. For example, the consumables and equipment for each group can be set up at each group's lab station before class begins, or one member from each group can collect them from a table or a cart when needed during class.

Safety Precautions and Laboratory Waste Disposal

Remind students to follow all normal lab safety rules. In addition, tell students to take the following safety precautions:

1. Wear sanitized indirectly vented chemical-splash goggles and chemical-resistant nonlatex gloves and aprons during lab setup, hands-on activity, and takedown.
2. Never put consumables in their mouth.
3. Use caution when working with hot plates, because they can burn skin and cause fires.
4. Hot plates also need to be kept away from water and other liquids.
5. Use only GFCI-protected electrical receptacles for hot plates.
6. Clean up any spilled water immediately to avoid a slip or fall hazard.

7. Be careful when working with hot water, because it can burn skin.
8. Handle all glassware with care.
9. Handle glass thermometers with care. They are fragile and can break, causing a sharp hazard that can cut or puncture skin.
10. Never return the consumables to stock bottles.
11. Wash their hands with soap and water when they are done collecting the data.

Water can be poured down any standard drain.

TABLE 1.1

Materials list for Lab 1

Item	Quantity
Consumable	
Water	As needed per group
Equipment and other materials	
Safety glasses or goggles	1 per student
Chemical-resistant apron	1 per student
Nonlatex gloves	1 pair per student
Beaker, 50 ml	1 per group
Beaker, 250 ml	1 per group
Graduated cylinder, 10 ml	1 per group
Graduated cylinder, 25 ml	1 per group
Graduated cylinder, 100 ml	1 per group
Electronic or triple beam balance	1 per group
Hot plate	1 per group
Thermometer (or temperature probe with interface)	1 per group
Stopwatch	1 per group
Investigation Proposal C (optional)	1 per group
Whiteboard, 2' × 3'*	1 per group
Lab Handout	1 per student
Peer-review guide	1 per student
Checkout Questions	1 per student

* As an alternative, students can use a computer and presentation software, such as Microsoft PowerPoint or Apple Keynote, to create their arguments.

Topics for the Explicit and Reflective Discussion

Concepts That Can Be Used to Justify the Evidence

To provide an adequate justification of their evidence, students must explain why they included the evidence in their arguments and make the assumptions underlying their analysis and interpretation of the data explicit. In this investigation, students can use the following concepts to help justify their evidence:

- All matter is made up of atoms.
- All atoms are constantly in motion.
- All atoms and molecules have kinetic energy.
- The state of matter depends on the kinetic energy of the molecules in the substance.
- Adding heat to a substance can increase the kinetic energy of the molecules in that substance.

We recommend that you review these concepts during the explicit and reflective discussion to help students make this connection.

How to Design Better Investigations

It is important for students to reflect on the strengths and weaknesses of the investigation they designed during the explicit and reflective discussion. Students should therefore be encouraged to discuss ways to eliminate potential flaws, measurement errors, or sources of bias in their investigations. To help students be more reflective about the design of their investigation, you can ask the following questions:

1. What were some of the strengths of your investigation? What made it scientific?
2. What were some of the weaknesses of your investigation? What made it less scientific?
3. If you were to do this investigation again, what would you do to address the weaknesses in your investigation? What could you do to make it more scientific?

Crosscutting Concepts

This investigation is well aligned with three crosscutting concepts found in *A Framework for K–12 Science Education,* and you should review these concepts during the explicit and reflective discussion.

- *Scale, proportion, and quantity:* It is critical for scientists to be able to recognize what is relevant at different sizes, time frames, and scales. Scientists must also be able to recognize proportional relationships between categories or quantities. In this lab students will explain how molecular processes influence changes in a substance.

- *Systems and system models:* Defining a system under study and making a model of it are tools for developing a better understanding of natural phenomena in science. In this lab students will model a substance as it undergoes physical changes.
- *Energy and matter: Flows, cycles, and conservation:* In science it is important to track how energy and matter move into, out of, and within systems. In this lab students will track energy as it moves between a system and the surroundings when a substance experiences a phase change.

The Nature of Science and the Nature of Scientific Inquiry

This investigation is well aligned with two important concepts related to the *nature of science* (NOS) and the *nature of scientific inquiry* (NOSI), and you should review these concepts during the explicit and reflective discussion.

- *The difference between laws and theories in science:* A scientific law describes the behavior of a natural phenomenon or a generalized relationship under certain conditions; a scientific theory is a well-substantiated explanation of some aspect of the natural world. Theories do not become laws even with additional evidence; they explain laws. However, not all scientific laws have an accompanying explanatory theory. It is also important for students to understand that scientists do not discover laws or theories; the scientific community develops them over time.
- *The importance of imagination and creativity in science:* Students should learn that developing explanations for or models of natural phenomena and then figuring out how they can be put to the test of reality is as creative as writing poetry, composing music, or designing video games. Scientists must also use their imagination and creativity to figure out new ways to test ideas and collect or analyze data.

Hints for Implementing the Lab

- Allowing students to design their own procedures for collecting data gives students an opportunity to try, to fail, and to learn from their mistakes. However, you can scaffold students as they develop their procedure by having them fill out an investigation proposal. These proposals provide a way for you to offer students hints and suggestions without telling them how to do it. You can also check the proposals quickly during a class period. For this lab we suggest using Investigation Proposal C.
- Larger amounts of water will take longer to heat than small amounts. If time is a concern, you may want to provide students with multiple beakers of each size, hot plates, and stopwatches, so that they can test multiple samples at once.
- Allow the students to become familiar with the lab equipment as part of the tool talk before they begin to design their investigation. Particularly, you may want to make sure all students are familiar with how to use the thermometer or

temperature probe. This gives students a chance to see what they can and cannot do with the equipment.

- Be sure that students record actual values (e.g., temperature and time) during the data collection stage (stage 2).

Topic Connections

Table 1.2 provides an overview of the scientific practices, crosscutting concepts, disciplinary core ideas, and supporting ideas at the heart of this lab investigation. In addition, it lists the NOS and NOSI concepts for the explicit and reflective discussion. Finally, it lists literacy and mathematics skills (*CCSS ELA* and *CCSS Mathematics*) that are addressed during the investigation.

TABLE 1.2

Lab 1 alignment with standards

Scientific practices	• Asking questions and defining problems • Developing and using models • Planning and carrying out investigations • Analyzing and interpreting data • Constructing explanations and designing solutions • Engaging in argument from evidence • Obtaining, evaluating, and communicating information
Crosscutting concepts	• Scale, proportion, and quantity • Systems and system models • Energy and matter: Flows, cycles, and conservation
Core ideas	• PS1.A: Structure and properties of matter • PS3.A: Definitions of energy • PS3.B: Conservation of energy and energy transfer
Supporting ideas	• Matter and substances • Atoms • Phase changes • Kinetic theory of matter
NOS and NOSI concepts	• Scientific laws and theories • Imagination and creativity in science
Literacy connections (*CCSS ELA*)	• *Reading:* Key ideas and details, craft and structure, integration of knowledge and ideas • *Writing:* Text types and purposes, production and distribution of writing, research to build and present knowledge, range of writing • *Speaking and listening:* Comprehension and collaboration, presentation of knowledge and ideas
Mathematics connections (*CCSS Mathematics*)	• Reason abstractly and quantitatively • Construct viable arguments and critique the reasoning of others • Use appropriate tools strategically • Attend to precision

Lab Handout

Lab 1. Thermal Energy and Matter

What Happens at the Molecular Level When Thermal Energy Is Added to a Substance?

Introduction

Every substance in the universe is made up of matter. A substance can exist in three different states: solid, liquid, or gas. A substance such as water can easily transition from one state of matter to the other. For example, water transitions from a solid state to a liquid state when an ice cube melts (Figure L1.1). The ice cube is able to melt and transition from a solid to a liquid because it absorbs thermal energy. Thermal energy is a type of energy that is transferred between two objects because they have different temperatures. In the example of an ice cube melting, thermal energy is transferred to the ice cube from the warm air surrounding it. Thermal energy always moves from the warmer object to the colder object. Think about another example, such as a cold can of soda in your hand. In that case, thermal energy is transferring from your hand to the soda; eventually the cold soda will gain enough thermal energy that it becomes the same temperature as its surroundings.

All substances, regardless of whether they are a solid, a liquid, or a gas, are made up of atoms and molecules. Atoms and molecules are submicroscopic, meaning they are too small to be seen with our eyes and even too small to be seen with most microscopes. The atoms or molecules that make up a substance are constantly in motion. The composition of a substance will always be the same even though the substance can transition from one state to a different state. Water, for example, is always made of H_2O molecules even when it is in a solid (ice), liquid, or gaseous (steam) state.

FIGURE L1.1

Water undergoes a phase change from solid to liquid when ice melts.

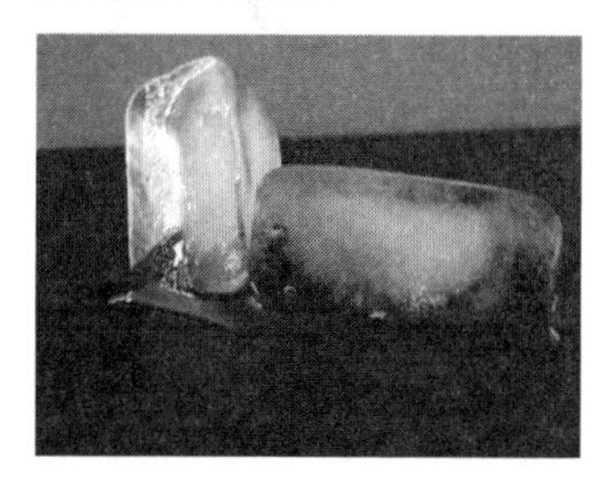

The difference between the solid, liquid, and gas states of a substance is due to the amount of kinetic energy the atoms or molecules have and how these particles are moving relative to each other. Kinetic energy is the energy of motion. Atoms or molecules that are moving quickly have more kinetic energy than atoms or molecules that move more slowly. For example, the molecules found within a sample of gaseous water move around quickly. These molecules therefore have a lot of kinetic energy. The molecules that are found in a sample of solid water, in contrast, move around slower and have less kinetic energy than the molecules in a sample of gaseous water. You can therefore measure the temperature of a substance to learn about the average kinetic energy of the molecules within that substance.

At this point, we have established several key ideas about the nature of matter. For example, we know that all matter can exist in three different states and all matter is composed of atoms or molecules that are really small. We also know that a substance has the same composition regardless of its state and that the atoms or molecule of a substance will have different amounts of kinetic energy at different temperatures. These ideas, when taken together, can serve as a foundation for the development of an explanatory model that can be used to illustrate what happens at the molecular level when thermal energy is added to a substance. This type of model is important to develop because explanatory models can help us predict the behavior of matter under different conditions. For example, we could use an explanatory model to help us predict how long it will take a substance to boil when it is heated on a hot plate or a stove. Your goal for this investigation will be to collect data about the behavior of a substance when thermal energy is added to it and then use what you learn to develop an explanatory model that describes what happens to the molecules that make up a substance when they are exposed to thermal energy.

Your Task

Develop a model that helps you explain what happens at the molecular level as thermal energy is added to a substance. The substance you will work with during this investigation is water. Your model should account for the mass of the substance, its temperature, and the amount of time that thermal energy is being added to the substance so that you can explain the relationship between the amount of water in a sample, the temperature at which the sample boils, and how long it takes to reach the boiling temperature. Your model, once fully developed, should enable you to make accurate predictions about the amount of time it will take for a particular sample of water to boil. Once you have developed your model, you will need to test it to determine if it leads to accurate predictions or not.

The guiding question of this investigation is, **What happens at the molecular level when thermal energy is added to a substance?**

Materials

You may use any of the following materials during your investigation:

Consumable
- Water

Equipment
- Beakers (various sizes)
- Graduated cylinders (various sizes)
- Electronic or triple beam balance
- Hot plate
- Thermometer or temperature probe
- Stopwatch
- Safety glasses or goggles
- Chemical-resistant apron
- Nonlatex gloves

Safety Precautions

Follow all normal lab safety rules. In addition, take the following safety precautions:

1. Wear sanitized indirectly vented chemical-splash goggles and chemical-resistant nonlatex gloves and aprons during lab setup, hands-on activity, and takedown.

2. Never put consumables in your mouth.
3. Use caution when working with hot plates, because they can burn skin and cause fires.
4. Hot plates also need to be kept away from water and other liquids.
5. Use only GFCI-protected electrical receptacles for hot plates.
6. Clean up any spilled water immediately to avoid a slip or fall hazard.
7. Be careful when working with hot water, because it can burn skin.
8. Handle all glassware with care.
9. Handle glass thermometers with care. They are fragile and can break, causing a sharp hazard that can cut or puncture skin.
10. Never return the consumables to stock bottles.
11. Wash hands with soap and water after completing the lab activity.

Investigation Proposal Required? ☐ Yes ☐ No

Getting Started

The first step in developing your model is to design and carry out an investigation to determine how long it takes for different samples of water to boil and the temperature at which each sample boils. To accomplish this task, you can heat a sample of water in a beaker on a hot plate (see Figure L1.2). Before you begin to heat different samples of water, you must determine what type of data you need to collect, how you will collect it, and how you will analyze it.

FIGURE L1.2

A sample of water can be heated on a hot plate.

To determine *what type of data you need to collect,* think about the following questions:

- What information do you need to make your model?
- What measurements will you take during your investigation?
- How will you know how much thermal energy has been transferred to your samples of water?

To determine *how you will collect the data,* think about the following questions:

- What equipment will you use to collect the data you need?
- How will you make sure that your data are of high quality (i.e., how will you reduce error)?
- How will you keep track of the data you collect?

- How will you organize your data?

To determine *how you will analyze the data,* think about the following questions:

- What type of calculations will you need to make?
- What type of table or graph could you create to help make sense of your data?

Once you have carried out your investigations, your group will need to develop a model that can be used to help explain what is happening at the molecular level when thermal energy is added to water. Your model must include the relationship between the amount of water being heated, the temperature at which that sample of water boils, and how long it takes the sample to reach the boiling temperature. Your model should also be able to account for any differences in the mass of water, differences in initial temperature between samples, and the amount of time that thermal energy is added to the substance.

The last step in this investigation is to test your model. To accomplish this goal, you can heat different amounts of water that you did not investigate to determine if your model leads to accurate predictions about the time it takes for each particular sample of water to boil. If you are able to use your model to make accurate predictions about the time it takes for different amounts of water to boil, then you will be able to generate the evidence you need to convince others that the model you developed is valid.

Connections to Crosscutting Concepts, the Nature of Science, and the Nature of Scientific Inquiry

As you work through your investigation, be sure to think about

- how scientists need to be able to recognize what is relevant at different scales;
- how scientists often need to track how energy moves into, out of, and within a system;
- the difference between laws and theories in science; and
- how scientists must use imagination and creativity when developing models and explanations.

Initial Argument

Once your group has finished collecting and analyzing your data, your group will need to develop an initial argument. Your argument needs to include a *claim, evidence* to support your claim, and a *justification* of the evidence. The claim is your group's answer to the guiding question. The evidence is an analysis and interpretation of your data. Finally, the justification of the evidence is why your group thinks the evidence matters. The justification of the evidence is important because scientists can use different kinds of evidence to support their claims. Your group will create your initial argument on a whiteboard.

Your whiteboard should include all the information shown in Figure L1.3.

FIGURE L1.3

Argument presentation on a whiteboard

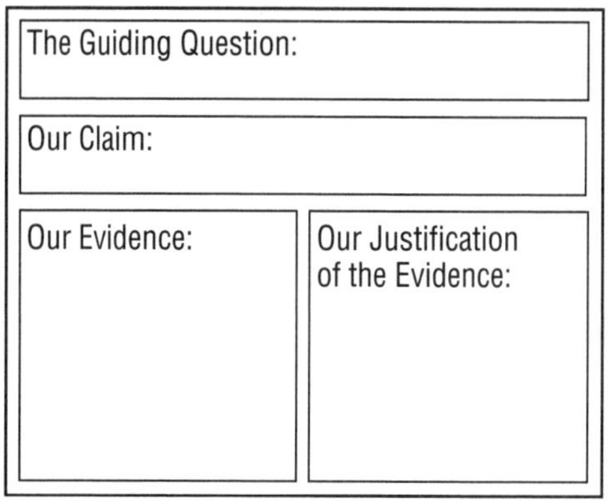

The Guiding Question:	
Our Claim:	
Our Evidence:	Our Justification of the Evidence:

Argumentation Session

The argumentation session allows all of the groups to share their arguments. One member of each group will stay at the lab station to share that group's argument, while the other members of the group go to the other lab stations to listen to and critique the arguments developed by their classmates. This is similar to how scientists present their arguments to other scientists at conferences. If you are responsible for critiquing your classmates' arguments, your goal is to look for mistakes so these mistakes can be fixed and they can make their argument better. The argumentation session is also a good time to think about ways you can make your initial argument better. Scientists must share and critique arguments like this to develop new ideas.

To critique an argument, you might need more information than what is included on the whiteboard. You will therefore need to ask the presenter lots of questions. Here are some good questions to ask:

- How did you collect your data? Why did you use that method? Why did you collect those data?
- What did you do to make sure the data you collected are reliable? What did you do to decrease measurement error?
- How did your group analyze the data? Why did you decide to do it that way? Did you check your calculations?
- Is that the only way to interpret the results of your analysis? How do you know that your interpretation of your analysis is appropriate?
- Why did your group decide to present your evidence in that way?
- What other claims did your group discuss before you decided on that one? Why did your group abandon those alternative ideas?
- How confident are you that your claim is valid? What could you do to increase your confidence?

Once the argumentation session is complete, you will have a chance to meet with your group and revise your initial argument. Your group might need to gather more data or design a way to test one or more alternative claims as part of this process. Remember, your goal at this stage of the investigation is to develop the most acceptable and valid answer to the research question!

Report

Once you have completed your research, you will need to prepare an *investigation report* that consists of three sections. Each section should provide an answer to the following questions:

1. What question were you trying to answer and why?
2. What did you do to answer your question and why?
3. What is your argument?

Your report should answer these questions in two pages or less. The report must be typed, and any diagrams, figures, or tables should be embedded into the document. Be sure to write in a persuasive style; you are trying to convince others that your claim is acceptable or valid!

Checkout Questions

Lab 1. Thermal Energy and Matter

What Happens at the Molecular Level When Thermal Energy Is Added to a Substance?

The image below shows three beakers filled with water. Each beaker is sitting on a different hot plate, and each hot plate is set to a specific temperature. Draw a model inside each circle that explains the behavior of the water molecules in each beaker.

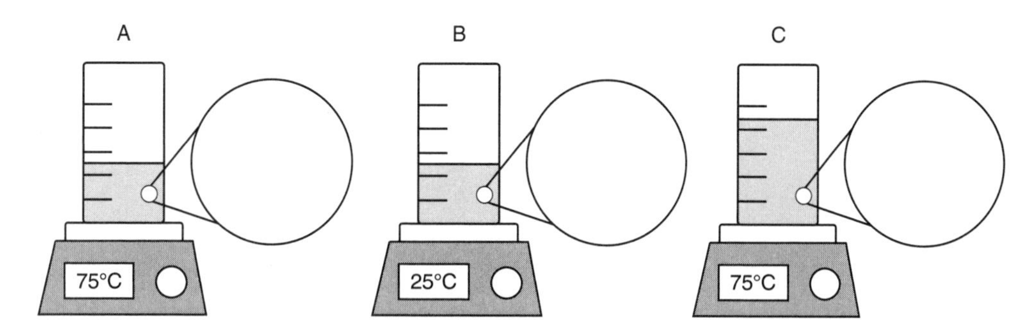

1. Explain your model. Why did you draw it that way?

2. A scientist has two beakers of water. As shown in the figure at right, one beaker has 50 ml of water at 75°C and the other beaker has 75 ml of water at 25°C. She then mixes the water together in a third beaker.

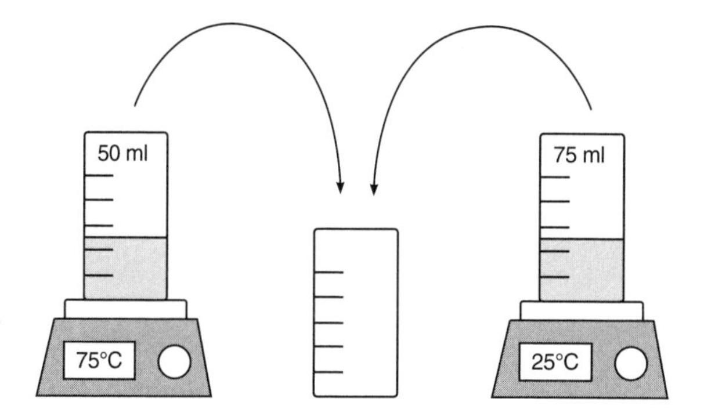

   What will be the temperature of the water after they are mixed?

   a. > 45°C
   b. 45°C
   c. < 45°C
   d. Unsure

Explain your answer. What rule did you use to make your decision?

3. A scientific law describes the behavior of a natural phenomenon or a generalized relationship under certain conditions; a scientific theory is a well-substantiated explanation.
   a. I agree with this statement.
   b. I disagree with this statement.

   Explain your answer, using an example from your investigation about thermal energy.

4. Scientists need to be creative and have a good imagination.
   a. I agree with this statement.
   b. I disagree with this statement.

   Explain your answer, using an example from your investigation about thermal energy.

5. Scientists often need to be able to recognize proportional relationships between groups or quantities. Explain why recognizing a proportional relationship is important, using an example from your investigation about thermal energy.

6. It is often important to track how matter flows into, out of, and within a system during an investigation. Explain why it is important to keep track of matter when studying a system, using an example from your investigation about thermal energy.

Teacher Notes

Lab 2. Chemical and Physical Changes

What Set of Rules Should We Use to Distinguish Between Chemical and Physical Changes in Matter?

Purpose

The purpose of this lab is to *introduce* students to the types of changes that matter can undergo. Through this activity, students will have an opportunity to explore the crosscutting concepts of the importance of identifying patterns in science and how changes affect the stability of chemical systems. Students will also learn about the difference between observations and inferences and how scientists use creativity and imagination.

The Content

Matter, the "stuff" of which the universe is composed, has two characteristics: it has mass and it occupies space. Matter can go through changes in both its physical and chemical properties. *Physical properties* of matter include qualities such as density, odor, color, melting point, boiling point, physical type (liquid, gas, or solid), and magnetism. These properties are often useful for identifying different substances. *Chemical properties* of matter involve the structure of the atoms or molecules that make up the matter.

All matter is composed of submicroscopic particles called atoms. The way an atom or molecule is shaped depends on the elements present. The structure of atoms affects how they react with other types of atoms. When atoms react with each other, they switch or share electrons. This can lead to forming ions or molecules. When molecules react with each other, they can trade off the atoms and become completely different molecules.

A *substance* is a sample of matter that has a constant composition. The physical and chemical properties of a substance, as a result, are the same throughout the sample. Scientists use atomic composition and specific physical or chemical properties to distinguish between different substances. The atomic composition of a substance refers to the different types of atoms found in it and the relative proportion of each type of atom. Water, for example, consists of atoms of hydrogen and oxygen in a ratio of two hydrogen atoms for every one oxygen atom. The physical and chemical properties of a substance refer to measurable or observable qualities or attributes that are used to distinguish between different substances. Scientists can identify substances based on their physical and chemical properties because every type of substance has a unique set of physical and chemical properties that reflect the unique atomic composition of that substance.

It is important to note, however, that the physical properties of a substance can change when environmental conditions change. Water at 25°C is a liquid, but when its temperature drops below 0°C it is a solid. This temperature change did not create a different substance, because the atomic composition of the water did not change. The only difference in liquid and solid water is how the molecules of water are oriented relative to each other. In liquid form, the water molecules tumble around each other, whereas water molecules are in a fixed position and can only vibrate when the water is in a solid form.

Water changing from a liquid to a solid when it is cooled to a temperature below 0°C is an example of a *physical change.* A physical change occurs when the appearance of a substance is changed but its atomic composition remains constant. Any change in a substance that involves a rearrangement of how the atoms within that substance are bonded together, in contrast, is called a *chemical change.* A chemical change results in a change in the atomic composition of a substance. It is therefore important for scientists to be aware of the difference between a physical and chemical change when they attempt to identify a substance based on its physical and chemical properties. It is also important for them to take into account environmental conditions, such as temperature or pressure, because the physical properties of a substance will differ based on environmental conditions.

Timeline

The instructional time needed to complete this lab investigation is 170–230 minutes. Appendix 2 (p. 411) provides options for implementing this lab investigation over several class periods. Option C (230 minutes) should be used if students are unfamiliar with scientific writing, because this option provides extra instructional time for scaffolding the writing process. You can scaffold the writing process by modeling, providing examples, and providing hints as students write each section of the report. Option D (170 minutes) should be used if students are familiar with scientific writing and have developed the skills needed to write an investigation report on their own. In option D, students complete stage 6 (writing the investigation report) and stage 8 (revising the investigation report) as homework.

Materials and Preparation

The materials needed to implement this investigation are listed in Table 2.1 (p. 46). Most of the consumable materials can be purchased from a grocery store. We recommend that you use a set routine for distributing and collecting the materials during the lab investigation. For example, the consumables and equipment for each group can be set up at each group's lab station before class begins, or one member from each group can collect them from a table or a cart when needed during class.

TABLE 2.1

Materials list for Lab 2

Item	Quantity
Consumables	
Water (in squirt bottle)	1 bottle per group
Baking soda	5 g per group
Calcium chloride	5 g per group
Sugar	5 g per group
Vinegar (acetic acid)	50 ml per group
Antacid tablets	4 per group
Steel wool	1 piece per group
Paper, 8½" × 11"	5 sheets per group
Candle (standard size such as might be used on a dinner table)	1 per group
Equipment and other materials	
Safety glasses or goggles	1 per student
Chemical-resistant apron	1 per student
Nonlatex gloves	1 pair per student
Electronic or triple beam balance	1 per group
Hot plate	1 per class
Beaker, 50 ml	1 per group
Beaker, 250 ml	1 per group
Beaker, 400 ml	1 per group
Graduated cylinder, 10 ml	1 per group
Graduated cylinder, 25 ml	1 per group
Graduated cylinder, 100 ml	1 per group
Disposable thin-stem pipettes, 4.0–7.0 ml capacity	3 per group
Spatulas	2 per group
Matches	1 box per group
Mortar and pestle	1 per group
pH paper	10 per group
Conductivity probe	1 per group
Well plate	1 per group
Investigation Proposal C (optional)	1 per group
Whiteboard, 2' × 3'*	1 per group
Lab Handout	1 per student
Peer-review guide	1 per student
Checkout Questions	1 per student

*As an alternative, students can use computer and presentation software such as Microsoft PowerPoint or Apple Keynote to create their arguments.

Safety Precautions and Laboratory Waste Disposal

Remind students to follow all normal lab safety rules. In addition, tell the students to take the following safety precautions:

1. Review salient information on safety data sheets for hazardous chemicals with students.
2. Wear sanitized indirectly vented chemical-splash goggles and chemical-resistant nonlatex gloves and aprons during lab setup, hands-on activity, and takedown.
3. Never put consumables in their mouth.
4. Use caution when working with hot plates or candles. They can burn skin and cause fires.
5. Hot plates also need to be kept away from water and other liquids.
6. Use only GFCI-protected electrical receptacles for hot plates.
7. Clean up any spilled liquid immediately to avoid a slip or fall hazard.
8. Use caution when working with hazardous chemicals in this lab that are corrosive and/or toxic.
9. Handle all glassware with care.
10. Use caution in handling steel wool, because it can cause metal slivers.
11. Never return the consumables to stock bottles.
12. Wash hands with soap and water after completing the lab activity.

All solutions may be rinsed down the drain with excess water according to Flinn suggested disposal method 26b. Information about laboratory waste disposal methods is included in the Flinn catalog and reference manual; you can request a free copy at *www.flinnsci.com/forms/please-request-a-catalog-most-appropriate-for-your-needs.aspx.*

Topics for the Explicit and Reflective Discussion

Concepts That Can Be Used to Justify the Evidence

To provide an adequate justification of their evidence, students must explain why they included the evidence in their arguments and make the assumptions underlying their analysis and interpretation of the data explicit. In this investigation, students can use the following concepts to help justify their evidence:

- Atomic theory of matter
- Physical and chemical properties

- Law of definite proportions
- Physical and chemical changes

We recommend that you review these concepts during the explicit and reflective discussion to help students make this connection.

How to Design Better Investigations

It is important for students to reflect on the strengths and weaknesses of the investigation they designed during the explicit and reflective discussion. Students should therefore be encouraged to discuss ways to eliminate potential flaws, measurement errors, or sources of bias in their investigations. To help students be more reflective about the design of their investigation, you can ask the following questions:

1. What were some of the strengths of your investigation? What made it scientific?
2. What were some of the weaknesses of your investigation? What made it less scientific?
3. If you were to do this investigation again, what would you do to address the weaknesses in your investigation? What could you do to make it more scientific?

Crosscutting Concepts

This investigation is well aligned with two crosscutting concepts found in *A Framework for K–12 Science Education*, and you should review these concepts during the explicit and reflective discussion.

- *Patterns*. Scientists look for patterns in nature and attempt to understand the underlying cause of these patterns. In this lab students will investigate patterns that can be used to identify chemical or physical changes.
- *Stability and change*. It is critical for scientists to understand what makes a system stable or unstable and what controls rates of change in system. In this lab students will investigate how systems change during chemical or physical processes.

The Nature of Science and the Nature of Scientific Inquiry

This investigation is well aligned with two important concepts related to the *nature of science* (NOS) and the *nature of scientific inquiry* (NOSI), and you should review these concepts during the explicit and reflective discussion.

- *The difference between observations and inferences in science:* An observation is a descriptive statement about a natural phenomenon, whereas an inference is an interpretation of an observation. Students should also understand that current scientific knowledge and the perspectives of individual scientists guide both observations and inferences. Thus, different scientists can have different but equally valid interpretations of the same observations due to differences in their perspectives and background knowledge.

- *The importance of imagination and creativity in science:* Students should learn that developing explanations for or models of natural phenomena and then figuring out how they can be put to the test of reality is as creative as writing poetry, composing music, or designing video games. Scientists must also use their imagination and creativity to figure out new ways to test ideas and collect or analyze data.

Hints for Implementing the Lab

- Allowing students to design their own procedures for collecting data gives students an opportunity to try, to fail, and to learn from their mistakes. However, you can scaffold students as they develop their procedure by having them fill out an investigation proposal. These proposals provide a way for you to offer students hints and suggestions without telling them how to do it. You can also check the proposals quickly during a class period. For this lab we suggest using Investigation Proposal C.
- Be sure to encourage groups to collect multiple types of data (e.g., mass, changes in color, presence of bubbles or gas produced) to develop a more complete set of evidence. Such work offers the opportunity to discuss the difference between data and evidence explicitly with students. Furthermore, using the type of analysis and comparison also provides opportunities to distinguish between observations (in this activity, the various data collected) and inferences (the rule set for changes is inferred through comparison of observations).
- Have students identify the set of changes they will be examining so you can review them and make sure they have a good mixture of both chemical and physical changes.
- Emphasize that students will need evidence to support the rules they develop. Robust evidence for such rules should involve more than one example of a change.

Topic Connections

Table 2.2 (p. 50) provides an overview of the scientific practices, crosscutting concepts, disciplinary core ideas, and supporting ideas at the heart of this lab investigation. In addition, it lists the NOS and NOSI concepts for the explicit and reflective discussion. Finally, it lists literacy and mathematics skills (*CCSS ELA* and *CCSS Mathematics*) that are addressed during the investigation.

TABLE 2.2

Lab 2 alignment with standards

Scientific practices	• Asking questions and defining problems • Planning and carrying out investigations • Analyzing and interpreting data • Using mathematics and computational thinking • Constructing explanations and designing solutions • Engaging in argument from evidence • Obtaining, evaluating, and communicating information
Crosscutting concepts	• Patterns • Stability and change
Core idea	• PS1.A: Structure and properties of matter
Supporting ideas	• Atomic theory of matter • Physical and chemical properties • Law of definite proportions • Physical and chemical changes
NOS and NOSI concepts	• Observations and inferences • Imagination and creativity in science
Literacy connections (*CCSS ELA*)	• *Reading:* Key ideas and details, craft and structure, integration of knowledge and ideas • *Writing:* Text types and purposes, production and distribution of writing, research to build and present knowledge, range of writing • *Speaking and listening:* Comprehension and collaboration, presentation of knowledge and ideas
Mathematics connections (*CCSS Mathematics*)	• Reason abstractly and quantitatively • Construct viable arguments and critique the reasoning of others • Use appropriate tools strategically • Attend to precision

Lab Handout

Lab 2. Chemical and Physical Changes

What Set of Rules Should We Use to Distinguish Between Chemical and Physical Changes in Matter?

Introduction

Matter has mass and occupies space. Although these two basic characteristics of matter are shared among all the different kinds of matter in the universe, most kinds have even more characteristics that are special to only certain groups or types. Scientists work to understand the special characteristics of the many kinds of matter so that they can also understand what happens when different kinds of matter interact with each other.

Every kind of matter, such as wood, steel, or water, has a unique set of physical and chemical properties. Physical properties of matter include qualities or attributes such as density, odor, color, melting point, boiling point, and magnetism. These properties are often useful for identifying different types of substances. Chemical properties of matter, in contrast, describe how matter interacts with other types of matter. For example, when a metal is added to an acid, it reacts with the acid to form a gas. A chemical property of metal, therefore, is reactivity with acids. All substances have a specific set of physical and chemical properties. Scientists can therefore use physical and chemical properties to identify an unknown substance.

The matter that is around us changes all of the time. Natural events, for example, can change matter in a variety of ways, such as when a forest fire turns wood into ash or when wind breaks down a rock to produce small particles of dust. Another example of a change in matter is when a solid turns into a liquid, which is what happens when a crayon melts (see Figure L2.1). Scientists classify changes in matter as either a chemical change or a physical change. A chemical change is defined as a change in the composition and properties of a substance. Chemical changes involve the rearrangement of molecules or atoms and result in the production of one or more new substances. A physical change is defined as a change in the state or energy of matter. A physical change does not result in the production of a new substance, although the starting and ending materials may look very different from each other.

FIGURE L2.1
A crayon melting

Matter will look very different after it goes through a chemical change because a chemical change transforms a substance into one or more new substances. The appearance of matter, however, can also change when it goes through a physical change. In addition, we cannot see what happens at the level of atoms as matter goes through a change. It is therefore often difficult to tell the difference between a chemical change and a physical change by just observing the appearance of a substance. Scientists, as a result, have developed several rules to help them classify changes in matter. In this investigation, you will have an opportunity to develop a set of rules that you can use to determine if a change in matter should be classified as a chemical change or a physical change.

Your Task

Use what you know about physical and chemical properties of matter, stability and change, and how to design and carry out an investigation to develop a set of rules you can use to distinguish between a chemical change and a physical change in matter.

The guiding question of this investigation is, **What set of rules should we use to distinguish between chemical and physical changes in matter?**

Materials

You may use any of the following materials during your investigation:

Consumables
- Water
- Baking soda
- Calcium chloride
- Sugar
- Vinegar
- Antacid tablets
- Steel wool
- Paper
- Candle

Equipment
- Well plate
- Beakers (various sizes)
- Electronic or triple beam balance
- Graduated cylinders (various sizes)
- Pipettes
- Spatulas
- Matches
- Mortar and pestle
- pH paper
- Conductivity probe
- Safety glasses or goggles
- Chemical-resistant apron
- Nonlatex gloves

Safety Precautions

Follow all normal lab safety rules. Your teacher will provide important information about working with the chemicals associated with this investigation. In addition, take the following safety precautions:

1. Follow safety precautions noted on safety data sheets for hazardous chemicals.
2. Wear sanitized indirectly vented chemical-splash goggles and chemical-resistant nonlatex gloves and aprons during lab setup, hands-on activity, and takedown.
3. Never put consumables in your mouth.

4. Use caution when working with hot plates or candles. They can burn skin and cause fires.
5. Hot plates also need to be kept away from water and other liquids.
6. Use only GFCI-protected electrical receptacles for hot plates.
7. Clean up any spilled liquid immediately to avoid a slip or fall hazard.
8. Use caution when working with hazardous chemicals in this lab that are corrosive and/or toxic.
9. Handle all glassware with care.
10. Use caution in handling steel wool, because it can cause metal slivers.
11. Never return the consumables to stock bottles.
12. Wash hands with soap and water after completing the lab activity.

Investigation Proposal Required? ☐ Yes ☐ No

Getting Started

The first step in this investigation is to identify all the various physical properties of matter that are possible to observe or measure using the available materials. Once you know what physical properties you can observe or measure during this investigation, you can then start collecting data about what happens to the physical properties of matter when it goes through a physical or a chemical change. Be sure to observe or measure several different physical properties of the matter you are using before and after the change takes place. Listed below are some examples of actions to cause physical and chemical changes that you can use to document the physical properties of the matter before and after it goes through a change.

Actions to cause a physical change

- Grind an antacid tablet.
- Cut a piece of steel wool to its smallest size.
- Put 1 g of candle wax in a test tube and heat the test tube in a hot-water bath.
- Add 1 g of sugar to 5 ml water.

Actions to cause a chemical change

- Add 1 g of baking soda to 5 ml of vinegar.
- Add 1 g of steel wool to 5 ml of vinegar.
- Add 1 g of antacid tablet to 5 ml water.
- Add 1 g of sugar in a test tube and then heat it using a candle.

The second step in this investigation is to develop a set of rules that you can use to determine if a change in matter is a physical one or a chemical one. Once you have a set of rules, you will need to test them to determine if they allow you to accurately identify a physical or chemical change in matter. It is important for you to test your rules because the results of your test will allow you to demonstrate that your rules are not only valid but also a useful way to identify a change in matter. Be sure to modify your rules as needed if they do not allow you to accurately classify a change in matter. To accomplish this final step of your investigation, you can test your rules using the following examples of actions to cause physical and chemical changes:

Actions to cause a physical change

- Add 1 ml of water to a balloon and tie the balloon so the water inside the balloon cannot escape. Then place the balloon in a microwave and heat the water (in the balloon) for 30 seconds.
- Mix 1 g of calcium chloride and 1 g of baking soda.
- Mix 1 g of crushed antacid and 1 g of sugar.

Actions to cause a chemical change

- Add 1 g of calcium chloride to 5 ml of water.
- Add 1 g of baking soda to 5 ml of water.
- Add 1 g of calcium chloride and 1 g of baking soda to 5 ml of water.

Remember, if you can use your rules to classify these changes in matter correctly, then you will be able to generate the evidence you need to convince others that the rules that you developed are valid and useful.

Connections to Crosscutting Concepts, the Nature of Science, and the Nature of Scientific Inquiry

As you work through your investigation, be sure to think about

- the importance of identifying patterns,
- stability and change in nature,
- the difference between observations and inferences, and
- the role of imagination and creativity in science.

Initial Argument

Once your group has finished collecting and analyzing your data, your group will need to develop an initial argument. Your argument needs to include a *claim*, *evidence* to support your claim, and a *justification* of the evidence. The claim is your group's answer to the guiding question. The evidence is an analysis and interpretation of your data. Finally, the justification of the evidence is why your group thinks the evidence matters. The justification of the evidence is important because scientists can use different kinds of evidence to support their claims. Your group will create your initial argument on a whiteboard. Your whiteboard should include all the information shown in Figure L2.2.

FIGURE L2.2

Argument presentation on a whiteboard

The Guiding Question:	
Our Claim:	
Our Evidence:	Our Justification of the Evidence:

Argumentation Session

The argumentation session allows all of the groups to share their arguments. One member of each group will stay at the lab station to share that group's argument, while the other members of the group go to the other lab stations to listen to and critique the arguments developed by their classmates. This is similar to how scientists present their arguments to other scientists at conferences. If you are responsible for critiquing your classmates' arguments, your goal is to look for mistakes so these mistakes can be fixed and they can make their argument better. The argumentation session is also a good time to think about ways you can make your initial argument better. Scientists must share and critique arguments like this to develop new ideas.

To critique an argument, you might need more information than what is included on the whiteboard. You will therefore need to ask the presenter lots of questions. Here are some good questions to ask:

- How did you collect your data? Why did you use that method? Why did you collect those data?
- What did you do to make sure the data you collected are reliable? What did you do to decrease measurement error?
- How did your group analyze the data? Why did you decide to do it that way? Did you check your calculations?
- Is that the only way to interpret the results of your analysis? How do you know that your interpretation of your analysis is appropriate?
- Why did your group decide to present your evidence in that way?
- What other claims did your group discuss before you decided on that one? Why did your group abandon those alternative ideas?
- How confident are you that your claim is valid? What could you do to increase your confidence?

Once the argumentation session is complete, you will have a chance to meet with your group and revise your initial argument. Your group might need to gather more data or design a way to test one or more alternative claims as part of this process. Remember, your goal at this stage of the investigation is to develop the most acceptable and valid answer to the research question!

Report

Once you have completed your research, you will need to prepare an *investigation report* that consists of three sections. Each section should provide an answer to the following questions:

1. What question were you trying to answer and why?
2. What did you do to answer your question and why?
3. What is your argument?

Your report should answer these questions in two pages or less. This report must be typed, and any diagrams, figures, or tables should be embedded into the document. Be sure to write in a persuasive style; you are trying to convince others that your claim is acceptable and valid!

Checkout Questions

Lab 2. Chemical and Physical Changes

What Set of Rules Should We Use to Distinguish Between Chemical and Physical Changes in Matter?

1. What is a physical change in matter?

2. What is a chemical change in matter?

3. A scientist has a collection of substances, both solids and liquids, that she mixes in different combinations. For each mixture, she puts a smaller amount of solid into a larger amount of liquid. She is trying to determine if mixing these substances produces any physical or chemical changes.

Mixture	Solid	Liquid	Observation after mixing
A	Sugar—white crystals	Water—clear	Clear solution
B	Sugar—white crystals	Vinegar—clear	Clear solution
C	Baking soda—white powder	Water—clear	Clear solution
D	Baking soda—white powder	Vinegar—clear	Clear solution with bubbles

a. Which mixture(s) involve only a physical change?

b. How do you know?

c. Which mixture(s) involve only a chemical change?

d. How do you know?

4. Scientists do not use creativity or imagination when they are investigating the physical world.

    a. I agree with this statement.
    b. I disagree with this statement.

    Explain your answer, using an example from your investigation about physical and chemical changes.

5. The result of mixing vinegar and baking soda is a chemical change.

    a. I agree with this statement.
    b. I disagree with this statement.

    Explain your answer, using an example from your investigation about physical and chemical changes.

6. Scientists often need to look for patterns that occur in the data they collect and analyze. Explain why identifying patterns is important, using an example from your investigation about physical and chemical changes.

7. Physical systems will become stable over time after experiencing a period of change. Explain why it is important to understand how a system stabilizes after experiencing change, using an example from your investigation about physical and chemical changes.

Application Labs

Teacher Notes

Lab 3. Physical Properties of Matter

What Are the Identities of the Unknown Substances?

Purpose

The purpose of this lab is for students to *apply* what they have learned about physical properties of matter to identify a set of unknown substances. Through this activity, students will have an opportunity to explore the role of patterns and of scale, proportion, and quantity in scientific investigations. Students will also learn about the difference between data and evidence and how scientists use different methods to answer different types of questions.

The Content

Matter is something that has mass and takes up space. All matter is composed of submicroscopic particles called atoms. A *substance* is a sample of matter that has a constant composition. The physical and chemical properties of a substance, as a result, are the same throughout the sample. Scientists use atomic composition and specific physical or chemical properties to distinguish between different substances.

The *atomic composition* of a substance refers to the different types of atoms found in it and the relative proportion of each type of atom. Water, for example, consists of atoms of hydrogen and oxygen in a ratio of two hydrogen atoms for every one oxygen atom. The physical and chemical properties of a substance refer to measurable or observable qualities or attributes that are used to distinguish between different substances. *Physical properties* are descriptive characteristics of matter. Examples of physical properties include color, density, conductivity, and malleability. *Chemical properties,* in contrast, describe how a substance interacts with other matter. For example, sodium and potassium react with water, but aluminum and gold do not. Scientists can identify substances based on their physical and chemical properties because every type of substance has a unique set of physical and chemical properties that reflect the unique atomic composition of that substance.

It is important to note, however, that the physical properties of a substance can change when environmental conditions change. Water at 25°C is a liquid, but when its temperature drops below 0°C it is a solid. This change did not create a different substance, because the atomic composition of the water did not change. The only difference between liquid and solid water is how the molecules of water are oriented relative to each other. In liquid form, the water molecules tumble around each other, whereas water molecules are in a fixed position and can only vibrate when the water is in a solid form. Water changing from a liquid to a solid when it is cooled to a temperature below 0°C is an example of a physical change.

A *physical change* occurs when the appearance of a substance is changed but its atomic composition remains constant. Any change in a substance that involves a rearrangement of how the atoms within that substance are bonded together, in contrast, is called a *chemical change.* A chemical change results in a change in the atomic composition of a substance. It is therefore important for scientists to be aware of the difference between a physical and chemical change when they attempt to identify a substance based on its physical and chemical properties. It is also important for them to take into account environmental conditions, such as temperature or pressure, because the physical properties of a substance will differ based on environmental conditions.

Timeline

The instructional time needed to complete this lab investigation is 200–230 minutes. Appendix 2 (p. 411) provides options for implementing this lab investigation over several class periods. Option C (230 minutes) should be used if students are unfamiliar with scientific writing, because this option provides extra instructional time for scaffolding the writing process. You can scaffold the writing process by modeling, providing examples, and providing hints as students write each section of the report. Option F (200 minutes) should be used if students are familiar with scientific writing and have developed the skills needed to write an investigation report on their own. In option F, students complete stage 6 (writing the investigation report) and stage 8 (revising the investigation report) as homework.

Materials and Preparation

The materials needed to implement this investigation are listed in Table 3.1 (p. 64). The consumables and equipment can be purchased from a science supply company such as Carolina, Flinn Scientific, or Ward's Science.

We recommend that the set of known substances include between 5 and 10 different materials (such as aluminum, steel, copper, brass, nylon, acrylic, pine, poplar, oak, and PVC) and the set of unknown substances include at least 4 materials (such as aluminum, steel, acrylic, and PVC). All of the materials in the set of unknown substances must be the same as materials that were included in the set of known substances. There are a number of options for creating the set of known substances and the set of unknown substances. You can purchase one or more different density sets from a science supply company and then use the materials from the kits to create your own sets. Good kits to use for this approach include the following (the shape of objects included in the kit and the item number are given in parentheses after the vendor name; item numbers are accurate at the time of writing this book but may change in the future):

- Deluxe Density Cube Set from Carolina (cubes; item 752476)
- Specific Gravity Set from Carolina (cylinders; item 752490)
- Study of Density Kit from Carolina (various shapes; item 840942)
- Density Rod Set from Carolina (cylinders; item 752480)

TABLE 3.1

Materials list for Lab 3

Item	Quantity
Consumables	
Water (in squirt bottle)	1 bottle per group
Set of known substances (should include 5–10 different materials such as aluminum, steel, copper, brass, nylon, acrylic, pine, poplar, oak, and PVC; label each material by name)	1 per group
Set of unknown substances (include 4 of the known materials but label as unknown A, B, C, and D)	1 per group
Equipment and other materials	
Safety glasses or goggles	1 per student
Chemical-resistant apron	1 per student
Nonlatex gloves	1 pair per student
Beaker, 250 ml	1 per group
Beaker, 400 ml	1 per group
Graduated cylinder, 10 ml	1 per group
Graduated cylinder, 25 ml	1 per group
Graduated cylinder, 100 ml	1 per group
Disposable thin-stem pipettes, 4.0–7.0 ml capacity	3 per group
Metric ruler	1 per group
Wire, 10 cm	3 per group
Size D battery	1 per group
Mini lightbulb	1 per group
Mini lightbulb holder	1 per group
Investigation Proposal C (optional)	1 per group
Whiteboard, 2' × 3'*	1 per group
Lab Handout	1 per student
Peer-review guide	1 per student
Checkout Questions	1 per student

*As an alternative, students can use computer and presentation software, such as Microsoft PowerPoint or Apple Keynote, to create their arguments.

- Specific Gravity Set, Square, from Carolina (rectangles; 752491)
- Measurement Challenge—a Density Super Value Guided-Inquiry Kit from Flinn Scientific (various shapes; item AP5939)
- Density Cube Set from Flinn Scientific (cubes; item AP6058)
- Specific Gravity Metal Specimens Set from Flinn Scientific (cylinders; item AP9234)
- Investigating Measurement and Density Kit from Ward's Science (various shapes; item 180222)
- Density Blocks Lab Activity from Ward's Science (cubes; item 366859)
- Specific Gravity Specimens Set from Ward's Science (various shapes; item 364001)

You can also purchase metal shot such as aluminum, brass, copper, tin, and zinc and some different plastic beads or blocks and then use these materials to create a set of known and unknown substances. Another option is to gather different substances from around the classroom to create your own sets.

Whichever approach you decide to use to create sets of known and unknown substances, it is important for the set of unknown substances to look different from the set of known substances. To make the sets look different from each other, simply change the amount or shape of each substance. For example, if the set of known substances consist of a set of cubes, then the set of unknown substances should include cylinders, rectangles, or shot.

We recommend that you use a set routine for distributing and collecting the materials during the lab investigation. For example, the consumables and equipment for each group can be set up at each group's lab station before class begins, or one member from each group can collect them from a table or a cart when needed during class.

Safety Precautions and Laboratory Waste Disposal

Remind students to follow all normal lab safety rules. In addition, tell students to take the following safety precautions:

1. Wear sanitized indirectly vented chemical-splash goggles and chemical-resistant nonlatex gloves and aprons during lab setup, hands-on activity, and takedown.
2. Clean up any spilled liquid immediately to avoid a slip or fall hazard.
3. Handle all glassware with care.
4. Lightbulbs are made of glass. Be careful handling them. If they break, clean them up immediately and place in a broken glass box.
5. Handle electrical wires with caution. They have sharp ends, which can cut or puncture skin.
6. Wash hands with soap and water after completing the lab activity.

The water can be disposed of down a drain if it is connected to a sanitation sewer system. The sets of known and unknown substances can be cleaned, dried, and stored for later use. Batteries, lightbulbs, and wire may be stored for future use. When batteries need replacing, dispose of old batteries according to manufacturer's recommendations. The box of broken glass should be discarded in accordance with local policies.

Topics for the Explicit and Reflective Discussion

Concepts That Can Be Used to Justify the Evidence

To provide an adequate justification of their evidence, students must explain why they included the evidence in their arguments and make the assumptions underlying their analysis and interpretation of the data explicit. In this investigation, students can use the following concepts to help justify their evidence.

- Matter and substances
- Atoms and atomic composition
- Physical and chemical properties
- Physical and chemical changes

We recommend that you review these concepts during the explicit and reflective discussion to help students make this connection.

How to Design Better Investigations

It is important for students to reflect on the strengths and weaknesses of the investigation they designed during the explicit and reflective discussion. Students should therefore be encouraged to discuss ways to eliminate potential flaws, measurement errors, or sources of bias in their investigations. To help students be more reflective about the design of their investigation, you can ask the following questions:

1. What were some of the strengths of your investigation? What made it scientific?
2. What were some of the weaknesses of your investigation? What made it less scientific?
3. If you were to do this investigation again, what would you do to address the weaknesses in your investigation? What could you do to make it more scientific?

Crosscutting Concepts

This investigation is well aligned with two crosscutting concepts found in *A Framework for K–12 Science Education*, and you should review these concepts during the explicit and reflective discussion.

- *Patterns:* Observed patterns guide the way scientists organize and classify substances and interactions. Scientists also explore the relationships between and the underlying causes of the patterns they observe in nature. In this lab students will identify substances by comparing observed patterns between known and unknown substances.
- *Scale, proportion, and quantity:* It is critical for scientists to be able to recognize what is relevant at different sizes, time frames, and scales. Scientists must also be able to recognize proportional relationships between categories or quantities. Physical properties often involve proportional relationships; in this lab students will use those relationships to identify unknown substances.

The Nature of Science and the Nature of Scientific Inquiry

This investigation is well aligned with two important concepts related to the *nature of science* (NOS) and the *nature of scientific inquiry* (NOSI), and you should review these concepts during the explicit and reflective discussion.

- *The difference between data and evidence in science*: Data are measurements, observations, and findings from other studies that are collected as part of an investigation. Evidence, in contrast, is analyzed data and an interpretation of the analysis.
- *Methods used in scientific investigations*: Examples of methods include experiments, systematic observations of a phenomenon, literature reviews, and analysis of existing data sets; the choice of method depends on the objectives of the research. There is no universal step-by step scientific method that all scientists follow; rather, different scientific disciplines (e.g., chemistry vs. physics) and fields within a discipline (e.g., organic vs. physical chemistry) use different types of methods, use different core theories, and rely on different standards to develop scientific knowledge.

Hints for Implementing the Lab

- Allowing students to design their own procedures for collecting data gives students an opportunity to try, to fail, and to learn from their mistakes. However, you can scaffold students as they develop their procedure by having them fill out an investigation proposal. These proposals provide a way for you to offer students hints and suggestions without telling them how to do it. You can also check the proposals quickly during a class period. For this lab we suggest using Investigation Proposal C.
- We recommend including 5–10 materials in the set of known substances and 4 materials in the set of unknown substances to foster higher-quality argumentation during the lab. The more identities that student groups have to determine, the

more opportunities there are for variation among groups that can lead to critical questioning and discussion during the argumentation session. However, if necessary for time or scheduling issues, the number of objects included in the set of known substances and the set of unknown substances can be decreased.

- The wire, battery, lightbulb, and lightbulb holder can be used to create a simple circuit. The students can then use the circuit to determine if a substance conducts electricity or not.
- Be sure to encourage groups to collect multiple types of data to develop a more complete set of evidence. Such work offers the opportunity to discuss the difference between data and evidence explicitly with students. Furthermore, using this type of analysis and comparison also provides opportunities to distinguish between observations (in this activity, the various physical property data collected) and inferences (the identity of the unknown is inferred through comparison of characteristics).
- We recommend that students use at least three different physical properties to identify the unknown objects, but you should not tell them which ones to use. The variation in approaches will foster better discussions during the argumentation session. We do recommend, however, that you insist that students be able to justify why they chose to use a specific physical property over another.
- Do not tell students how to determine the volume of the samples; this is a methodological challenge associated with the lab and will provide an ideal opportunity to discuss measurement error and how choice of method can influence the accuracy of a measurement during the explicit and reflective discussion.
- This is a good lab for students to make mistakes during the data collection stage. Students will quickly figure out what they did wrong during the argumentation session. It will also create an opportunity for students to reflect on and identify ways to improve the way they design investigations during the explicit and reflective discussion.
- Be sure to allow students to go back and re-collect data at the end of the argumentation session. Students often realize that they made numerous mistakes when they were collecting data as a result of their discussions during the argumentation session. The students, as a result, will want a chance to re-collect data, and the re-collection of data should be encouraged when time allows. This also offers an opportunity to discuss what scientists do when they realize a mistake is made inside the lab.

Topic Connections

Table 3.2 provides an overview of the scientific practices, crosscutting concepts, disciplinary core ideas, and supporting ideas at the heart of this lab investigation. In addition, it lists the NOS and NOSI concepts for the explicit and reflective discussion. Finally, it lists literacy and mathematics skills (*CCSS ELA* and *CCSS Mathematics*) that are addressed during the investigation.

TABLE 3.2

Lab 3 alignment with standards

Scientific practices	• Asking questions and defining problems • Planning and carrying out investigations • Analyzing and interpreting data • Using mathematics and computational thinking • Constructing explanations and designing solutions • Engaging in argument from evidence • Obtaining, evaluating, and communicating information
Crosscutting concepts	• Patterns • Scale, proportion, and quantity
Core idea	• PS1.A: Structure and properties of matter
Supporting ideas	• Matter and substances • Atoms and atomic composition • Physical and chemical properties • Physical and chemical changes
NOS and NOSI concepts	• Difference between data and evidence • Methods used in scientific investigations
Literacy connections (*CCSS ELA*)	• *Reading:* Key ideas and details, craft and structure, integration of knowledge and ideas • *Writing:* Text types and purposes, production and distribution of writing, research to build and present knowledge, range of writing • *Speaking and listening:* Comprehension and collaboration, presentation of knowledge and ideas
Mathematics connections (*CCSS Mathematics*)	• Reason abstractly and quantitatively • Construct viable arguments and critique the reasoning of others • Use appropriate tools strategically • Attend to precision

Lab Handout

Lab 3. Physical Properties of Matter

What Are the Identities of the Unknown Substances?

Introduction

Matter, the "stuff" of which the universe is composed, is all around us. Anything that we can touch, feel, or see is an example of matter. Matter can be defined as something that has mass and takes up space. All matter is composed of submicroscopic particles called atoms. A substance is a sample of matter that has a constant composition. Examples of substances include water, iron, plastic, and glass. On Earth, substances are found in one of three different states (i.e., solid, liquid, and gas), and it is common to see a substance change from one state to another. The types of atoms, the interactions that occur between atoms, and how the atoms are moving within a substance determine its state and its behavior under different conditions.

Scientists use atomic composition and specific chemical or physical properties to distinguish between different substances (see Figure L3.1). The atomic composition of a substance refers to the different types of atoms found in it and the relative proportion of each type of atom. Water, for example, is composed of hydrogen atoms and oxygen atoms in a ratio of two hydrogen atoms for every one oxygen atom. The chemical and physical properties of a substance refer to measurable or observable qualities or attributes that are used to distinguish between different substances. Chemical properties describe how a substance interacts with other matter. Sodium and potassium, for example, react with water, but aluminum and gold do not. Physical properties are descriptive characteristics of matter. Examples of physical properties include color, density, conductivity, and malleability. Every substance will have a unique set of chemical and physical properties that can be used to identify it, because every type of substance has a unique atomic composition.

FIGURE L3.1

How scientists distinguish between different substances

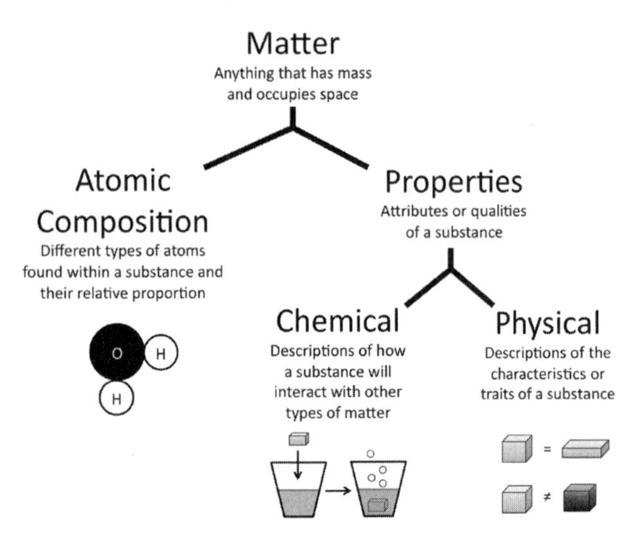

It is often challenging to determine the identity of an unknown substance based on its chemical and physical properties. A scientist, for example, may only have a small amount of a substance. As a result, the scientist may not be able to conduct all the different types of tests that he or she wants to because some tests may change the characteristics of the sample during the process (such as when a metal is mixed with an acid). It is also difficult to determine many of the physical properties of the sample, such as its density or its malleability, when there is only a small amount of the substance, because taking measurements is harder. To complicate matters further, an unknown substance may have an irregular shape, which can make it difficult to accurately measure its volume. Without knowing the mass and the volume of a substance, it is impossible to calculate its density.

In this investigation, you will have an opportunity to learn about some of the challenges scientists face when they need to identify an unknown substance based on its physical properties and why it is important to make accurate measurements inside the laboratory.

Your Task

You will be given a set of known substances. You will then document, measure, or calculate at least three different physical properties for each substance. From there, you will return the known substances to your teacher, who will then give you a set of unknown substances. The unknown substances will consist of one or more of the known substances. Your goal is to use what you know about the physical properties of matter, proportional relationships, and patterns to design and carry out an investigation that will enable you to collect the data you need to determine the identity of the unknown substances.

The guiding question of this investigation is, **What are the identities of the unknown substances?**

Materials

You may use any of the following materials during your investigation:

Consumables
- Water (in squirt bottles)
- Set of known substances
- Set of unknown substances

Equipment
- Electronic or triple beam balance
- Beakers (various sizes)
- Graduated cylinders (various sizes)
- Pipettes
- Metric ruler
- Wire
- Size D battery
- Mini lightbulb
- Mini lightbulb holder
- Safety glasses or goggles
- Chemical-resistant apron
- Nonlatex gloves

Safety Precautions

Follow all normal lab safety rules. In addition, take the following safety precautions:

1. Wear sanitized indirectly vented chemical-splash goggles and chemical-resistant nonlatex gloves and aprons during lab setup, hands-on activity, and takedown.
2. Clean up any spilled liquid immediately to avoid a slip or fall hazard.
3. Handle all glassware with care.

4. Lightbulbs are made of glass. Be careful handling them. If they break, clean them up immediately and place in a broken glass box.
5. Handle electrical wires with caution. They have sharp ends, which can cut or puncture skin.
6. Wash hands with soap and water after completing the lab activity.

Investigation Proposal Required? ☐ Yes ☐ No

Getting Started

To answer the guiding question, you will need to make several systematic observations of the known and unknown substances. To accomplish this task, you must determine what type of data you need to collect, how you will collect it, and how you will analyze it.

To determine *what type of data you need to collect,* think about the following questions:

- Which three physical properties (e.g., color, density, conductivity, malleability, luster) will you focus on as you make your systematic observations?
- What information do you need to determine or calculate each of the physical properties?

To determine *how you will collect the data,* think about the following questions:

- What equipment will you need to collect the data you need?
- How will you make sure that your data are of high quality (i.e., how will you reduce error)?
- How will you keep track of the data you collect?
- How will you organize your data?

To determine *how you will analyze the data,* think about the following questions:

- What type of calculations will you need to make?
- What patterns do you need to look for in your data?
- What type of table or graph could you create to help make sense of your data?
- How will you determine if the physical properties of the various objects are the same or different?

Connections to Crosscutting Concepts, the Nature of Science, and the Nature of Scientific Inquiry

As you work through your investigation, be sure to think about

- how scientists use patterns as a basis for classification systems;
- how scientists need to be able to recognize proportional relationships between categories, groups, or quantities;
- the difference between data and evidence in science; and
- how scientists use different types of methods to answer different types of questions.

Initial Argument

Once your group has finished collecting and analyzing your data, your group will need to develop an initial argument. Your initial argument needs to include a *claim*, *evidence* to support your claim, and a *justification* of the evidence. The claim is your group's answer to the guiding question. The evidence is an analysis and interpretation of your data. Finally, the justification of the evidence is why your group thinks the evidence matters. The justification of the evidence is important because scientists can use different kinds of evidence to support their claims. Your group will create your initial argument on a whiteboard. Your whiteboard should include all the information shown in Figure L3.2.

FIGURE L3.2

Argument presentation on a whiteboard

<table>
<tr><td colspan="2">The Guiding Question:</td></tr>
<tr><td colspan="2">Our Claim:</td></tr>
<tr><td>Our Evidence:</td><td>Our Justification of the Evidence:</td></tr>
</table>

Argumentation Session

The argumentation session allows all of the groups to share their arguments. One member of each group will stay at the lab station to share that group's argument, while the other members of the group go to the other lab stations to listen to and critique the arguments developed by their classmates. This is similar to how scientists present their arguments to other scientists at conferences. If you are responsible for critiquing your classmates' arguments, your goal is to look for mistakes so these mistakes can be fixed and they can make their argument better. The argumentation session is also a good time to think about ways you can make your initial argument better. Scientists must share and critique arguments like this to develop new ideas.

To critique an argument, you might need more information than what is included on the whiteboard. You will therefore need to ask the presenter lots of questions. Here are some good questions to ask:

- How did you collect your data? Why did you use that method? Why did you collect those data?

- What did you do to make sure the data you collected are reliable? What did you do to decrease measurement error?
- How did your group analyze the data? Why did you decide to do it that way? Did you check your calculations?
- Is that the only way to interpret the results of your analysis? How do you know that your interpretation of your analysis is appropriate?
- Why did your group decide to present your evidence in that way?
- What other claims did your group discuss before you decided on that one? Why did your group abandon those alternative ideas?
- How confident are you that your claim is valid? What could you do to increase your confidence?

Once the argumentation session is complete, you will have a chance to meet with your group and revise your initial argument. Your group might need to gather more data or design a way to test one or more alternative claims as part of this process. Remember, your goal at this stage of the investigation is to develop the most acceptable and valid answer to the research question!

Report

Once you have completed your research, you will need to prepare an *investigation report* that consists of three sections. Each section should provide an answer to the following questions:

1. What question were you trying to answer and why?
2. What did you do to answer your question and why?
3. What is your argument?

Your report should answer these questions in two pages or less. This report must be typed, and any diagrams, figures, or tables should be embedded into the document. Be sure to write in a persuasive style; you are trying to convince others that your claim is acceptable and valid!

Checkout Questions

Lab 3. Physical Properties of Matter

What Are the Identities of the Unknown Substances?

1. What is the difference between a physical and a chemical property of matter?

2. Why do substances have unique physical and chemical properties?

3. A scientist is given four different objects. She collects the following information about them.

Object	Mass	Volume	Conducts electricity?	Malleable?
A	24.5 g	2.5 cm^3	Yes	Yes
B	10.1 g	1.5 cm^3	Yes	No
C	4.6 g	2.2 cm^3	No	No
D	27.7 g	3.8 cm^3	Yes	Yes

a. Are any of the objects the same substance?

b. How do you know?

4. Scientists can use an experiment to answer any type of research question.

   a. I agree with this statement.
   b. I disagree with this statement.

   Explain your answer, using an example from your investigation about identification of an unknown based on physical properties.

5. "The freezing points of solutions A and B are both –1°C" is an example of evidence.

   a. I agree with this statement.
   b. I disagree with this statement.

   Explain your answer, using an example from your investigation about identification of an unknown based on physical properties.

6. Scientists often need to look for proportional relationships between different quantities during an investigation. Explain what a proportional relationship is and why these relationships are important, using an example from your investigation about identification of an unknown based on physical properties.

7. It is often important for scientists to identify patterns during an investigation. Explain why it is important to identify patterns when studying a system, using an example from your investigation about identification of an unknown based on physical properties.

Teacher Notes

Lab 4. Conservation of Mass

How Does the Total Mass of the Substances Formed as a Result of a Chemical Change Compare With the Total Mass of the Original Substances?

Purpose

The purpose of this lab is for students to *apply* what they have learned about atoms and chemical change to determine if mass is conserved during a chemical reaction. Through this activity, students will have an opportunity to explore the crosscutting concepts of systems and the flow, cycles, and conservation of matter. Students will also learn about the difference between data and evidence and the important role that creativity and imagination play in science.

The Content

A *substance* is a sample of matter that has a constant composition. The physical and chemical properties of a substance, as a result, are the same throughout any sample of that substance. A substance has qualities or attributes that make it different from other substances. These qualities or attributes are called physical and chemical properties. *Physical properties* are descriptive characteristics of a substance. *Chemical properties,* in contrast, describe how a substance interacts with other matter. Scientists can identify substances based on their physical and chemical properties because every type of substance has a unique set of physical and chemical properties that reflect the unique atomic composition of that substance.

A substance can go through both physical and chemical changes. A *physical change* is simply a change in the appearance of a substance. In a physical change, the atoms or molecules that make up a substance stay the same but the position and relative motion of these atoms or molecules within the substance change in some way. A *chemical change,* in contrast, involves a rearrangement of how the atoms within that substance are bonded together. The properties of the substance or substances that are formed as a result of a chemical change will have different properties because the new substance or substances will have a different atomic or molecular composition. A chemical change therefore causes one or more substances to be transformed into one or more different substances.

A chemical change is often described as a *chemical reaction.* The original substance or substances involved in the chemical reaction are called *reactants,* and the new substance or substances are called *products.* The mass of the reactants does not change as they turn into products during a chemical reaction because atoms are not created or destroyed during the process. This phenomenon is often described as the *law of conservation of mass.* This law can be used, along with knowledge of the chemical properties of particular elements, to describe and predict the outcomes of different types of reactions.

Timeline

The instructional time needed to complete this lab investigation is 200–230 minutes. Appendix 2 (p. 411) provides options for implementing this lab investigation over several class periods. Option C (230 minutes) should be used if students are unfamiliar with scientific writing, because this option provides extra instructional time for scaffolding the writing process. You can scaffold the writing process by modeling, providing examples, and providing hints as students write each section of the report. Option F (200 minutes) should be used if students are familiar with scientific writing and have developed the skills needed to write an investigation report on their own. In option F, students complete stage 6 (writing the investigation report) and stage 8 (revising the investigation report) as homework.

Materials and Preparation

The materials needed to implement this investigation are listed in Table 4.1 (p. 80). The consumables and equipment can be purchased from a science supply company such as Carolina, Flinn Scientific, or Ward's Science. You can purchase the solutions from these companies or prepare your own. You will need about 1 L of each solution for a class of 30. For best results, prepare all the solutions with analytical precision using an analytical balance and volumetric flasks. Prepare the solutions for this investigation as follows:

- *1 M solution of acetic acid, $C_2H_4O_2$*: Add 57.1 ml of 17.4 M solution of $C_2H_4O_2$ to 500 ml of distilled water in a 1 L volumetric flask. Mix well. Dilute to the 1 L mark with distilled water. *Caution: Be sure to add the acid to the water (do not add the water to the acid).*
- *1 M solution of hydrochloric acid, HCl*: Add 83.5 ml of 11.6 M solution of HCl to 500 ml of distilled water in a 1 L volumetric flask. Mix well. Dilute to the 1 L mark with distilled water. *Caution: Be sure to add the acid to the water (do not add the water to the acid).*
- *0.1 M solution of aluminum nitrate, $Al(NO_3)_3$*: Add 37.5 g of $Al(NO_3)_3$ to 500 ml of distilled water in a 1 L volumetric flask. Mix well. Dilute to the 1 L mark with distilled water.
- *0.1 M solution of sodium hydroxide, NaOH*: Add 4.0 g of NaOH to 500 ml of distilled water in a 1 L volumetric flask. Mix well. Dilute to the 1 L mark with distilled water.
- *0.1 M solution of copper(II) nitrate, $Cu(NO_3)_2$*: Add 24.2 g of $Cu(NO_3)_2$ to 500 ml of distilled water in a 1 L volumetric flask. Mix well. Dilute to the 1 L mark with distilled water.

We recommend that you use a set routine for distributing and collecting the materials during the lab investigation. For example, the consumables and equipment for each group can be set up at each group's lab station before class begins, or one member from each group can collect them from a table or a cart when needed during class.

TABLE 4.1

Materials list for Lab 4

Item	Quantity
Consumables	
Sodium bicarbonate, $NaHCO_3$	50 g per group
Magnesium (Mg) metal ribbon	10 cm per group
1 M acetic acid, $C_2H_4O_2$	100 ml per group
1 M hydrochloric acid, HCl	100 ml per group
0.1 M aluminum nitrate, $Al(NO_3)_3$	100 ml per group
0.1 M sodium hydroxide, NaOH	100 ml per group
0.1 M copper(II) nitrate, $Cu(NO_3)_2$	100 ml per group
Equipment and other materials	
Safety glasses or goggles	1 per student
Chemical-resistant apron	1 per student
Nonlatex gloves	1 pair per student
Erlenmeyer flask, 125 ml	2 per group
Erlenmeyer flask, 250 ml	2 per group
Test tubes, 5.0 ml	2 per group
Rubber stoppers (to seal the Erlenmeyer flasks)	4 per group
Balloons (that will fit over the opening of the Erlenmeyer flasks)	4 per group
Weighing dishes or paper	4 per group
Electronic or triple beam balance	1 per group
Investigation Proposal B (optional)	1 per group
Whiteboard, 2' × 3'*	1 per group
Lab Handout	1 per student
Peer-review guide	1 per student
Checkout Questions	1 per student

*As an alternative, students can use computer and presentation software, such as Microsoft PowerPoint or Apple Keynote, to create their arguments.

Safety Precautions and Laboratory Waste Disposal

Remind students to follow all normal lab safety rules. Acetic acid, hydrochloric acid, and sodium hydroxide are corrosive to eyes, skin, and other body tissues. Aluminum nitrate, copper(II) nitrate, and sodium hydroxide are toxic by ingestion. Be sure to warn students about the chemical and then explain how to safely work with them. In addition, tell the students to take the following safety precautions:

1. Review salient information on safety data sheets for hazardous chemicals with students.
2. Wear sanitized indirectly vented chemical-splash goggles and chemical-resistant nonlatex gloves and aprons during lab setup, hands-on activity, and takedown.
3. Never put consumables in their mouth.
4. Clean up any spilled liquid immediately to avoid a slip or fall hazard.
5. Use caution when working with hazardous chemicals in this lab that are corrosive and/or toxic.
6. Handle all glassware with care.
7. Never return the consumables to stock bottles.
8. Follow proper procedure for disposal of chemicals and solutions.
9. Wash hands with soap and water after completing the lab activity.

We recommend following Flinn laboratory waste disposal methods, as follows: use method 10 for the 0.1 M NaOH solution, method 24b for the 1 M HCl solution and the 1 M $C_2H_4O_2$ solution, method 26a for the 0.1 M $Cu(NO_3)_2$ solution, and method 26b for the 0.1 M $Al(NO_3)_3$ solution. Information about laboratory waste disposal methods is included in the Flinn catalog and reference manual; you can request a free copy at *www.flinnsci.com/forms/please-request-a-catalog-most-appropriate-for-your-needs.aspx*.

Topics for the Explicit and Reflective Discussion

Concepts That Can Be Used to Justify the Evidence

To provide an adequate justification of their evidence, students must explain why they included the evidence in their arguments and make the assumptions underlying their analysis and interpretation of the data explicit. In this investigation, students can use the following concepts to help justify their evidence:

- Characteristics of matter and substances

- Atomic theory of matter
- Physical and chemical properties of matter
- Physical and chemical changes

We recommend that you review these concepts during the explicit and reflective discussion to help students make this connection.

How to Design Better Investigations

It is important for students to reflect on the strengths and weaknesses of the investigation they designed during the explicit and reflective discussion. Students should therefore be encouraged to discuss ways to eliminate potential flaws, measurement errors, or sources of bias in their investigations. To help students be more reflective about the design of their investigation, you can ask the following questions:

1. What were some of the strengths of your investigation? What made it scientific?
2. What were some of the weaknesses of your investigation? What made it less scientific?
3. If you were to do this investigation again, what would you do to address the weaknesses in your investigation? What could you do to make it more scientific?

Crosscutting Concepts

This investigation is well aligned with two crosscutting concepts found in *A Framework for K–12 Science Education*, and you should review these concepts during the explicit and reflective discussion.

- *Systems and system models:* Defining a system under study and making a model of it are tools for developing a better understanding of natural phenomena in science. In this lab students will investigate potential changes within a system as a result of a chemical reaction.
- *Energy and matter: Flows, cycles, and conservation*. In science it is important to track how energy and matter move into, out of, and within systems. In this lab students will track potential changes in matter that occur during chemical reactions.

The Nature of Science and the Nature of Scientific Inquiry

This investigation is well aligned with two important concepts related to the *nature of science* (NOS) and the *nature of scientific inquiry* (NOSI), and you should review these concepts during the explicit and reflective discussion.

- *The difference between data and evidence in science*: Data are measurements, observations, and findings from other studies that are collected as part of an investigation. Evidence, in contrast, is analyzed data and an interpretation of the analysis.

- *The importance of imagination and creativity in science:* Students should learn that developing explanations for or models of natural phenomena and then figuring out how they can be put to the test of reality is as creative as writing poetry, composing music, or designing video games. Scientists must also use their imagination and creativity to figure out new ways to test ideas and collect or analyze data.

Hints for Implementing the Lab

- Allowing students to design their own procedures for collecting data gives students an opportunity to try, to fail, and to learn from their mistakes. However, you can scaffold students as they develop their procedure by having them fill out an investigation proposal. These proposals provide a way for you to offer students hints and suggestions without telling them how to do it. You can also check the proposals quickly during a class period. For this lab we suggest using Investigation Proposal B.
- Many students have alternative conceptions about the properties of matter that will affect their thinking during the lab or how they go about attempting to answer the guiding question. For example, many students believe that gases do not have mass or that solids have more mass than liquids. Many of the students in your class, as a result, will think that it will not be important to capture the gas that is produced during a reaction or that the appearance of a solid precipitate will increase the mass of a system.
- Some of the reactions produce a gas. Students will need to determine a way to ensure that the reactants do not mix before they can determine the total mass of system before the reaction begins. Students will also need a way to capture all the gas that is produced once the reactants are mixed. Students can use a balloon to hold one of the reactants that is then stretched over the top of an Erlenmeyer flask. This strategy will keep the reactants separate and help capture the gas produced.
- Some of the reactions produce a precipitate. Students will need to determine a way to ensure that the reactants do not mix before they can determine the total mass of system before the reaction. Students can use small test tubes to keep the reactants separate in an Erlenmeyer flask or a beaker.
- Do not tell students how to determine if the mass of the products is different from the mass of the reactants. This methodological challenge will provide an ideal opportunity to discuss measurement error, how the choice of method can influence the accuracy of a measurement, and the role of imagination or creativity in science during the explicit and reflective discussion.
- This is a good lab for students to make mistakes during the data collection stage. Students will quickly figure out what they did wrong during the argumentation session. It will also create an opportunity for students to reflect on and identify ways to improve the way they design investigations during the explicit and reflective discussion.

- Be sure to allow students to go back and re-collect data at the end of the argumentation session. Students often realize that they made numerous mistakes when they were collecting data as a result of their discussions during the argumentation session. The students, as a result, will want a chance to re-collect data, and the re-collection of data should be encouraged when time allows. This also offers an opportunity to discuss what scientists do when they realize a mistake is made inside the lab.

Topic Connections

Table 4.2 provides an overview of the scientific practices, crosscutting concepts, disciplinary core ideas, and supporting ideas at the heart of this lab investigation. In addition, it lists the NOS and NOSI concepts for the explicit and reflective discussion. Finally, it lists literacy and mathematics skills (*CCSS ELA* and *CCSS Mathematics*) that are addressed during the investigation.

TABLE 4.2

Lab 4 alignment with standards

Scientific practices	• Asking questions and defining problems • Planning and carrying out investigations • Analyzing and interpreting data • Using mathematics and computational thinking • Constructing explanations and designing solutions • Engaging in argument from evidence • Obtaining, evaluating, and communicating information
Crosscutting concepts	• System and system models • Energy and matter: Flows, cycles, and conservation
Core ideas	• PS1.A: Structure and properties of matter • PS1.B: Chemical reactions
Supporting ideas	• Characteristics of matter and substances • Atomic theory of matter • Physical and chemical properties • Physical and chemical changes
NOS and NOSI concepts	• Difference between data and evidence • Imagination and creativity in science
Literacy connections (*CCSS ELA*)	• *Reading:* Key ideas and details, craft and structure, integration of knowledge and ideas • *Writing:* Text types and purposes, production and distribution of writing, research to build and present knowledge, range of writing • *Speaking and listening:* Comprehension and collaboration, presentation of knowledge and ideas
Mathematics connections (*CCSS Mathematics*)	• Reason abstractly and quantitatively • Construct viable arguments and critique the reasoning of others • Use appropriate tools strategically • Attend to precision

Lab Handout

Lab 4. Conservation of Mass

How Does the Total Mass of the Substances Formed as a Result of a Chemical Change Compare With the Total Mass of the Original Substances?

Introduction

Matter is defined as anything that has mass and takes up space. Matter is composed of submicroscopic particles called atoms. To date, we know of 118 different types of atoms. All atoms share the same basic structure. At the center of an atom is a nucleus, which is composed of even smaller particles called protons and neutrons. Atoms are also composed of a third type of particle called electrons, which are found in specific regions around the nucleus. These regions are called orbitals. Scientists use the number of protons found in the nucleus of an atom to distinguish between the 118 different types of atoms. For example, there is 1 proton in the nucleus of a hydrogen atom and 30 protons in the nucleus of a zinc atom. Each type of atom also has a specific mass that reflects the composition of its nucleus.

Atoms can be bonded together in different combinations to create different types of molecules. Atoms or molecules can be combined to create different types of substances. A substance is a sample of matter that has a constant composition. Substances that consist of a single type of atom, such as gold or tin, are called elements. Substances that consist of a single type of molecule, such as water or sugar, are called compounds. A substance has qualities or attributes that distinguish it from other substances. These qualities or attributes are called physical and chemical properties. Physical properties are observable or measurable characteristics of a substance. Examples of physical properties include such things as density, melting point, and boiling point. Chemical properties, in contrast, describe how a substance interacts with other substances. For example, zinc reacts with hydrochloric acid but not with water. Scientists can identify a substance by examining its physical and chemical properties because every type of substance has a unique set of physical and chemical properties that reflect its unique atomic or molecular composition.

A substance can go through both chemical and physical changes. Any change in a substance that involves a rearrangement of how the atoms within that substance are bonded together is called a chemical change. A chemical change causes one or more substances to be transformed into one or more different substances. This process is often described as a chemical reaction. The original substance or substances involved in the chemical reaction are called reactants and the new substance or substances are called products. A physical change in matter, in contrast, does not involve a rearrangement of how the atoms within that substance are bonded together. A physical change is simply a change in the appearance of a substance. Examples of a physical change include a liquid turning into a solid or a solid turning into a liquid and a substance being broken or cut into smaller pieces. Figure L4.1 (p. 86) illustrates what happens at the submicroscopic level when a

FIGURE L4.1

Difference between a chemical change and physical change at the submicroscopic level

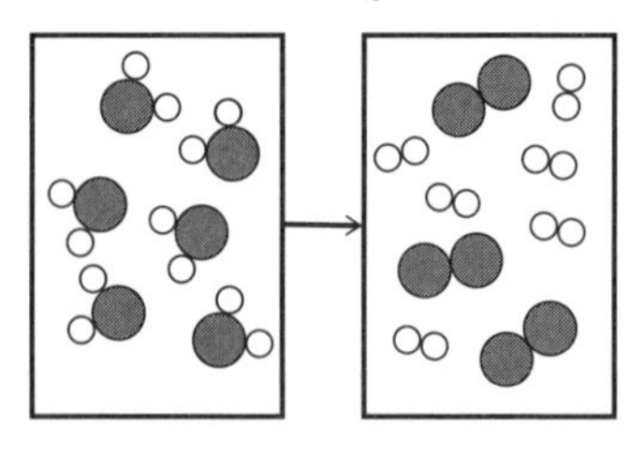

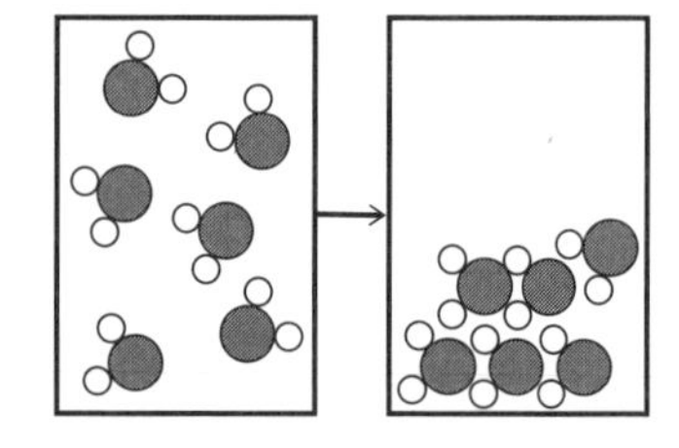

substance, such as water, goes through a chemical or physical change.

Many substances will react with other substances in predictable ways. Take the reaction of zinc and hydrochloric acid as an example. When zinc and hydrochloric acid are mixed together, the resulting products of the reaction will be hydrogen gas and a solution of zinc chloride. Another example is the reaction that takes place between a solution of silver nitrate and a solution of sodium chloride. When these two clear solutions are mixed together, the atoms in each substance interact and then rearrange to produce a different solution containing sodium nitrate and a solid substance called silver chloride. The properties of the products that are formed as a result of a chemical reaction are different than the properties of the reactants because the atoms in the original substances were broken apart and then rearranged and combined in a new way. The new configuration of atoms results in substances that have a different atomic or molecular composition. The unique atomic or molecular composition of a substance, as noted earlier, gives a substance its unique chemical and physical properties.

The chemical and physical properties of the reactants and the products of a chemical reaction are often very different even when the reactants and the products are composed of the same types of atoms. To illustrate, consider what happens when zinc (a metal) and hydrochloric acid (HCl molecules dissolved in water) are mixed. Zinc and hydrochloric acid, as noted earlier, react to produce hydrogen (a gas) and a solution of zinc chloride ($ZnCl_2$ molecules dissolved in water). Table L4.1 shows the composition of these four substances and some of the physical properties of each one. As can be seen in the table, zinc and hydrochloric acid (the reactants) have very different physical properties than hydrogen and zinc chloride (the products), even though the reactants and the products of this reaction are composed of the same three types of atoms.

TABLE L4.1

Formulas and some physical properties of zinc, hydrochloric acid, hydrogen, and zinc chloride

Substance	Formula	Physical properties			
		Density (g/cm^3)	Phase (at 23°C)	Melting point (°C)	Boiling point (°C)
Zinc	Zn	7.14	Solid	419	907
Hydrochloric acid	HCl	1.2	Liquid	–26	48
Hydrogen	H_2	0.00009	Gas	–259	–253
Zinc chloride	$ZnCl_2$	2.9	Solid	290	732

At this point, we have established several fundamental ideas about the nature of matter. We know that all matter has mass, that matter is composed of atoms, and that each type of atom has a specific mass. We also know that the reactants and the products of a reaction contain the same types of atoms, because a chemical change is just a rearrangement of atoms. These fundamental ideas, when taken together, suggest that the total mass of the reactants should be the same as the total mass of the products left at the end of a chemical reaction. This claim, however, seems highly unlikely, because the substances that are left at the end of a reaction often have very different physical properties than the substances at the start of the reaction. Your goal for this investigation will be to test the validity or the acceptability of this hypothesis.

Your Task

Use what you to know about atoms, chemical reactions, systems, and how to track the movement of matter to design and carry out an investigation to determine if the total mass of the reactants is same as the total mass of the products left at the end of a chemical reaction.

The guiding question of this investigation is, **How does the total mass of the substances formed as a result of a chemical change compare with the total mass of the original substances?**

Materials

You may use any of the following materials during your investigation:

Consumables
- Sodium bicarbonate, $NaHCO_3$
- Magnesium (Mg) metal ribbon
- 1 M acetic acid, $C_2H_4O_2$
- 1 M hydrochloric acid, HCl
- 0.1 M aluminum nitrate, $Al(NO_3)_3$
- 0.1 M sodium hydroxide, NaOH
- 0.1 M copper(II) nitrate, $Cu(NO_3)_2$

Equipment
- 4 Beakers (various sizes)
- 4 Erlenmeyer flasks (various sizes)
- 2 5.0 ml Test tubes
- 4 Rubber stoppers
- 4 Balloons
- Weighing dishes or paper
- Electronic or triple beam balance
- Safety glasses or goggles
- Chemical-resistant apron
- Nonlatex gloves

Safety Precautions

Follow all normal lab safety rules. Acetic acid, hydrochloric acid, and sodium hydroxide are corrosive to eyes, skin, and other body tissues. Aluminum nitrate, copper(II) nitrate, and sodium hydroxide are toxic by ingestion. Your teacher will explain relevant and important information about working with the chemicals associated with this investigation. In addition, take the following safety precautions:

1. Follow safety precautions noted on safety data sheets for hazardous chemicals.

2. Wear sanitized indirectly vented chemical-splash goggles and chemical-resistant nonlatex gloves and aprons during lab setup, hands-on activity, and takedown.
3. Never put consumables in your mouth.
4. Clean up any spilled liquid immediately to avoid a slip or fall hazard.
5. Use caution when working with hazardous chemicals in this lab that are corrosive and/or toxic.
6. Handle all glassware with care.
7. Never return the consumables to stock bottles.
8. Follow proper procedure for disposal of chemicals and solutions.
9. Wash hands with soap and water after completing the lab activity.

Investigation Proposal Required? ☐ Yes ☐ No

Getting Started

To answer the guiding question, you will investigate four different chemical reactions. The reactants and products for each chemical reaction are provided in Table L4.2. Your goal is to determine if the total mass of the reactants that you use in each reaction is the same or different than the total mass of the products.

TABLE L4.2

Reactants and products of the four chemical reactions

Reaction	Reactants	Products
1	Sodium bicarbonate (s) and acetic acid (aq)	Carbon dioxide (g), sodium acetate (aq), and water (l)
2	Magnesium (s) and hydrochloric acid (aq)	Magnesium chloride (aq) and hydrogen (g)
3	Aluminum nitrate (aq) and sodium hydroxide (aq)	Aluminum hydroxide (s) and sodium nitrate (aq)
4	Copper(II) nitrate (aq) and sodium hydroxide (aq)	Copper hydroxide (s) and sodium nitrate (aq)

Note: aq = aqueous solution (solid dissolved in water); g = gas ; l = liquid; s = solid.

Some of products that you will produce during your investigation will be solids, some will be liquids, and some will be gases. Your challenge will be to find a way to ensure that none of the substances that you create when you mix the reactants together escape from the container you are using to hold them during the reaction or once the reaction is complete. You will only be given a limited amount of each reactant, so it is important to find a way to create a closed system before you mix any of the reactants together. You will also need to determine what type of data you need to collect, how you will collect it, and how you will analyze it before you begin your investigation.

To determine *what type of data you need to collect,* think about the following questions:

- What observations (color change, production of gas, etc.) will you need to make during your investigation?
- What measurements (mass of the reactants, mass of the containers, etc.) will you need to make during your investigation?

To determine *how you will collect the data,* think about the following questions:

- How will you ensure that none of the substances that you create when you mix the reactants together escape during the reaction or once the reaction is complete?
- How will you take into account the mass of the containers?
- When will you need to make your observations or measurements?
- What equipment will you need to collect the data?
- How will you make sure that your data are of high quality (i.e., how will you reduce error)?
- How will you keep track of the data you collect?
- How will you organize your data?

To determine *how you will analyze the data,* think about the following questions:

- What type of calculations will you need to make?
- What type of table or graph could you create to help make sense of your data?
- How will you determine if the total mass of the reactants and the products is the same or different?

Connections to Crosscutting Concepts, the Nature of Science, and the Nature of Scientific Inquiry

As you work through your investigation, be sure to think about

- the importance of defining a system under study;
- how scientists often need track how matter moves into, out of, and within a system;
- the difference between data and evidence in science; and
- how testing explanations requires imagination and creativity.

Initial Argument

Once your group has finished collecting and analyzing your data, your group will need to develop an initial argument. Your initial argument needs to include a *claim*, *evidence* to support your claim, and a *justification* of the evidence. The claim is your group's answer to the guiding question. The evidence is an analysis and interpretation of your data. Finally, the justification of the evidence is why your group thinks the evidence matters. The justification of the evidence is important because scientists can use different kinds of evidence to support their claims. Your group will create your initial argument on a whiteboard. Your whiteboard should include all the information shown in Figure L4.2.

FIGURE L4.2

Argument presentation on a whiteboard

The Guiding Question:	
Our Claim:	
Our Evidence:	Our Justification of the Evidence:

Argumentation Session

The argumentation session allows all of the groups to share their arguments. One member of each group will stay at the lab station to share that group's argument, while the other members of the group go to the other lab stations to listen to and critique the arguments developed by their classmates. This is similar to how scientists present their arguments to other scientists at conferences. If you are responsible for critiquing your classmates' arguments, your goal is to look for mistakes so these mistakes can be fixed and they can make their argument better. The argumentation session is also a good time to think about ways you can make your initial argument better. Scientists must share and critique arguments like this to develop new ideas.

To critique an argument, you might need more information than what is included on the whiteboard. You will therefore need to ask the presenter lots of questions. Here are some good questions to ask:

- How did you collect your data? Why did you use that method? Why did you collect those data?
- What did you do to make sure the data you collected are reliable? What did you do to decrease measurement error?
- How did your group analyze the data? Why did you decide to do it that way? Did you check your calculations?
- Is that the only way to interpret the results of your analysis? How do you know that your interpretation of your analysis is appropriate?
- Why did your group decide to present your evidence in that way?
- What other claims did your group discuss before you decided on that one? Why did your group abandon those alternative ideas?

- How confident are you that your claim is valid? What could you do to increase your confidence?

Once the argumentation session is complete, you will have a chance to meet with your group and revise your initial argument. Your group might need to gather more data or design a way to test one or more alternative claims as part of this process. Remember, your goal at this stage of the investigation is to develop the most acceptable and valid answer to the research question!

Report

Once you have completed your research, you will need to prepare an *investigation report* that consists of three sections. Each section should provide an answer to the following questions:

1. What question were you trying to answer and why?
2. What did you do to answer your question and why?
3. What is your argument?

Your report should answer these questions in two pages or less. This report must be typed, and any diagrams, figures, or tables should be embedded into the document. Be sure to write in a persuasive style; you are trying to convince others that your claim is acceptable and valid!

Checkout Questions

Lab 4. Conservation of Mass

How Does the Total Mass of the Substances Formed as a Result of a Chemical Change Compare With the Total Mass of the Original Substances?

1. The figure below shows a submicroscopic view of matter going through either a physical or chemical change.

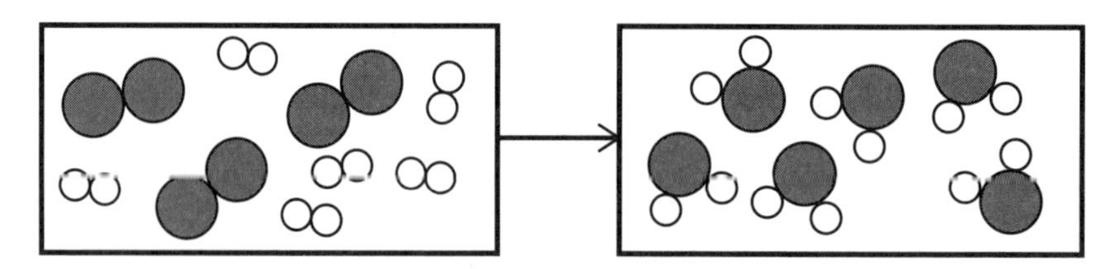

What type of change is illustrated in this figure?

a. A physical change
b. A chemical change
c. Unsure

Explain your answer. What rule did you use to make your decision?

2. A scientist takes the mass of two liquids. As shown in the figure below, the total mass of the two liquids and the containers holding them is 245 g. She then mixes the two liquids and a chemical reaction take place. A precipitate is produced as a result of the reaction.

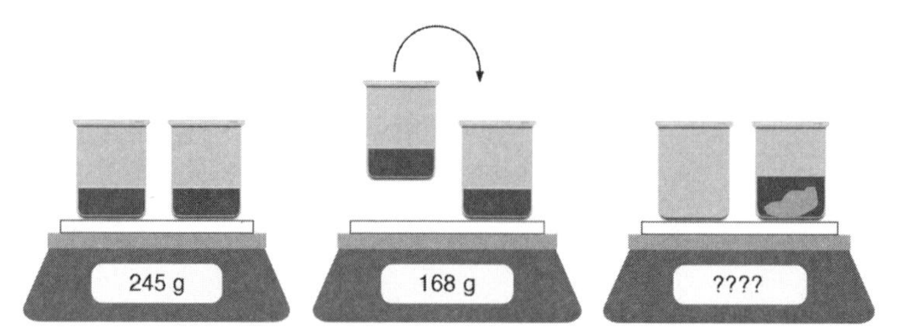

What will be the total mass of the products and the containers holding them?

a. > 245 g
b. 245 g
c. < 245 g
d. Unsure

Explain your answer. What rule did you use to make your decision?

3. Scientists do not need to have a good imagination or to be creative when testing ideas.

   a. I agree with this statement.
   b. I disagree with this statement.

   Explain your answer, using an example from your investigation about the conservation of mass.

4. "The mass of the reactants is 245 grams" is an example of evidence.

   a. I agree with this statement.
   b. I disagree with this statement.

   Explain your answer, using an example from your investigation about the conservation of mass.

5. Scientists often need to define the system under study during an investigation. Explain why defining the system under study is important, using an example from your investigation about the conservation of mass.

6. It is often important to track how matter flows into, out of, and within a system during an investigation. Explain why it is important to keep track of matter when studying a system, using an example from your investigation about the conservation of mass.

Teacher Notes

Lab 5. Design Challenge

Which Design Will Cool a Soda the Best?

Purpose

The purpose of this lab is for students to *apply* what they know about endothermic and exothermic chemical processes and the flow of energy into and out of systems. Through this activity, students will have an opportunity to investigate the structure and function of objects and how energy moves between different objects. Students will also learn about the different methods scientists and engineers use during investigations and how imagination and creativity play a role in designing solutions to problems.

The Content

Endothermic and exothermic processes are the result of a system absorbing energy from its surroundings or releasing energy to the surroundings. When ionic compounds such as those used during this investigation are dissolved into water, there are several steps that take place. Each of these steps involves a change in energy, and the overall process is described as endothermic or exothermic based on the sum of the changes in energy for each step in the entire process.

Figure 5.1 describes the flow of energy during the process of dissolving an ionic compound. *Enthalpy* is the measure of heat content of a system. During the first two steps of dissolving, energy is absorbed by the system from the surroundings; this energy is needed to break the attraction between the ions of the solute and to break the attractive forces between the water molecules. During the third step in the dissolving process, the individual ions of the solute are attracted to the water molecules; the formation of these attractive bonds releases energy from the system to the surroundings. If the energy absorbed in steps 1 and 2 is greater than the energy released in step 3, then the overall process is endothermic. In contrast, if more energy is released in step 3 than is absorbed in steps 1 and 2, the process is exothermic. When a process is endothermic, it will register a decrease in temperature; when a process is exothermic, it will register an increase in temperature.

In this investigation students are designing a device that will use a chemical process to cool another substance (e.g., a can of soda). When the chemicals that are being dissolved are viewed as the system, the can of soda serves as the surroundings. In this way, the system can absorb energy from the surroundings, thereby reducing the total energy in the surroundings and effectively reducing the temperature of the surroundings. When objects of different temperatures are in contact with each other, the flow of heat energy will always be from the warmer object toward the colder object. Energy is transferred in this manner due to the natural movement of particles. All particles are in constant motion, and the kinetic energy of the particles in a substance is directly related to the temperature of that substance. The more kinetic energy particles have, the greater the temperature of that

FIGURE 5.1

The dissociation process and hydration process that take place when an ionic compound dissolves in water

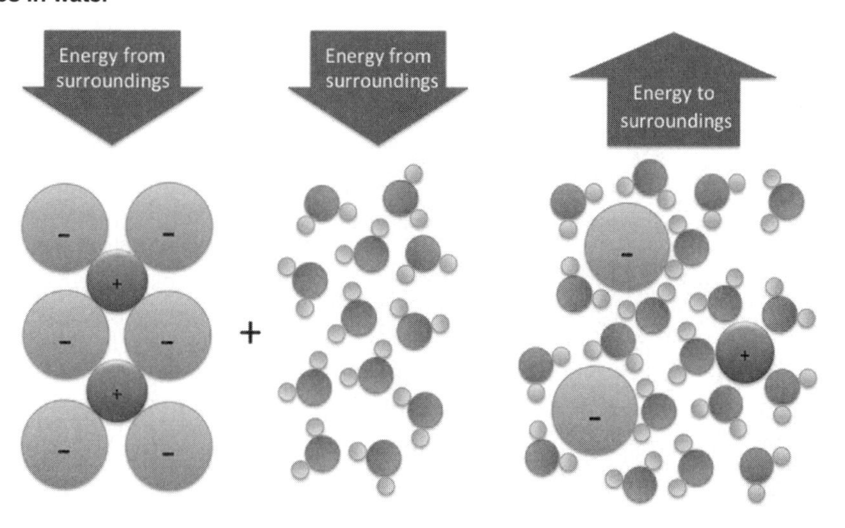

substance. When warm and cold objects come into contact, the kinetic energy of the particles in the warmer substance is in part transferred to the particles in the cooler substance. In this way, the warmer substance loses energy and cools and the cooler substance gains energy and warms.

Whenever two objects of different temperature come into contact with each other, heat energy will be transferred from the warmer object to the cooler object. However, the rate at which heat energy transfers from one object to the next depends on several factors, one being the materials involved. Metals are good conductors of heat energy, meaning that heat energy transfers into and out of metals easily. Other substances, such as Styrofoam, are poor conductors of heat energy, meaning that heat energy does not transfer through the substance easily. Poor conductors of heat energy are called *insulators.*

Engineering Connection

This investigation engages students in a design challenge to construct a device that both cools a beverage using a chemical process and insulates the user's hand. Unlike a typical scientific investigation, in this lab students are not trying to explain a natural phenomenon; rather, they are determining the best solution to a problem. To determine the best solution, students will need to develop multiple solutions and test them using an iterative process of design, test, refine, and optimize. During this iterative cycle of design, students' solutions are tested and refined based on data related to how well the design helps solve the given problem. The optimization stage of the design cycle is where the final design is improved based on trading off more or less important features of the design given the constraints of the task. Design constraints may include the size of the device, cost of the materials, the ratio of cost to outcome, or acceptable margin of error for a specified outcome. We suggest that you and your students work collaboratively to determine what the constraints will be for this design challenge.

The outcome for this investigation will be a design solution that meets the constraints identified for the given problem. At the conclusion of the design cycle, students will still generate an argument that includes a claim, evidence, and justification; however, the arguments will be slightly modified with respect to a more typical argument-driven inquiry investigation. Students may make a general claim about the features of a successful design, or their claim may be the specific design developed by their group. The evidence portion of the argument will include data that have been analyzed and interpreted to support the success of their design. Finally, the justification of their evidence will include a connection to scientific ideas like a typical scientific argument, with the addition of how their design also addresses the constraints of the problem they are solving.

Timeline

The instructional time needed to complete this lab investigation is 200–280 minutes. Appendix 2 (p. 411) provides options for implementing this lab investigation over several class periods. Option G (280 minutes) should be used if students are unfamiliar with scientific writing, because this option provides extra instructional time for scaffolding the writing process. You can scaffold the writing process by modeling, providing examples, and providing hints as students write each section of the report. Option H (200 minutes) should be used if students are familiar with scientific writing and have developed the skills needed to write an investigation report on their own. In option H, students complete stage 6 (writing the investigation report) and stage 8 (revising the investigation report) as homework.

Materials and Preparation

The materials needed to implement this investigation are listed in Table 5.1. The consumables and equipment can be purchased from a science supply company such as Carolina, Flinn Scientific, or Ward's Science. We recommend that you use a set routine for distributing and collecting the materials during the lab investigation. For example, the consumables and equipment for each group can be set up at each group's lab station before class begins, or one member from each group can collect them from a table or a cart when needed during class.

TABLE 5.1

Materials list for Lab 5

Item	Quantity
Consumables	
Ammonium chloride, NH_4Cl	10 g per group
Ammonium nitrate, NH_4NO_3	10 g per group
Calcium chloride, $CaCl_2$	10 g per group
Magnesium sulfate, $MgSO_4$	10 g per group
Sodium bicarbonate, $NaHCO_3$	10 g per group
Sodium chloride, NaCl	10 g per group
Sodium thiosulfate, $Na_2S_2O_3$	10 g per group
Paper cups	2–3 per group
Plastic baggies	2–3 per group
Water (for dissolving the salt)	500 ml per group
Soda (water or any beverage may be used)	300 ml per group
Equipment and other materials	
Safety glasses or goggles	1 per student
Chemical-resistant apron	1 per student
Nonlatex gloves	1 pair per student
Graduated cylinder, 25 ml	1 per group
Assorted insulating materials	As needed
Electronic or triple beam balance	1 per group
Thermometer (or temperature probe with interface)	1 per group
Chemical spatula	1 per group
Scissors	1 per group
Tape	As needed
Rubber bands	As needed
Investigation Proposal C (optional)	1 per group
Whiteboard, 2' × 3'*	1 per group
Lab Handout	1 per student
Peer-review guide	1 per student
Checkout Questions	1 per student

* As an alternative, students can use computer and presentation software, such as Microsoft PowerPoint or Apple Keynote, to create their arguments.

During this investigation, students will need to test each salt to determine which will be the best for cooling their sample beverage (or water). A small amount of salt can be used to determine which salts result in an endothermic or exothermic process; we recommend 1 g of salt dissolved into 10 ml of water. A larger amount of salt will be needed to actually cool the sample beverage. The actual amount of salt depends on the amount of beverage and its starting temperature. We recommend conducting a test before doing the lab with students, to determine the amount of salt needed for the amount of beverage you will provide to your student groups.

Additionally, this investigation requires that students design a device to cool their sample beverage without cooling their hand. In other words, the students will need to use some sort of insulator in their design. The materials list includes the vague description of "assorted insulating materials" to allow for variety in the items that are provided. Examples of such items include cotton or wool fabric pieces, polystyrene or cornstarch packing material, and foam or neoprene from manufactured drink coolers.

Safety Precautions and Laboratory Waste Disposal

Remind students to follow all normal lab safety rules. Ammonium chloride, ammonium nitrate, sodium thiosulfate, and magnesium sulfate are all moderately toxic by ingestion and are tissue irritants. You will therefore need to explain the potential hazards of working with ammonium chloride, ammonium nitrate, sodium thiosulfate, and magnesium sulfate and how to work with hazardous chemicals. In addition, tell students to take the following safety precautions:

1. Review salient information on safety data sheets for hazardous chemicals with students.
2. Wear sanitized indirectly vented chemical-splash goggles and chemical-resistant nonlatex gloves and aprons during lab setup, hands-on activity, and takedown.
3. Never put consumables in their mouth.
4. Clean up any spilled liquid immediately to avoid a slip or fall hazard.
5. Use caution when working with hazardous chemicals in this lab that are corrosive, toxic, and/or irritant.
6. Handle all glassware with care.
7. Handle glass thermometers with care. They are fragile and can break, causing a sharp hazard that can cut or puncture skin.
8. Never return the consumables to stock bottles.
9. Follow proper procedure for disposal of chemicals and solutions.
10. Wash hands with soap and water after completing the lab activity.

The aqueous calcium chloride and sodium chloride salt solutions can be disposed of down a drain if the drain is connected to a sanitation sewer system. We recommend following Flinn laboratory waste disposal method 26a or 26b to dispose of the materials from this investigation. Information about laboratory waste disposal methods is included in the Flinn catalog and reference manual; you can request a free copy at *www.flinnsci.com/forms/please-request-a-catalog-most-appropriate-for-your-needs.aspx.*

Topics for the Explicit and Reflective Discussion

Concepts That Can Be Used to Justify the Evidence

To provide an adequate justification of their evidence, students must explain why they included the evidence in their arguments and make the assumptions underlying their analysis and interpretation of the data explicit. In this investigation, students can use the following concepts to help justify their evidence:

- Endothermic and exothermic processes
- The difference between heat and temperature
- Heat transfer
- Insulators versus conductors

We recommend that you review these concepts during the explicit and reflective discussion to help students make this connection.

How to Design Better Investigations

It is important for students to reflect on the strengths and weaknesses of the investigation they designed during the explicit and reflective discussion. Students should therefore be encouraged to discuss ways to eliminate potential flaws, measurement errors, or sources of bias in their investigations. To help students be more reflective about the design of their investigation, you can ask the following questions:

- What were some of the strengths of your investigation? What made it scientific?
- What were some of the weaknesses of your investigation? What made it less scientific?
- If you were to do this investigation again, what would you do to address the weaknesses in your investigation? What could you do to make it more scientific?
- Did you meet the goal of the design challenge?
- Did you ensure that your solution is consistent with the design parameters?

Crosscutting Concepts

This investigation is well aligned with two crosscutting concepts found in *A Framework for K–12 Science Education,* and you should review these concepts during the explicit and reflective discussion.

- *Energy and matter: Flows, cycles, and conservation:* In science it is important to track how energy and matter move into, out of, and within systems. In this lab, heat energy is transferred between the chemical solution and the sample beverage. Additionally, students use insulators to prevent the transfer of heat energy from their hand to the cooling device they design.
- *Structure and function:* The way an object is shaped or structured determines many of its properties and functions. In this lab, the relationship between the various components of the cooling device directly influences how the sample beverage is cooled and if the device is able to accomplish the desired task.

The Nature of Science and the Nature of Scientific Inquiry

This investigation is well aligned with two important concepts related to the *nature of science* (NOS) and the *nature of scientific inquiry* (NOSI), and you should review these concepts during the explicit and reflective discussion.

- *Methods used in scientific investigations*: Examples of methods include experiments, systematic observations of a phenomenon, literature reviews, and analysis of existing data sets; the choice of method depends on the objectives of the research. There is no universal step-by-step scientific method that all scientists follow; rather, different scientific disciplines (e.g., chemistry vs. physics) and fields within a discipline (e.g., organic vs. physical chemistry) use different types of methods, use different core theories, and rely on different standards to develop scientific knowledge.
- *The importance of imagination and creativity in science:* Students should learn that developing explanations for or models of natural phenomena and then figuring out how they can be put to the test of reality is as creative as writing poetry, composing music, or designing video games. Scientists must also use their imagination and creativity to figure out new ways to test ideas and collect or analyze data.

Hints for Implementing the Lab

- We recommend conducting a test before doing the lab with students, to ensure that you provide students with access to enough salt to cool the desired amount of sample beverage. It is easier to cool a small sample than a large sample, such as a whole can of soda.

- Allowing students to design their own procedures for collecting data gives students an opportunity to try, to fail, and to learn from their mistakes. However, you can scaffold students as they develop their procedure by having them fill out an investigation proposal. These proposals provide a way for you to offer students hints and suggestions without telling them how to do it. You can also check the proposals quickly during a class period. For this lab we suggest using Investigation Proposal C.
- Encourage students to test each salt to determine if dissolving is an endothermic or exothermic process for that salt.
- Allowing students to test each salt in separate plastic baggies or having students use plastic baggies when they design their cooling device will aid in cleanup at the conclusion of the design challenge.

Topic Connections

Table 5.2 (p. 104) provides an overview of the scientific practices, crosscutting concepts, disciplinary core ideas, and supporting ideas at the heart of this lab investigation. In addition, it lists NOS and NOSI concepts for the explicit and reflective discussion. Finally, it lists literacy and mathematics skills (*CCSS ELA* and *CCSS Mathematics*) that are addressed during the investigation.

TABLE 5.2

Lab 5 alignment with standards

Scientific practices	• Asking questions and defining problems • Planning and carrying out investigations • Analyzing and interpreting data • Using mathematics and computational thinking • Constructing explanations and designing solutions • Engaging in argument from evidence • Obtaining, evaluating, and communicating information
Crosscutting concepts	• Energy and matter: Flows, cycles, and conservation • Structure and function
Core ideas	• PS1.B: Chemical reactions • PS3.A: Definitions of energy • ETS1.A: Defining and delimiting an engineering problem • ETS1.B: Developing possible solutions
Supporting ideas	• Endothermic and exothermic processes • Heat and temperature • Heat transfer • Insulators versus conductors
NOS and NOSI concepts	• Methods used in scientific investigations • Imagination and creativity in science
Literacy connections (*CCSS ELA*)	• *Reading:* Key ideas and details, craft and structure, integration of knowledge and ideas • *Writing:* Text types and purposes, production and distribution of writing, research to build and present knowledge, range of writing • *Speaking and listening:* Comprehension and collaboration, presentation of knowledge and ideas
Mathematics connections (*CCSS Mathematics*)	• Reason abstractly and quantitatively • Construct viable arguments and critique the reasoning of others

Lab Handout

Lab 5. Design Challenge

Which Design Will Cool a Soda the Best?

Introduction

Many chemical processes are accompanied by a change in temperature of the substances involved. These temperature changes occur because energy is either released or absorbed during the process. When energy is released during a chemical process, the process is exothermic. Think about a hand warmer pack that you might place in a glove on a cold day. A chemical reaction is taking place inside the pack, which releases energy that is absorbed by your hand and warms them up. Other chemical processes need to absorb energy to occur; these processes are endothermic. Cold packs inside a first aid kit are a good example of an endothermic process. The pack is normally at room temperature, but shaking it up activates it, and the chemicals inside are allowed to mix together and they get very cold. The cold pack feels cold because it must absorb energy from its surroundings to help the chemicals dissolve together. The energy the cold pack absorbs is coming from your hand and the air, and because energy is leaving your hand, your hand gets cold.

FIGURE L5.1

Cardboard sleeves on coffee cups act as insulators to slow the transfer of energy from the hot beverage to our hand.

Endothermic and exothermic processes are common, but sometimes it is necessary to control the flow of energy during these processes. Substances called insulators slow down the transfer of heat energy. Insulation in the walls and ceiling of homes is used to keep a home warm for longer on a cold winter day. The same insulation also helps keep the inside of the home cool on a hot summer day. Some materials work very well as insulators, but others do not. Styrofoam is a good insulator and is often used to keep hot drinks warm, such as coffee (see Figure L5.1), but it is also used to keep other drinks cold, such as soda. Engineers often have to investigate how much insulation to use for different products so that they can regulate the amount of thermal energy that transfers into or out of a system, so they can design a quality product.

Your Task

Use what you know about chemical reactions, thermal energy, and how energy flows into and out of a system to design a product that will help to cool a can of soda (or other beverage) by using a chemical reaction. The structure and function of your design should be related in a way that causes the drink to cool down but also ensures that your hand does not get cold if you are holding the drink.

The guiding question of this investigation is, **Which design will cool a soda the best?**

Materials

You may use any of the following materials during your investigation:

Consumables

- Ammonium chloride, NH_4Cl
- Ammonium nitrate, NH_4NO_3
- Calcium chloride, $CaCl_2$
- Magnesium sulfate, $MgSO_4$
- Sodium bicarbonate, $NaHCO_3$
- Sodium chloride, NaCl
- Sodium thiosulfate, $Na_2S_2O_3$
- Paper cups
- Plastic baggies
- Water or soda

Equipment

- Polystyrene cups
- Beaker (50 ml)
- Graduated cylinder (25 ml)
- Insulators
- Electronic or triple beam balance
- Thermometer or temperature probe
- Chemical spatula
- Scissors
- Tape
- Rubber bands
- Safety glasses or goggles
- Chemical-resistant apron
- Nonlatex gloves

Safety Precautions

Follow all normal lab safety rules. Ammonium chloride, ammonium nitrate, sodium thiosulfate, and magnesium sulfate are all moderately toxic by ingestion and tissue irritants. Your teacher will explain relevant and important information about working with the chemicals associated with this investigation.

1. Follow safety precautions noted on safety data sheets for hazardous chemicals.
2. Wear sanitized indirectly vented chemical-splash goggles and chemical-resistant nonlatex gloves and aprons during lab setup, hands-on activity, and takedown.
3. Never put consumables in your mouth.
4. Clean up any spilled liquid immediately to avoid a slip or fall hazard.
5. Use caution when working with hazardous chemicals in this lab that are corrosive, toxic, and/or irritant.
6. Handle all glassware with care.
7. Handle glass thermometers with care. They are fragile and can break, causing a sharp hazard that can cut or puncture skin.
8. Never return the consumables to stock bottles.

9. Follow proper procedure for disposal of chemicals and solutions.
10. Wash hands with soap and water after completing the lab activity.

Investigation Proposal Required? ☐ Yes ☐ No

Getting Started

To answer the guiding question, you will need to make systematic observations of the various salts. To accomplish this task, you must determine what type of data you need to collect, how you will collect it, and how you will analyze it.

To determine *what type of data you need to collect,* think about the following questions:

- How will you know if a chemical will help make the drink sample cold?
- What information do you need to determine if your design is successful?

To determine *how you will collect your data,* think about the following questions:

- What equipment will you use to collect the data you need?
- How will you make sure that your data are of high quality (i.e., how will you reduce error)?
- How will you keep track of the data you collect?
- How will you organize your data?

To determine *how you will analyze your data,* think about the following questions:

- What type of calculations will you need to make?
- What type of table or graph could you create to help make sense of your data?
- How will you determine if your design is successful or not?

Connections to Crosscutting Concepts, the Nature of Science, and the Nature of Scientific Inquiry

As you work through your investigation, be sure to think about

- how scientists often need to track how energy moves into, out of, and within a system;
- the relationship between the structure and function of an object;
- how scientists and engineers use a variety of methods to answer questions and solve problems; and
- the role of imagination and creativity when scientists and engineers attempt to solve problems.

Initial Argument

Once your group has finished collecting and analyzing your data, your group will need to develop an initial argument. Your initial argument needs to include a *claim, evidence* to support your claim, and a *justification* of the evidence. The claim is your group's answer to the guiding question. The evidence is an analysis and interpretation of your data. Finally, the justification of the evidence is why your group thinks the evidence matters. The justification of the evidence is important because scientists can use different kinds of evidence to support their claims. Your group will create your initial argument on a whiteboard. Your whiteboard should include all the information shown in Figure L5.2.

FIGURE L5.2

Argument presentation on a whiteboard

The Guiding Question:	
Our Claim:	
Our Evidence:	Our Justification of the Evidence:

Argumentation Session

The argumentation session allows all of the groups to share their arguments. One member of each group will stay at the lab station to share that group's argument, while the other members of the group go to the other lab stations to listen to and critique the arguments developed by their classmates. This is similar to how scientists present their arguments to other scientists at conferences. If you are responsible for critiquing your classmates' arguments, your goal is to look for mistakes so these mistakes can be fixed and they can make their argument better. The argumentation session is also a good time to think about ways you can make your initial argument better. Scientists must share and critique arguments like this to develop new ideas.

To critique an argument, you might need more information than what is included on the whiteboard. You will therefore need to ask the presenter lots of questions. Here are some good questions to ask:

- How did you collect your data? Why did you use that method? Why did you collect those data?
- What did you do to make sure the data you collected are reliable? What did you do to decrease measurement error?
- How did your group analyze the data? Why did you decide to do it that way? Did you check your calculations?
- Is that the only way to interpret the results of your analysis? How do you know that your interpretation of your analysis is appropriate?
- Why did your group decide to present your evidence in that way?
- What other claims did your group discuss before you decided on that one? Why did your group abandon those alternative ideas?

- How confident are you that your claim is valid? What could you do to increase your confidence?

Once the argumentation session is complete, you will have a chance to meet with your group and revise your initial argument. Your group might need to gather more data or design a way to test one or more alternative claims as part of this process. Remember, your goal at this stage of the investigation is to develop the most acceptable and valid answer to the research question!

Report

Once you have completed your research, you will need to prepare an *investigation report* that consists of three sections. Each section should provide an answer to the following questions:

1. What question were you trying to answer and why?
2. What did you do to answer your question and why?
3. What is your argument?

Your report should answer these questions in two pages or less. This report must be typed, and any diagrams, figures, or tables should be embedded into the document. Be sure to write in a persuasive style; you are trying to convince others that your claim is acceptable and valid!

Checkout Questions

Lab 5. Design Challenge

Which Design Will Cool a Soda the Best?

1. Describe what you know about materials that act as insulators and conductors when it comes to heat energy.

2. Insulated coffee mugs are popular because they keep coffee warm for much longer than a simple paper cup. Vacuum-style containers, such as the one shown below, are made with empty space between the walls.

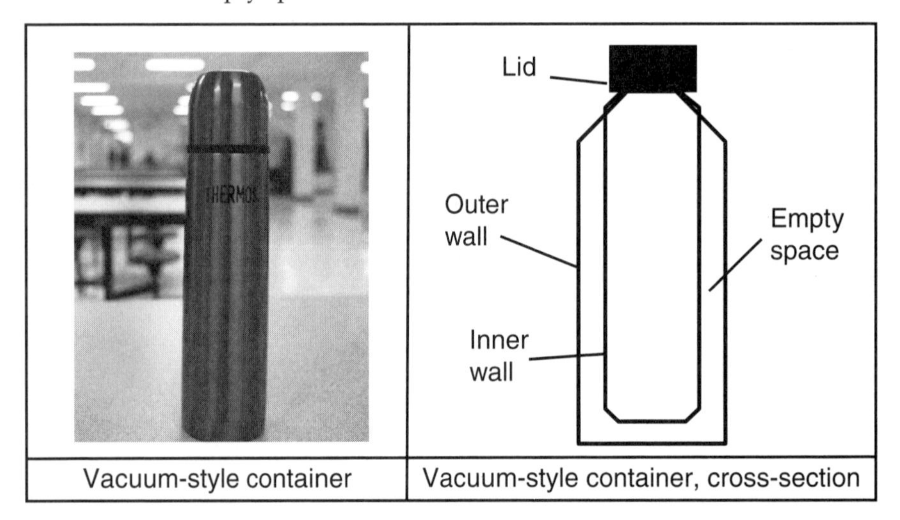

Vacuum-style container | Vacuum-style container, cross-section

Use what you know about insulators and conductors to explain why this style of container is good for keeping liquids hot or cold.

3. Scientists use the same procedures to answer all the questions they investigate in their work.

   a. I agree with this statement.
   b. I disagree with this statement.

   Explain your answer, using an example from your investigation about thermal energy.

4. Scientists do not need to be creative and have a good imagination.

   a. I agree with this statement.
   b. I disagree with this statement.

   Explain your answer, using an example from your investigation about thermal energy.

5. Scientists often study the various relationships between the structure of an object and its functions. Explain why recognizing these relationships is important, using an example from your investigation about thermal energy.

6. It is often important to track how matter flows into, out of, and within system during an investigation. Explain why it is important to keep track of matter when studying a system, using an example from your investigation about thermal energy.

SECTION 3

Physical Science
Core Idea 2

Motion and Stability: Forces and Interactions

Introduction Labs

Teacher Notes

Lab 6. Strength of Gravitational Force

How Does the Gravitational Force That Exists Between Two Objects Relate to Their Masses and the Distance Between Them?

Purpose

The purpose of this lab is to *introduce* students to the relationship between mass, distance, and the strength of a gravitational force. Through this activity, students will have an opportunity to explore the role of patterns and of scale, proportion, and quantity in scientific investigations. Students will also learn about how scientific knowledge changes over time and the culture of science.

The Content

In science, one of the fundamental questions to be answered is related to the forces that act on and between different objects. One of the forces that we are most familiar with is the *gravitational force*. The gravitational force is responsible for balls rolling down hills, a pen falling to the ground when you drop it, the Moon revolving around Earth, and Earth revolving around the Sun. Of all his many achievements, the idea that gravity is responsible for all of these things may have been the most important contribution Isaac Newton made to science.

The gravitational force between two objects is a function of three things. The first two are the mass of each object, respectively. The third factor is the distance between the two objects. Newton's formula for the force of gravity between any two objects, called the *law of universal gravitation*, is $F = G\,(m_1 m_2)/r^2$ where F is the gravitational force between the two objects, m_1 is the mass of the first object, m_2 is the mass of the second object, and r is the distance between them. The unit for force is newtons (N), for mass is kilograms (kg), and for distance is meters (m).

$$F = G\,\frac{(m_1 m_2)}{r^2}$$

There are a few important things to realize about this equation. First, the gravitational force acts between two objects. This means that if the gravitational force of Earth on a person (using the example of someone who weighs about 115 pounds) is 500 N, then the force of that person on Earth is also 500 N. The person pulls on Earth with the same force that Earth pulls on a person. The reason we don't see Earth move due to this force is because the mass of Earth is more than one trillion times larger than the mass of a person.

The second important thing to realize is that if you increase the mass of one object, the gravitational force between them will increase in a linear way. That means if you double the mass of one object, you will double the gravitational force between the two objects if you leave the mass of the second object and the distance between them unchanged. If you have students graph the force of gravity between two objects as a function of the increased mass of one of the objects, the graph will form a straight line with a positive slope.

The final important thing to recognize about Newton's law of universal gravitation is that when the distance between the two objects increases, the force of gravity between the two objects decreases, and the decrease is not linear. Instead, the relationship forms a decreasing curve (see Figure 6.1). For example, when the distance between the two objects is doubled, the gravitational force between the two objects is quartered.

FIGURE 6.1

Graph of the relationship between the magnitude of a gravitational force and the distance between two objects

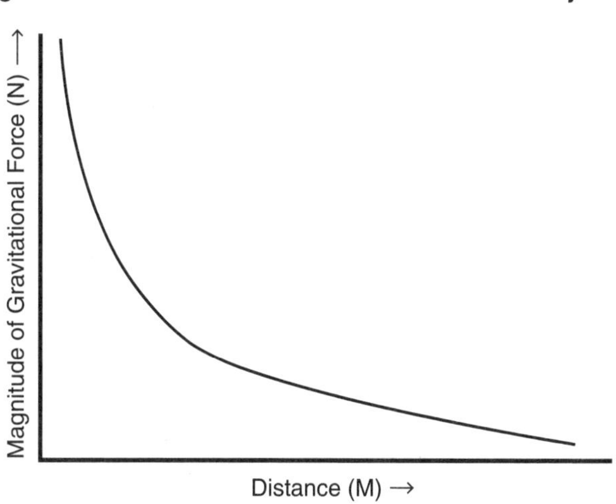

Timeline

The instructional time needed to complete this lab investigation is 170–230 minutes. Appendix 2 (p. 411) provides options for implementing this lab investigation over several class periods. Option C (230 minutes) should be used if students are unfamiliar with scientific writing, because this option provides extra instructional time for scaffolding the writing process. You can scaffold the writing process by modeling, providing examples, and providing hints as students write each section of the report. Option D (170 minutes) should be used if students are familiar with scientific writing and have developed the skills needed to write an investigation report on their own. In option D, students complete stage 6 (writing the investigation report) and stage 8 (revising the investigation report) as homework.

Materials and Preparation

The materials needed to implement this investigation are listed in Table 6.1 (p. 118). The *Gravity Force Lab* simulation was developed by PhET Interactive Simulations, University of Colorado (*http://phet.colorado.edu*), and is available at *http://phet.colorado.edu/en/simulation/gravity-force-lab*. It is free to use and can be run online using an internet browser on a school computer or tablet. You should access the website and learn how the simulation works before beginning the lab investigation. In addition, it is important to check if students can access and use the simulation from a school computer, because some schools have set up firewalls and other restrictions on web browsing.

TABLE 6.1

Materials list for Lab 6

Item	Quantity
Computer (or tablet) with internet access	1 per group
Investigation Proposal C (optional)	1 per group
Whiteboard, 2' × 3'*	1 per group
Lab Handout	1 per student
Peer-review guide	1 per student
Checkout Questions	1 per student

* As an alternative, students can use computer and presentation software, such as Microsoft PowerPoint or Apple Keynote, to create their arguments.

Safety Precautions and Laboratory Waste Disposal

Remind students to follow all normal lab safety rules. There is no laboratory waste associated with this activity

Topics for the Explicit and Reflective Discussion

Concepts That Can Be Used to Justify the Evidence

To provide an adequate justification of their evidence, students must explain why they included the evidence in their arguments and make the assumptions underlying their analysis and interpretation of the data explicit. In this investigation, students can use the following concepts to help justify their evidence:

- Forces and motion
- Gravity as a non-contact force of attraction
- Mass and inertia

We recommend that you review these concepts during the explicit and reflective discussion to help students make this connection.

How to Design Better Investigations

It is important for students to reflect on the strengths and weaknesses of the investigation they designed during the explicit and reflective discussion. Students should therefore be encouraged to discuss ways to eliminate potential flaws, measurement errors, or sources of bias in their investigations. To help students be more reflective about the design of their investigation, you can ask the following questions:

1. What were some of the strengths of your investigation? What made it scientific?
2. What were some of the weaknesses of your investigation? What made it less scientific?
3. If you were to do this investigation again, what would you do to address the weaknesses in your investigation? What could you do to make it more scientific?

Crosscutting Concepts

This investigation is well aligned with two crosscutting concepts found in *A Framework for K–12 Science Education*, and you should review these concepts during the explicit and reflective discussion.

- *Patterns:* Scientists look for patterns in nature and attempt to understand the underlying cause of these patterns. Scientists, for example, often collect data and then look for patterns to identify a relationship between two variables, a trend over time, or a difference between groups. In this lab students will investigate patterns associated with the force of gravity between different objects.
- *Scale, proportion, and quantity:* It is critical for scientists to be able to recognize what is relevant at different sizes, time frames, and scales. Scientists must also be able to recognize proportional relationships between categories, groups, or quantities. In this lab students will investigate the proportional relationship of mass, distance between objects, and the resulting force due to gravity.

The Nature of Science and the Nature of Scientific Inquiry

This investigation is well aligned with two important concepts related to the *nature of science* (NOS) and the *nature of scientific inquiry* (NOSI), and you should review these concepts during the explicit and reflective discussion.

- *Changes in scientific knowledge over time:* A person can have confidence in the validity of scientific knowledge but must also accept that scientific knowledge may be abandoned or modified in light of new evidence or because existing evidence has been reconceptualized by scientists. There are many examples in the history of science of both evolutionary changes (i.e., the slow or gradual refinement of ideas) and revolutionary changes (i.e., the rapid abandonment of a well-established idea) in scientific knowledge.
- *Science as a culture*: Scientists share a set of values, norms, and commitments that shape what counts as knowing, how to represent or communicate information, and how to interact with other scientists. The culture of science affects who gets to do science, what scientists choose to investigate, how investigations are conducted, how research findings are interpreted, and what people see as implications. People also view some research as being more important than others because of cultural values and current events.

Hints for Implementing the Lab

- Learn how to use the online simulation before the lab begins. It is important for you to know how to use the simulation so you can help students when they get stuck or confused.
- Allowing students to design their own procedures for collecting data gives students an opportunity to try, to fail, and to learn from their mistakes. However, you can scaffold students as they develop their procedure by having them fill out an investigation proposal. These proposals provide a way for you to offer students hints and suggestions without telling them how to do it. You can also check the proposals quickly during a class period. For this lab we suggest using Investigation Proposal C.
- A group of three students per computer or tablet tends to work well.
- Allow the students to play with the simulation as part of the tool talk before they begin to design their investigation. This gives students a chance to see what they can and cannot do with the simulation. You can also show them the full-color version of the simulation screenshot that appears in black and white in the Lab Handout; the full-color version is available at *www.nsta.org/adi-physicalscience.*
- Be sure that students record actual values (e.g., gravitational force in newtons, mass of objects, distance between objects) when they use the simulation, rather than just attempting to describe what they see on the computer screen (e.g., "the force was bigger").
- Encourage student to focus on changing only one factor at time (mass of the red object, mass of the blue object, distance between the objects) so it is easier to determine a relationship.
- We recommend that they fill out an investigation proposal for each experiment that they do.
- Encourage students to sketch graphs for each prediction based on each hypothesis. You can then encourage students to think about how well their result fits with each prediction as they analyze their data.
- This is a good lab for students to make mistakes during the data collection stage. Students will quickly figure out what they did wrong during the argumentation session, and it will only take them a short period of time to re-collect data. It will also create an opportunity for students to reflect on and identify ways to improve the way they design investigations (especially how they attempt to control variables as part of an experiment) during the explicit and reflective discussion.
- This lab also provides an excellent opportunity to discuss how scientists identify a signal (a pattern or trend) from the noise (measurement error) in their data during the explicit and reflective discussion. Be sure to use this activity as a concrete example.

Topic Connections

Table 6.2 provides an overview of the scientific practices, crosscutting concepts, disciplinary core ideas, and supporting ideas at the heart of this lab investigation. In addition, it lists the NOS and NOSI concepts for the explicit and reflective discussion. Finally, it lists literacy and mathematics skills (*CCSS ELA* and *CCSS Mathematics*) that are addressed during the investigation.

TABLE 6.2

Lab 6 alignment with standards

Scientific practices	• Asking questions and defining problems • Developing and using models • Planning and carrying out investigations • Analyzing and interpreting data • Using mathematics and computational thinking • Constructing explanations and designing solutions • Engaging in argument from evidence • Obtaining, evaluating, and communicating information
Crosscutting concepts	• Patterns • Scale, proportion, and quantity
Core ideas	• PS2.A: Forces and motion • PS2.B: Types of interactions
Supporting ideas	• Forces and motion • Gravity as a non-contact force • Mass and inertia
NOS and NOSI concepts	• Changes in scientific knowledge over time • Culture of science
Literacy connections (*CCSS ELA*)	• *Reading*: Key ideas and details, craft and structure, integration of knowledge and ideas • *Writing*: Text types and purposes, production and distribution of writing, research to build and present knowledge, range of writing • *Speaking and listening*: Comprehension and collaboration, presentation of knowledge and ideas
Mathematics connections (*CCSS Mathematics*)	• Reason abstractly and quantitatively • Construct viable arguments and critique the reasoning of others • Use appropriate tools strategically • Attend to precision

LAB 6

Lab Handout

Lab 6. Strength of Gravitational Force

How Does the Gravitational Force That Exists Between Two Objects Relate to Their Masses and the Distance Between Them?

Introduction

The motion of an object is the result of all the different forces that are acting on the object. If you pull on the handle of a drawer, the drawer will move in the direction you pulled it. If a ball is rolling down a driveway and hits a curb, the force of the curb will cause the ball to stop. Applying a pull or push to an object is an example of a contact force, where one object applies a force to another object through direct contact. There are other types of forces that can act on objects that do not involve objects touching. For example, the magnetic force produced by a magnet can make a paper clip move toward it or make another magnet move away from it without touching them. Another example is static electricity. Static electricity in a rubber balloon can cause a person's hair to stand up without the balloon actually touching any of his or her hair. Magnetic forces and electrical forces are therefore called non-contact forces because they can act on objects at a distance. Perhaps the most common non-contact force is gravity. Gravity is a force of attraction between two objects; the force due to gravity always works to bring objects closer together.

Any two objects, as long as they have some mass, will have a gravitational force of attraction between them. The force of gravity that exists between any two objects is influenced by the masses of those two objects. The mass of an object refers to the amount of matter that is contained by the object. Mass is also a measure of inertia, which is the resistance an object has to a change in its state of motion. All objects resist changes in their state of motion; however, the more massive an object, the more it will resist changes in its state of motion. The distance between any two objects will also influence the force of gravity that exists between them because the distance between any two objects can, and does, change. The exact relationship between these factors, however, was not well understood until 1687.

The first person to determine how mass and distance affect the strength of the gravitational force that exists between two objects was Isaac Newton. Newton described the relationship between these three factors in the book Philosophiae Naturalis Principia Mathematica (Newton 1687). The ability to describe the relationship between mass, distance, and the strength of a gravitational force was a major milestone in physics. It not only explained why objects fall toward the center of Earth (see Figure L6.1) but also explained why the planets move around the Sun, which was established by Copernicus in 1543 (see Figure L6.2). Before Newton put forth his revolutionary ideas about gravity, many people thought objects on Earth and objects in the sky moved because of different forces. Newton was the first person to suggest that the force of gravity is universal.

In this investigation, you will have an opportunity to explore the relationship between mass, distance, and the strength of the gravitational force that exists between two objects in order to learn more about the behavior of gravity. This type of investigation can be difficult, however, because identifying the exact nature of the relationship that exists between several different factors is challenging. Take mass as an example. There are many potential ways that the strength of a gravitational force between two objects can be related to the mass of those two objects. The strength of the gravitational force between two objects may depend on the mass of the larger object or the mass of the smaller object. The strength of the gravitational force could also be related to the total mass of the two objects. In addition to mass, there are many different ways that the strength of a gravitational force between two objects can be related to the distance between the two objects. The strength of a gravitational force may increase as the distance between the two objects increases, or it may decrease as the distance between the two objects increases. It may also increase or decrease exponentially as the distance between the two objects changes. All of these different relationships are possible (along with many others). Your goal is to figure out the actual relationship.

Your Task

Use what you know about forces, motion, patterns, and proportional relationships to design and carry out an investigation using a simulation to determine the relationship between mass, distance, and the strength of a gravitational force.

The guiding question of this investigation is, **How does the gravitational force that exists between two objects relate to their masses and the distance between them?**

FIGURE L6.1

Objects fall toward the center of Earth.

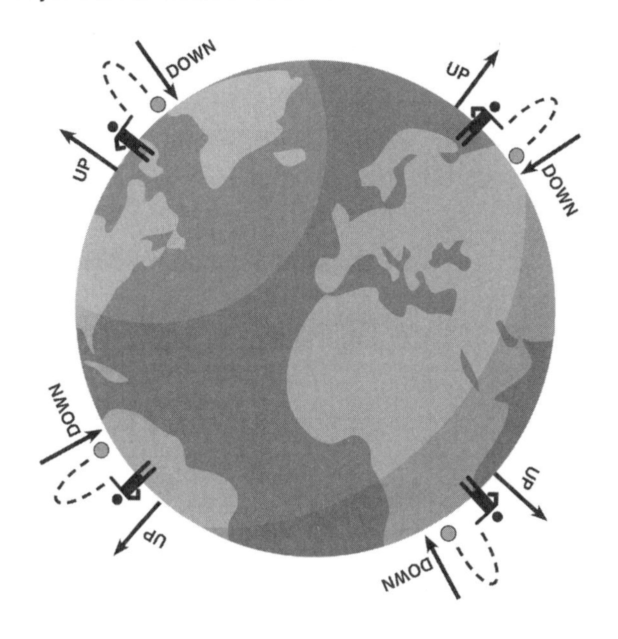

FIGURE L6.2

The heliocentric universe proposed by Copernicus in 1543

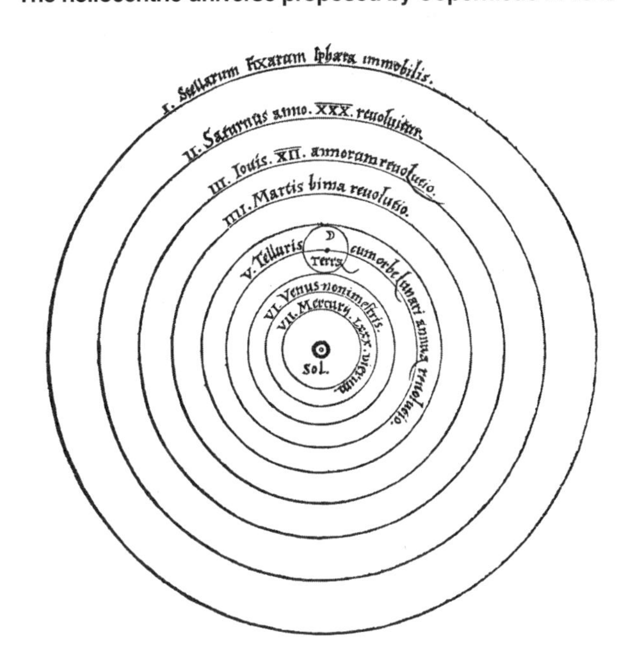

LAB 6

Materials

You will use an online simulation called *Gravity Force Lab* to conduct your investigation. You can access the simulation by going to the following website: *http://phet.colorado.edu/en/simulation/gravity-force-lab.*

Safety Precautions

Follow all normal lab safety rules.

Investigation Proposal Required? ☐ Yes ☐ No

Getting Started

The *Gravity Force Lab* simulation (see screen shot in Figure L6.3) enables you to measure the amount of gravitational force that two objects exert on each other. You can adjust the mass of the two different objects in the simulation and the amount of distance between them. As you change mass and distance, you will be able to see the amount of force, in newtons, that each object exerts on the other one. To use this simulation, start by clicking on the "Show values" box in the lower-right corner of the window. This will allow you to see the amount of force exerted by the blue object (m1) on the red object (m2) and the amount of force exerted by the red object (m2) on the blue object (m1). You can change the masses of the blue and red objects by using the sliders at the bottom of the window. To change the distance between the red and blue objects, you can simply drag and drop each one to a different spot above the ruler. This simulation is useful because it allows you to measure the force of gravity under different conditions, and perhaps more important, it provides a way for you to design and carry out controlled experiments so you can focus on one factor at a time.

You will need to design and carry out at least three different experiments using the *Gravity Force Lab* simulation in order to determine the relationship between mass, distance, and gravitational force. You will need to conduct three different experiments because you will need to be able to answer three specific questions before you will be able to develop an answer to the guiding question:

1. How does changing the mass of the red object (m2) affect the amount of gravitational force?
2. How does changing the mass of the blue object (m1) affect the amount of gravitational force?
3. How does changing the distance between the two objects affect the amount of gravitational force?

FIGURE L6.3

A screen shot of the *Gravity Force Lab* simulation

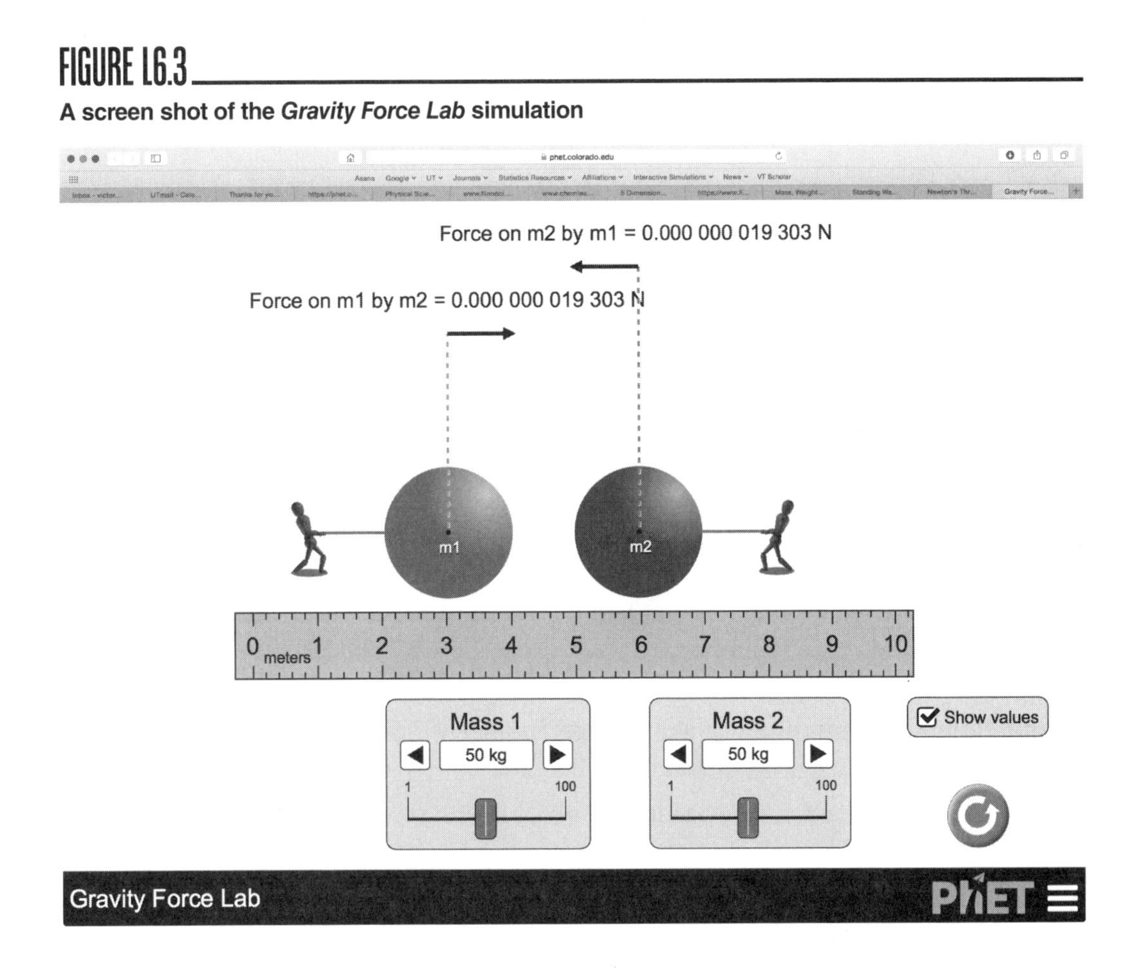

You will also need to determine what type of data you need to collect, how you will collect it, and how you will analyze the data for each experiment, because each experiment is slightly different.

To determine *what type of data you need to collect,* think about the following questions:

- What type of measurements will you need to record during each experiment?
- When will you need to make these measurements or observations?

To determine *how you will collect the data* using the simulation, think about the following questions:

- What will serve as your dependent variable for each experiment?
- What will serve as your independent variable for each experiment?
- How will you vary the independent variable during each experiment?
- What will you do to hold the other variables constant during each experiment?

- What types of comparisons will you need to make using the simulation?
- How many comparisons will you need to make to determine a trend or a relationship?
- How will you keep track of the data you collect and how will you organize it?

To determine *how you will analyze the data*, think about the following questions:

- What type of calculations will you need to make?
- What type of graph could you create to help make sense of your data?

Once you have carried out all your different experiments, your group will need to use your findings to develop an answer to the guiding question for this investigation. Your answer to the guiding question will need to be able to explain how the gravitational force that exists between two objects is related to the masses of the objects and distance between them. For your claim to be sufficient, your answer will need to be based on findings from all three of your experiments. You can then transform the data you collected during each experiment into evidence to support the validity of your overall explanation.

Connections to Crosscutting Concepts, the Nature of Science, and the Nature of Scientific Inquiry

As you work through your investigation, be sure to think about

- the importance of looking for and understanding patterns in data,
- the importance of understanding proportional relationships in science,
- how scientific knowledge can change over time, and
- the culture of science and how it influences the work of scientists.

Initial Argument

Once your group has finished collecting and analyzing your data, your group will need to develop an initial argument. Your initial argument needs to include a *claim, evidence* to support your claim, and a *justification* of the evidence. The claim is your group's answer to the guiding question. The evidence is an analysis and interpretation of your data. Finally, the justification of the evidence is why your group thinks the evidence matters. The justification of the evidence is important because scientists can use different kinds of evidence to support their claims. Your group will create your initial argument on a whiteboard. Your whiteboard should include all the information shown in Figure L6.4.

Argumentation Session

The argumentation session allows all of the groups to share their arguments. One member of each group will stay at the lab station to share that group's argument, while the other members of the group go to the other lab stations to listen to and critique the arguments developed by their classmates. This is similar to how scientists present their arguments to other scientists at conferences. If you are responsible for critiquing your classmates' arguments, your goal is to look for mistakes so these mistakes can be fixed and they can make their argument better. The argumentation session is also a good time to think about ways you can make your initial argument better. Scientists must share and critique arguments like this to develop new ideas.

FIGURE L6.4

Argument presentation on a whiteboard

The Guiding Question:	
Our Claim:	
Our Evidence:	Our Justification of the Evidence:

To critique an argument, you might need more information than what is included on the whiteboard. You will therefore need to ask the presenter lots of questions. Here are some good questions to ask:

- How did you collect your data? Why did you use that method? Why did you collect those data?
- What did you do to make sure the data you collected are reliable? What did you do to decrease measurement error?
- How did your group analyze the data? Why did you decide to do it that way? Did you check your calculations?
- Is that the only way to interpret the results of your analysis? How do you know that your interpretation of your analysis is appropriate?
- Why did your group decide to present your evidence in that way?
- What other claims did your group discuss before you decided on that one? Why did your group abandon those alternative ideas?
- How confident are you that your claim is valid? What could you do to increase your confidence?

Once the argumentation session is complete, you will have a chance to meet with your group and revise your initial argument. Your group might need to gather more data or design a way to test one or more alternative claims as part of this process. Remember, your goal at this stage of the investigation is to develop the most acceptable and valid answer to the research question!

Report

Once you have completed your research, you will need to prepare an *investigation report* that consists of three sections. Each section should provide an answer to the following questions:

1. What question were you trying to answer and why?
2. What did you do to answer your question and why?
3. What is your argument?

Your report should answer these questions in two pages or less. This report must be typed, and any diagrams, figures, or tables should be embedded into the document. Be sure to write in a persuasive style; you are trying to convince others that your claim is acceptable and valid!

References

Copernicus, N. 1543. De revolutionibus orbium coelestium [On the revolutions of heavenly spheres]. Johannes Petreius.

Newton, I. 1687. Philosophiae naturalis principia mathematica [Mathematical principles of natural philosophy].

Checkout Questions

Lab 6. Strength of Gravitational Force

How Does the Gravitational Force That Exists Between Two Objects Relate to Their Masses and the Distance Between Them?

1. The diagrams below show two objects and the distance between them.

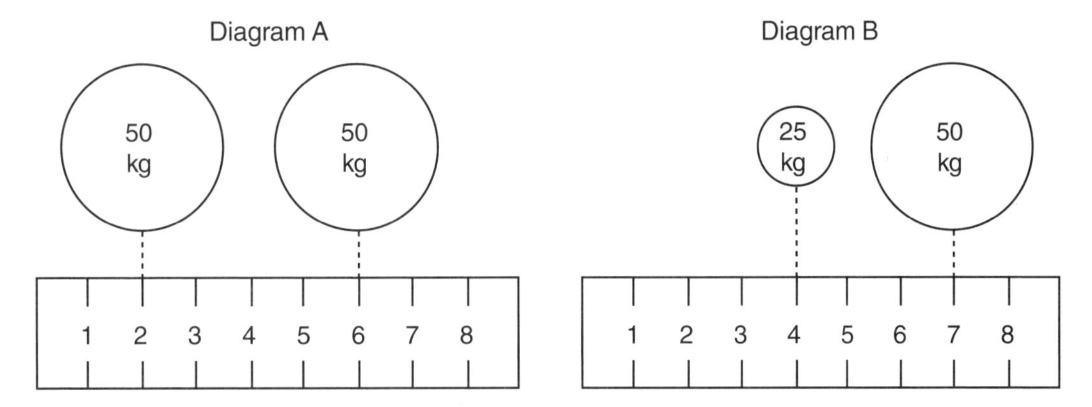

Which two objects have the greater gravitational attraction between them?

a. The objects in diagram A
b. The objects in diagram B
c. The gravitational attraction between the objects is the same in diagrams A and B
d. Unsure

How do you know?

2. The diagrams below show two objects and the distance between them.

Diagram A

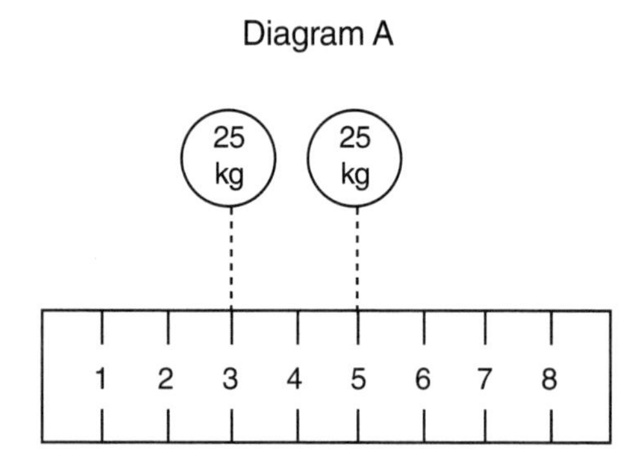

Diagram B

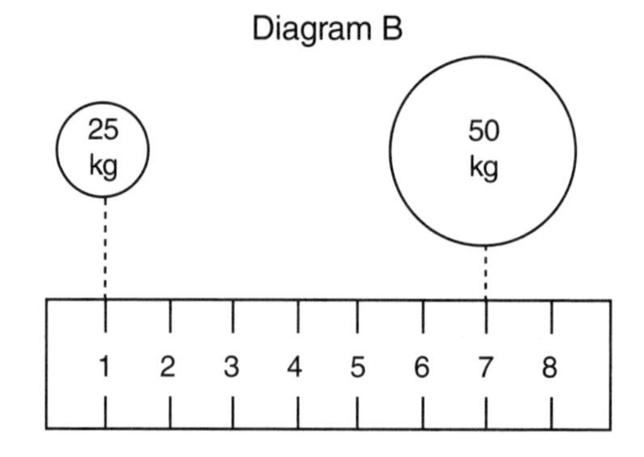

Which two objects have the greater gravitational attraction between them?

a. The objects in diagram A
b. The objects in diagram B
c. The gravitational attraction between the objects is the same in diagrams A and B.
d. Unsure

How do you know?

3. Once a scientist or a team of scientists develop a scientific law, it will never change.

a. I agree with this statement.
b. I disagree with this statement.

Explain your answer, using an example from your investigation about gravitational force.

4. Cultural values and expectations determine who gets to do science, what scientists choose to investigate, how investigations are conducted, and how research findings are interpreted.

   a. I agree with this statement.
   b. I disagree with this statement.

   Explain your answer, using an example from your investigation about gravitational force.

5. Scientists often need to look for and understand the underlying cause of patterns in data. Explain why it is important to be able to identify and understand patterns in data in science, using an example from your investigation about gravitational force.

6. Scientists often look for proportional relationships between two or more quantities in science. Explain what proportional relationships are and why they are important in science, using an example from your investigation about gravitational force.

Teacher Notes

Lab 7. Mass and Free Fall

How Does Mass Affect the Amount of Time It Takes for an Object to Fall to the Ground?

Purpose

The purpose of this lab is to *introduce* students to the relationship between gravitational force and motion of an object. Through this activity, students will have an opportunity to explore the role of patterns and of scale, proportion, and quantity in scientific investigations. Students will also learn about the difference between theories and laws and between data and evidence in science.

The Content

Imagine that a bowling ball and a feather are dropped at the same time from a height of 20 m in a large vacuum, so there is gravity but no air resistance. In a vacuum, the bowling ball and the feather will hit the ground at the same time because the only force acting upon the two objects is the force of gravity. The force of gravity that acts on the bowling ball and the feather, however, is not the same because they are not the same mass, and the magnitude of the force of gravity experienced by an object is dependent on the mass of that object. Earth therefore pulls downward upon the bowling ball with more force than it pulls downward upon the feather. Newton's second law of motion, however, indicates that the *acceleration* of an object is directly related to the net force and inversely related to its mass. The greater mass of the bowling ball therefore offsets the greater downward force that is acting on it. The acceleration of a falling object in the absence of air resistance, as a result, is determined by a ratio of force to mass, and this ratio is a constant. A simple rule to remember is that all objects, regardless of mass, experience the same acceleration when in a state of free fall. When the only force is gravity, the acceleration is the same value for all objects. On Earth, this acceleration value is 9.8 m/s^2. This is such an important value in physics that it is given a specific name (acceleration due to gravity) and symbol (g).

When there is air resistance, however, the bowling ball will hit the ground long before the feather. In this scenario, the bowling ball and the feather are being pulled downward due to the force of gravity just as before. Both objects, however, will also encounter the upward force of air resistance as they fall. *Air resistance* is the result of an object colliding with air molecules. The amount of air resistance that an object experiences as it falls is dependent on its speed and its surface area. According to Newton's second law, an object will accelerate if the forces acting upon it are unbalanced and the amount of acceleration is directly proportional to the amount of net force acting on it. Falling objects initially accelerate because there is no force big enough to balance the downward force of gravity. But as an object gains speed, it encounters an increasing amount of upward air resistance force. In fact, an object will continue to accelerate as it falls, until the air resistance force increases to a large enough value to balance the downward force of gravity. Once the downward force of gravity and the upward force of air resistance are balanced, the object will cease to accelerate. At this point, the object has reached its *terminal velocity.* The object will therefore continue to fall to the ground at this velocity. In the case of the bowling ball and the feather, the bowling ball has a much greater terminal velocity than the feather so it will take less time to reach the ground. A video that shows both of these scenarios can be found at *www.youtube.com/watch?v=E43-CfukEgs.*

Timeline

The instructional time needed to complete this lab investigation is 170–230 minutes. Appendix 2 (p. 411) provides options for implementing this lab investigation over several class periods. Option C (230 minutes) should be used if students are unfamiliar with scientific writing, because this option provides extra instructional time for scaffolding the writing process. You can scaffold the writing process by modeling, providing examples, and providing hints as students write each section of the report. Option D (170 minutes) should be used if students are familiar with scientific writing and have developed the skills needed to write an investigation report on their own. In option D, students complete stage 6 (writing the investigation report) and stage 8 (revising the investigation report) as homework.

Materials and Preparation

The materials needed to implement this investigation are listed in Table 7.1 (p. 134). Most of the equipment can be purchased from a science supply company such as Carolina, Flinn Scientific, or Ward's Science. You will, however, need to prepare your own beanbags before the first class period for this investigation. The beanbags must all be the same size, but each beanbag must be a different mass. The bags should be about 12 cm × 12 cm in size. You can make your own beanbags by sewing two pieces of fabric together or by using sandwich bags. We recommend filling the beanbags with either plastic beads or rice. The plastic beads can be purchased from a retail craft store such as Michaels or Hobby Lobby. Prepare the beanbags for this investigation as follows:

- Beanbag A: Add 60 ml (¼ cup) of plastic beads (or rice) and seal the bag.
- Beanbag B: Add 120 ml (½ cup) of plastic beads (or rice) and seal the bag.
- Beanbag C: Add 240 ml (1 cup) of plastic beads (or rice) and seal the bag.

TABLE 7.1

Materials list for Lab 7

Item	Quantity
Safety glasses or goggles	1 per student
Beanbag A	1 per group
Beanbag B	1 per group
Beanbag C	1 per group
Meterstick	1 per group
Stopwatch	1 per group
Electronic or triple beam balance	1 per group
Masking tape	As needed
Investigation Proposal B (optional)	1 per group
Whiteboard, 2' × 3'*	1 per group
Lab Handout	1 per student
Peer-review guide	1 per student
Checkout Questions	1 per student

*As an alternative, students can use computer and presentation software, such as Microsoft PowerPoint or Apple Keynote, to create their arguments.

We recommend that you use a set routine for distributing and collecting the materials during the lab investigation. For example, the equipment for each group can be set up at each group's lab station before class begins, or one member from each group can collect them from a table or a cart when needed during class.

Safety Precautions and Laboratory Waste Disposal

Remind students to follow all normal lab safety rules. In addition, tell the students to take the following safety precautions:

1. Wear sanitized safety glasses or goggles during lab setup, hands-on activity, and takedown.
2. Do not throw the beanbags.
3. Do not stand on tables or chairs.
4. Wash hands with soap and water after completing the lab activity.

There is no laboratory waste associated with this activity.

Topics for the Explicit and Reflective Discussion

Concepts That Can Be Used to Justify the Evidence

To provide an adequate justification of their evidence, students must explain why they included the evidence in their arguments and make the assumptions underlying their analysis and interpretation of the data explicit. In this investigation, students can use the following concepts to help justify their evidence:

- Gravitational force and mass
- Speed, velocity, and acceleration
- Free fall and air resistance

We recommend that you review these concepts during the explicit and reflective discussion to help students make this connection.

How to Design Better Investigations

It is important for students to reflect on the strengths and weaknesses of the investigation they designed during the explicit and reflective discussion. Students should therefore be encouraged to discuss ways to eliminate potential flaws, measurement errors, or sources of bias in their investigations. To help students be more reflective about the design of their investigation, you can ask the following questions:

1. What were some of the strengths of your investigation? What made it scientific?
2. What were some of the weaknesses of your investigation? What made it less scientific?
3. If you were to do this investigation again, what would you do to address the weaknesses in your investigation? What could you do to make it more scientific?

Crosscutting Concepts

This investigation is well aligned with two crosscutting concepts found in *A Framework for K–12 Science Education*, and you should review these concepts during the explicit and reflective discussion.

- *Patterns:* Scientists look for patterns in nature and attempt to understand the underlying cause of these patterns. Scientists, for example, often collect data and then look for patterns to identify a relationship between two variables, a trend over time, or a difference between groups. In this lab students will investigate trends and patterns in the amount of time it takes for objects with different masses to fall to the ground.

- *Scale, proportion, and quantity:* It is critical for scientists to be able to recognize what is relevant at different sizes, times, and scales. Scientists must also be able to recognize proportional relationships between categories, groups, or quantities. In this lab students will investigate any proportional relationships between the mass of an object and how long it takes that object to fall to the ground.

The Nature of Science and the Nature of Scientific Inquiry

This investigation is well aligned with two important concepts related to the *nature of science* (NOS) and the *nature of scientific inquiry* (NOSI), and you should review these concepts during the explicit and reflective discussion.

- *The difference between laws and theories in science:* A scientific law describes the behavior of a natural phenomenon or a generalized relationship under certain conditions; a scientific theory is a well-substantiated explanation of some aspect of the natural world. Theories do not become laws even with additional evidence; they explain laws. However, not all scientific laws have an accompanying explanatory theory. It is also important for students to understand that scientists do not discover laws or theories; the scientific community develops them over time.
- *The difference between data and evidence in science*: Data are measurements, observations, and findings from other studies that are collected as part of an investigation. Evidence, in contrast, is analyzed data and an interpretation of the analysis.

Hints for Implementing the Lab

- Allowing students to design their own procedures for collecting data gives students an opportunity to try, to fail, and to learn from their mistakes. However, you can scaffold students as they develop their procedure by having them fill out an investigation proposal. These proposals provide a way for you to offer students hints and suggestions without telling them how to do it. You can also check the proposals quickly during a class period. For this lab we suggest using Investigation Proposal B.
- It is important to use the same size beanbags in this investigation to control for the effect of air resistance. Each bag should have the same surface area so the amount of air resistance acting on the bags as they fall will be the same.
- Do not tell students how to determine the time it takes for a beanbag to fall to the ground. This methodological challenge will provide an ideal opportunity to discuss measurement error and how the choice of method can influence the accuracy of a measurement.

- Students will often try to make the data they collect fit their initial ideas about how mass affects the time it takes for an object to fall to the ground. To help address this issue, encourage students to fill out an investigation proposal and sketch graphs for each prediction. You can then encourage students to think about how well their result fits with each prediction.
- This is a good lab for students to make mistakes during the data collection stage. Students will quickly figure out what they did wrong during the argumentation session. It will also create an opportunity for students to reflect on and identify ways to improve the way they design investigations during the explicit and reflective discussion.
- This lab also provides an excellent opportunity to discuss how scientists identify a signal (a pattern or trend) from the noise (measurement error) in their data during the explicit and reflective discussion. Be sure to use this activity as a concrete example.
- Be sure to allow students to go back and re-collect data at the end of the argumentation session. Students often realize that they made numerous mistakes when they were collecting data as a result of their discussions during the argumentation session. The students, as a result, will want a chance to re-collect data, and the re-collection of data should be encouraged when time allows. This also offers an opportunity to discuss what scientists do when they realize a mistake is made inside the lab.
- After the lab is complete and the students have turned in their investigation reports, show them the video of a bowling ball and a feather falling after being dropped in a vacuum.

Topic Connections

Table 7.2 (p. 138) provides an overview of the scientific practices, crosscutting concepts, disciplinary core ideas, and supporting ideas at the heart of this lab investigation. In addition, it lists the NOS and NOSI concepts for the explicit and reflective discussion. Finally, it lists literacy and mathematics skills (*CCSS ELA* and *CCSS Mathematics*) that are addressed during the investigation.

TABLE 7.2

Lab 7 alignment with standards

Scientific practices	• Asking questions and defining problems • Planning and carrying out investigations • Analyzing and interpreting data • Using mathematics and computational thinking • Constructing explanations and designing solutions • Engaging in argument from evidence • Obtaining, evaluating, and communicating information
Crosscutting concepts	• Patterns • Scale, proportion, and quantity
Core ideas	• PS2.A: Forces and motion • PS2.B: Types of interactions
Supporting ideas	• Gravitational force and mass • Speed, velocity, and acceleration • Free fall and air resistance
NOS and NOSI concepts	• Scientific laws and theories • Difference between data and evidence
Literacy connections (*CCSS ELA*)	• *Reading*: Key ideas and details, craft and structure, integration of knowledge and ideas • *Writing*: Text types and purposes, production and distribution of writing, research to build and present knowledge, range of writing • *Speaking and listening*: Comprehension and collaboration, presentation of knowledge and ideas
Mathematics connections (*CCSS Mathematics*)	• Reason abstractly and quantitatively • Construct viable arguments and critique the reasoning of others • Use appropriate tools strategically • Attend to precision

Lab Handout

Lab 7. Mass and Free Fall

How Does Mass Affect the Amount of Time It Takes for an Object to Fall to the Ground?

Introduction

The motion of an object is the result of all the different forces that are acting on the object. If you push a toy car across the floor, it moves in the direction you pushed it. If the car then hits a wall, the force of the wall causes the car to stop. Applying a push or a pull to an object is an example of a contact force, where one object applies a force to another object through direct contact. There are other types of forces that can act on objects that do not involve objects touching. For example, a strong magnet can pull on a paper clip and make it move without ever actually touching the paper clip. Another example is static electricity. Static electricity in a rubber balloon can cause a person's hair to stand up without the balloon actually touching any of his or her hair. Magnetic forces and electrical forces are therefore called non-contact forces. Perhaps the most common non-contact force is gravity. Gravity is a force of attraction between two objects; the force due to gravity always works to bring objects closer together.

Any two objects that have mass will also have a gravitational force of attraction between them. Consider the Sun, Earth, and Moon as examples. Earth and the Moon are very large and have a lot of mass; the force of gravity between Earth and the Moon is strong enough to keep the Moon orbiting Earth even though they are very far apart. Similarly, the force of gravity between the Sun and Earth is strong enough to keep Earth in orbit around the Sun, despite Earth and the Sun being millions of miles apart. The force of gravity between two objects depends on the amount of mass of each object and how far apart they are. Objects that are more massive produce a greater gravitational force. The force of gravity between two objects also weakens as the distance between the two objects increases. So even though Earth and the Sun are very far apart from each other (which means less gravity), the fact that they are both very massive (which means more gravity) results in a gravitational force that is strong enough to keep Earth in orbit.

FIGURE L7.1

These skydivers are falling toward Earth because of gravity.

The gravitational force that acts between two objects, as noted earlier, can cause one of those objects to move. The skydivers in Figure L7.1, for example, are moving toward the center of Earth because of the force of gravity. Scientists describe the motion of an object by describing its speed, velocity, and acceleration. Speed is the distance an object travels in a specific amount of time. Velocity is the speed of an object in a given direction. Acceleration is the change in velocity divided by time. The amount of force required to change the motion of an object depends on the mass of that object. Therefore, as the mass of an object increases, so does the amount of force that is needed to change its motion.

In this investigation you will have an opportunity to explore the relationship between mass and the time it takes for an object to fall to the ground. Many people think that heavier objects fall to the ground faster than lighter ones because gravity will pull on heavier objects with more force, so the heavier object will accelerate faster. Other people, however, think that heavier objects have more inertia (the tendency of an object to stay still if it is still or keep moving if it is currently moving) so heavier objects will be less responsive to the force of gravity and take longer to accelerate. Unfortunately, it is challenging to determine which of these two explanations is the most valid because objects encounter air resistance as they fall. Air resistance is the result of an object moving through a layer of air and colliding with air molecules. The more air molecules that an object collides with, the greater the air resistance force. Air resistance is therefore dependent on the speed of the falling object and the surface area of the falling object. Since massive objects are often larger than less massive ones (consider a bowling ball and a marble as an example), it is often difficult to design a fair test of these two explanations. To determine the relationship between mass and the time it takes an object to fall to the ground, you will therefore need to design an experiment that will allow you to control for the influence of air resistance.

Your Task

Use what you know about forces and motion, patterns, and rates of change to design and carry out an experiment to determine the relationship between mass and the time it takes an object to fall to the ground.

The guiding question of this investigation is, **How does mass affect the amount of time it takes for an object to fall to the ground?**

Materials

You may use any of the following materials during your investigation:

- Beanbag A
- Beanbag B
- Beanbag C
- Meterstick
- Stopwatch
- Electronic or triple beam balance
- Masking tape
- Safety glasses or goggles

Safety Precautions

Follow all normal lab safety rules. In addition, take the following safety precautions:

1. Wear sanitized safety glasses or goggles during lab setup, hands-on activity, and takedown.
2. Do not throw the beanbags.

3. Do not stand on tables or chairs.
4. Wash hands with soap and water after completing the lab activity.

Investigation Proposal Required? ☐ Yes ☐ No

Getting Started

To answer the guiding question, you will need to design and conduct an experiment as part of your investigation. To accomplish this task, you must determine what type of data you need to collect, how you will collect it, and how you will analyze it.

To determine *what type of data you need to collect,* think about the following questions:

- What will serve as your independent variable?
- What will serve as your dependent variable?
- What measurements will you need to determine the rate of a falling object?

To determine *how you will collect your data,* think about the following questions:

- What variables will need to be controlled and how will you control them?
- How many tests will you need to run to have reliable data (to make sure it is consistent)?
- How will you make sure that your data are of high quality (i.e., how will you reduce error)?
- How will you keep track of the data you collect, and how will you organize it?

To determine *how you will analyze your data,* think about the following questions:

- How will you calculate the rate of a falling object?
- What type of calculations will you need to make to take into account multiple trials?
- What types of graphs or tables could you create to help make sense of your data?

Connections to Crosscutting Concepts, the Nature of Science, and the Nature of Scientific Inquiry

As you work through your investigation, be sure to think about

- the importance of looking for and understanding patterns in data,
- the importance of understanding proportional relationships in science,
- the difference between laws and theories in science, and
- the difference between data and evidence in science.

Initial Argument

Once your group has finished collecting and analyzing your data, your group will need to develop an initial argument. Your initial argument needs to include a *claim, evidence* to support your claim, and a *justification* of the evidence. The claim is your group's answer to the guiding question. The evidence is an analysis and interpretation of your data. Finally, the justification of the evidence is why your group thinks the evidence matters. The justification of the evidence is important because scientists can use different kinds of evidence to support their claims. Your group will create your initial argument on a whiteboard. Your whiteboard should include all the information shown in Figure L7.2.

FIGURE L7.2

Argument presentation on a whiteboard

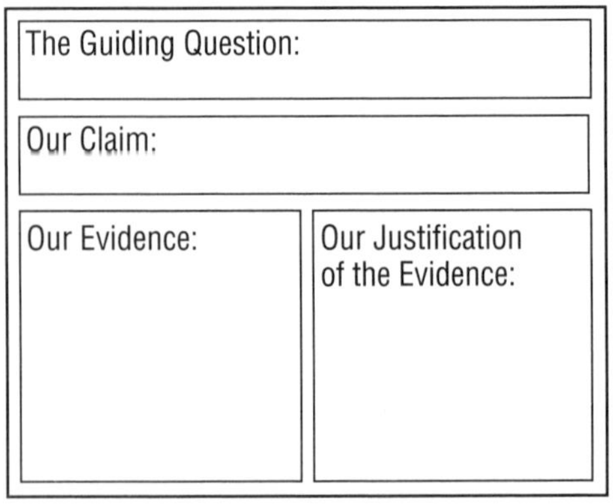

The Guiding Question:	
Our Claim:	
Our Evidence:	Our Justification of the Evidence:

Argumentation Session

The argumentation session allows all of the groups to share their arguments. One member of each group will stay at the lab station to share that group's argument, while the other members of the group go to the other lab stations to listen to and critique the arguments developed by their classmates. This is similar to how scientists present their arguments to other scientists at conferences. If you are responsible for critiquing your classmates' arguments, your goal is to look for mistakes so these mistakes can be fixed and they can make their argument better. The argumentation session is also a good time to think about ways you can make your initial argument better. Scientists must share and critique arguments like this to develop new ideas.

To critique an argument, you might need more information than what is included on the whiteboard. You will therefore need to ask the presenter lots of questions. Here are some good questions to ask:

- How did you collect your data? Why did you use that method? Why did you collect those data?
- What did you do to make sure the data you collected are reliable? What did you do to decrease measurement error?
- How did your group analyze the data? Why did you decide to do it that way? Did you check your calculations?
- Is that the only way to interpret the results of your analysis? How do you know that your interpretation of your analysis is appropriate?
- Why did your group decide to present your evidence in that way?
- What other claims did your group discuss before you decided on that one? Why did your group abandon those alternative ideas?

- How confident are you that your claim is valid? What could you do to increase your confidence?

Once the argumentation session is complete, you will have a chance to meet with your group and revise your initial argument. Your group might need to gather more data or design a way to test one or more alternative claims as part of this process. Remember, your goal at this stage of the investigation is to develop the most acceptable and valid answer to the research question!

Report

Once you have completed your research, you will need to prepare an *investigation report* that consists of three sections. Each section should provide an answer to the following questions:

1. What question were you trying to answer and why?
2. What did you do to answer your question and why?
3. What is your argument?

Your report should answer these questions in two pages or less. This report must be typed, and any diagrams, figures, or tables should be embedded into the document. Be sure to write in a persuasive style; you are trying to convince others that your claim is acceptable and valid!

Checkout Questions

Lab 7. Mass and Free Fall

How Does Mass Affect the Amount of Time It Takes for an Object to Fall to the Ground?

1. A group of students is investigating the relationship between mass and the time it takes for an object to reach the ground. They have six different cubes. Each cube is the same size but a different mass. They label the cubes in order of their relative mass. Cube A is the heaviest and cube F is the lightest. They then drop each cube from a height of 5 meters and time how long it takes for each cube to hit the ground. Use Tables 1–3 to answer the question below.

Table 1

Cube	Time (seconds)
A	1.3
B	1.2
C	1.1
D	1.0
E	0.9
F	0.8

Table 2

Cube	Time (seconds)
A	1.0
B	1.1
C	1.0
D	0.9
E	1.0
F	1.0

Table 3

Cube	Time (seconds)
A	0.8
B	0.9
C	1.0
D	1.1
E	1.2
F	1.3

Which table do you think best represents the data that the students would have collected?

a. Table 1
b. Table 2
c. Table 3
d. Unsure

How do you know?

2. "The force of gravitational attraction is directly dependent on the masses of both objects and inversely proportional to the square of the distance that separates their centers" is an example of a scientific theory.

    a. I agree with this statement.
    b. I disagree with this statement.

    Explain your answer, using an example from your investigation about mass and free fall.

3. "It took the cube 1.1 seconds to reach the ground" is an example of evidence.

    a. I agree with this statement.
    b. I disagree with this statement.

    Explain your answer, using an example from your investigation about mass and free fall.

4. Scientists often need to look for and understand the underlying cause of patterns in data. Explain why it important to be able to identify and understand patterns in data in science, using an example from your investigation about mass and free fall.

5. Scientists often look for proportional relationships between quantities in science. Explain what proportional relationships are and why they are important in science, using an example from your investigation about mass and free fall.

Teacher Notes

Lab 8. Force and Motion

How Do Changes in Pulling Force Affect the Motion of an Object?

Purpose

The purpose of this lab is to *introduce* students to the relationship between the force acting on an object and the resulting motion of that object. This lab gives students an opportunity to investigate what makes a system stable and what causes changes within a system under investigation. Students will also learn about the different methods that scientists use to answer different questions and about the nature and role of experiments in science.

The Content

This lab is intended to scaffold students' introduction to *Newton's second law of motion,* which describes that the acceleration of an object is directly proportional to the sum of the forces acting on that object and inversely proportional to the mass of the object. Newton's second law of motion is often written as follows:

$$\Sigma F = ma, \text{ where}$$

- ΣF is the sum of the forces (F) or net force on the object,
- measured in newtons (1 N = 1 kg*m/s^2);
- m is the mass of the object, measured in kilograms (kg); and
- a is the acceleration of the object, or the change in velocity for the object, measured in meters per second squared (m/s^2)

This equation describes that the sum of the forces acting on an object (ΣF) is equal to the product of the mass of the object (m) and the acceleration of the object (a). For an object with a constant mass, the acceleration of the object is directly related to the net force acting on it. If the net force on the object increases, then the acceleration of the object will increase. Likewise, if the net force on the object decreases, then the acceleration of the object will decrease.

It is important to remember that Newton's second law of motion describes the motion of an object when taking into account the sum of all the forces acting on that object. Therefore, it is important to remember two conditions related to the motion of an object; whether the forces acting on that object are *balanced* or *unbalanced.* When the forces acting on an object are balanced, that means that for every push or pull, there is a push or pull in the other direction that is equal strength. Consider the tug-of-war example from the Lab Handout.

If the teams pull with equal strength, the forces are balanced and there is no acceleration in either direction. A common misconception is that when the forces acting on an object are balanced, then there is no movement. This is not correct. When the forces acting on an object are balanced the object experiences zero acceleration, which simply means its velocity is not changing. Therefore, an object under the influence of balanced forces may have a speed of 0 m/s (i.e., sitting still with no movement) or it may have a constant speed of some value (e.g., 10 m/s or 50 mph). A car traveling down the road at a constant 50 mph is under the influence of balanced forces. Air resistance and friction from the tires and road balance the force from the engine; the car is moving but there is no acceleration. If the driver presses the gas pedal, which changes the strength of the force from the engine, the car will accelerate to a new speed; this represents unbalanced forces. When the forces (i.e., all the pushes and pulls) acting on an object do not add up to zero, the forces are unbalanced, which means that there is a net force in some direction. In this case the object will accelerate in the direction of that net force.

When physicists study the motion of objects, they often reduce the scenario to its most basic components and draw a diagram. These diagrams are called free-body diagrams and include a representation of the object under study (usually drawn as a dot or a box) and arrows indicating the different forces acting on the object, which are drawn in the direction that the force is acting on the object. Consider the scenario in the free-body diagram in Figure 8.1.

FIGURE 8.1

Sample free-body diagram

A 5 kg box sitting on the floor is being pulled to the right with a force of 12 N.
There is a friction force between the box and the floor of 2 N.
What is the acceleration of the box?

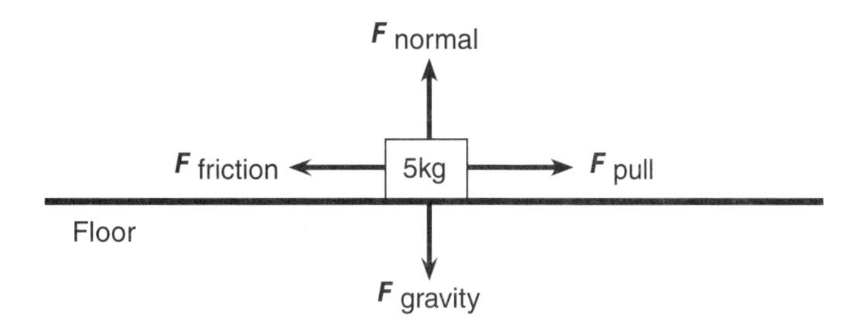

In this scenario, there are four forces displayed: F_{pull}, the force of the pull on the box (1 N); $F_{friction}$, the force of friction for the box sliding across the floor (2 N); $F_{gravity}$, the force of gravity pulling the box down; and F_{normal}, the force of the floor pushing up on the box. To determine the acceleration of the box, it is necessary to know the sum of all the forces acting on the box. The pulling force and the force due to friction are both given in the scenario. The force due to gravity is determined by multiplying the mass of the box (5 kg) by the acceleration the box experiences due to the force of gravity, which is a constant on Earth of 9.8 m/s^2. As a result, the force of gravity acting on the box is 49 N [$F = ma$, where F = (5 kg)*(9.8 m/s^2)]. In this example, the normal force is equal in strength to the force of gravity but in the opposite direction. Therefore, these two forces are considered balanced and cancel each other out, and there is no acceleration up or down due to either of these forces. In this scenario the pulling force is greater than the force of friction, meaning those forces are unbalanced, so intuitively we can predict that the box will accelerate to the right in the direction of the pull. For this example we can simply subtract the force due to friction from the force of the pull to determine that the net force is 10 N [$\Sigma F = F_{pull} - F_{friction}$ = 12 N – 2 N = 10 N].

Knowing the net force applied to the box and the mass of the box allows us to solve for the acceleration of the box using Newton's second law of motion as follows:

$$\Sigma F = ma, \text{ where } \Sigma F = 10 \text{ N and } m = 5 \text{ kg}$$

$$\Sigma F = ma$$

$$10 \text{ N} = (5 \text{ kg})*a$$

$$10 \text{ N}/5 \text{ kg} = a$$

$$a = 2 \text{ m/s}^2$$

If all other factors remain the same, changing the pulling force will result in a greater acceleration. Also, in many contexts the forces due to friction and air resistance are very small and therefore are not taken into account in calculations. This can result in small errors between predicted and observed values.

Timeline

The instructional time needed to complete this lab investigation is 200–280 minutes. Appendix 2 (p. 411) provides options for implementing this lab investigation over several class periods. Option A (280 minutes) should be used if students are unfamiliar with scientific writing, because this option provides extra instructional time for scaffolding the writing process. You can scaffold the writing process by modeling, providing examples, and providing hints as students write each section of the report. Option F (200 minutes) should be used if students are familiar with scientific writing and have developed the skills needed to write an investigation report on their own. In option F, students complete stage 6 (writing the investigation report) and stage 8 (revising the investigation report) as homework.

Materials and Preparation

The materials needed to implement this investigation are listed in Table 8.1.

TABLE 8.1

Materials list for Lab 8

Item	Quantity
Safety glasses or goggles	1 per student
Pull cart	1 per group
Pull cart track (or flat table)	1 per group
Pulley	1 per group
Pulley clamp	1 per group
String (approximately 1 m long)	1 per group
Hanging weight set	1 per group
Meterstick	1 per group
Electronic or triple beam balance	1 per group
Motion sensor with interface	1 per group
Investigation Proposal A (optional)	1 per group
Whiteboard, 2' × 3'*	1 per group
Lab Handout	1 per student
Peer-review guide	1 per student
Checkout Questions	1 per student

* As an alternative, students can use computer and presentation software, such as Microsoft PowerPoint or Apple Keynote, to create their arguments.

If you choose to use motion sensors for this investigation, that equipment, along with carts and accompanying tracks, can be purchased from vendors such as Pasco or Vernier. A sample equipment setup for using motion sensors is shown in Figure 8.2. We recommend using motion sensors for this lab because students will have the opportunity to collect data related to the cart's position, velocity, and acceleration simultaneously, which provides a variety of data that can be incorporated into their scientific arguments. Suitable motion sensor and cart equipment from these vendors include the following (item numbers are accurate at the time of writing this book but may change in the future):

FIGURE 8.2

Setting up the motion sensor, cart, and pulley

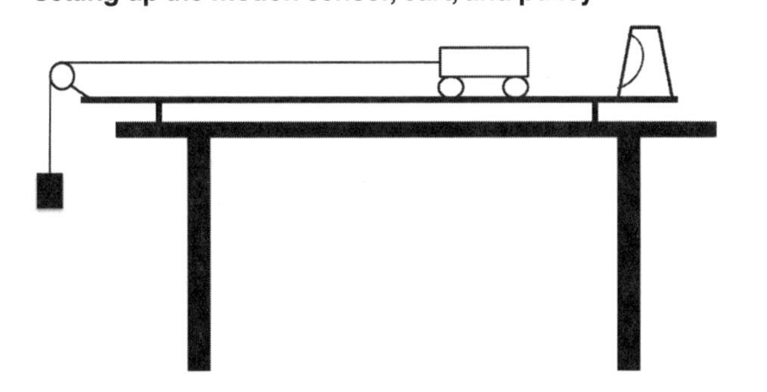

- 1.2 m PAScar Dynamics Systems from Pasco (item ME-6955A)
- PASPORT Motion Sensor from Pasco (item PS-2103A)
- Dynamics Cart and Track System from Vernier (item DTS)
- Go!Motion motion detector from Vernier (item GO-MOT)

Alternatively, basic tabletop pull carts, table clamps, and pulleys can be purchased from a science supply company such as Carolina, Flinn Scientific, or Ward's Science. Tabletop pull carts can be coupled with a motion sensor without a track system or can be used with a stand-alone timing device such as a spark timer. The data generated from a spark timer will require additional measurements and calculations to determine the velocity and/or acceleration of the cart. However, the marks on several strips of spark timer paper can also be compared side by side, in a qualitative manner, to generate evidence related to the cart's motion under different scenarios. Spark timers can be purchased from a variety of vendors.

We recommend that you use a set routine for distributing and collecting the materials during the lab investigation. For example, equipment for each group can be set up at each group's lab station before class begins, or one member from each group can collect them from a table or a cart when needed during class.

Safety Precautions and Laboratory Waste Disposal

Remind students to follow all normal lab safety rules. In addition, tell students to take the following safety precautions:

1. Wear sanitized safety glasses or goggles during lab setup, hands-on activity, and takedown.
2. Use caution when working with the moving pulleys, cart, and weights.
3. Keep their fingers and toes out of the way of the moving objects.
4. Wash hands with soap and water after completing the lab activity.

There is no laboratory waste associated with this activity.

Topics for the Explicit and Reflective Discussion

Concepts That Can Be Used to Justify the Evidence

To provide an adequate justification of their evidence, students must explain why they included the evidence in their arguments and make the assumptions underlying their analysis and interpretation of the data explicit. In this investigation, students can use the following concepts to help justify their evidence:

- Velocity
- Acceleration

- Balanced and unbalanced forces
- Newton's second law of motion

We recommend that you review these concepts during the explicit and reflective discussion to help students make this connection.

How to Design Better Investigations

It is important for students to reflect on the strengths and weaknesses of the investigation they designed during the explicit and reflective discussion. Students should therefore be encouraged to discuss ways to eliminate potential flaws, measurement errors, or sources of bias in their investigations. To help students be more reflective about the design of their investigation, you can ask the following questions:

1. What were some of the strengths of your investigation? What made it scientific?
2. What were some of the weaknesses of your investigation? What made it less scientific?
3. If you were to do this investigation again, what would you do to address the weaknesses in your investigation? What could you do to make it more scientific?

Crosscutting Concepts

This investigation is well aligned with two crosscutting concepts found in *A Framework for K–12 Science Education,* and you should review these concepts during the explicit and reflective discussion.

- *Systems and system models:* It is critical for scientists to be able to define the system under study and then make a model of it to understand it. Models can be physical, conceptual, or mathematical. In this lab students will observe motion and then have the opportunity to model that motion with graphs and diagrams to look for relationships.
- *Stability and change:* It is critical for scientists to understand what makes a system stable or unstable and what controls rates of change in a system. In this lab students will investigate how balanced or unbalanced forces influence the stability of a system.

The Nature of Science and the Nature of Scientific Inquiry

This investigation is well aligned with two important concepts related to the *nature of science* (NOS) and the *nature of scientific inquiry* (NOSI), and you should review these concepts during the explicit and reflective discussion.

- *Methods used in scientific investigations:* Examples of methods include experiments, systematic observations of a phenomenon, literature reviews, and analysis of existing data sets; the choice of method depends on the objectives of the research. There is no universal step-by step scientific method that all scientists follow; rather, different scientific disciplines (e.g., chemistry vs. biology) and fields within a discipline (e.g., organic vs. physical chemistry) use different types of methods, use different core theories, and rely on different standards to develop scientific knowledge.
- *The nature and role of experiments:* Scientists use experiments to test the validity of a hypothesis (i.e., a tentative explanation) for an observed phenomenon. Experiments include a test and the formulation of predictions (expected results) if the test is conducted and the hypothesis is valid. The experiment is then carried out and the predictions are compared with the observed results of the experiment. If the observed results match the predictions, then the hypothesis is supported. If the observed results do not match the predictions, then the hypothesis is not supported. A signature feature of an experiment is the control of variables to help eliminate alternative explanations for observed results.

Hints for Implementing the Lab

- Learn how to use the motion sensors and interface software before the lab begins. It is important for you to know how to use the equipment so you can help students when they get stuck or confused or if technical issues arise.
- Allowing students to design their own procedures for collecting data gives students an opportunity to try, to fail, and to learn from their mistakes. However, you can scaffold students as they develop their procedure by having them fill out an investigation proposal. These proposals provide a way for you to offer students hints and suggestions without telling them how to do it. You can also check the proposals quickly during a class period. For this lab we suggest using Investigation Proposal A.
- Allow the students to become familiar with the motion sensor and interface as part of the tool talk before they begin to design their investigation. This gives students a chance to see what they can and cannot do with the equipment.
- Be sure that students record actual values (e.g., mass, velocity, acceleration, or save/print graphs) and do not just attempt to hand draw what they see on the computer screen.
- Depending on how you set up the equipment for this lab, the hanging weights can sometimes fall to the floor during trials. It can be helpful if you provide each group with a cushioned landing spot for their hanging weights. A cardboard box lined with a towel or paper towels works well for this.

Topic Connections

Table 8.2 provides an overview of the scientific practices, crosscutting concepts, disciplinary core ideas, and supporting ideas at the heart of this lab investigation. In addition, it lists NOS and NOSI concepts for the explicit and reflective discussion. Finally, it lists literacy and mathematics skills (*CCSS ELA* and *CCSS Mathematics*) that are addressed during the investigation.

TABLE 8.2

Lab 8 alignment with standards

Scientific practices	• Asking questions and defining problems • Developing and using models • Planning and carrying out investigations • Analyzing and interpreting data • Using mathematics and computational thinking • Constructing explanations and designing solutions • Engaging in argument from evidence • Obtaining, evaluating, and communicating information
Crosscutting concepts	• Systems and system models • Stability and change
Core idea	• PS2.A: Forces and motion
Supporting ideas	• Velocity • Acceleration • Balanced and unbalanced forces • Newton's second law of motion
NOS and NOSI concepts	• Methods used in scientific investigations • Nature and role of experiments
Literacy connections (*CCSS ELA*)	• *Reading*: Key ideas and details, craft and structure, integration of knowledge and ideas • *Writing*: Text types and purposes, production and distribution of writing, research to build and present knowledge, range of writing • *Speaking and listening*: Comprehension and collaboration, presentation of knowledge and ideas
Mathematics connections (*CCSS Mathematics*)	• Reason abstractly and quantitatively • Construct viable arguments and critique the reasoning of others • Attend to precision • Look for and make use of structure

Lab Handout

Lab 8. Force and Motion

How Do Changes in Pulling Force Affect the Motion of an Object?

Introduction

A force can be described simply as a push or pull that acts on an object. For example, when you push or pull on a doorknob, you are applying a force that moves the door. In addition to a push or a pull, forces can be described as contact or non-contact. Pushing a box across the floor is an example of a contact force; the force to move the box is being applied by your hands, which are in contact with the box. However, non-contact forces can act on objects without having to actually touch the object. For example, a magnet can push or pull another magnet without the two ever touching each other. Similarly, gravity is a non-contact force that pulls objects closer together, such as when something falls toward Earth.

When you apply a force to an object, that object often will move. Sometimes, however, when you apply a force to an object it doesn't move. It is relatively easy to apply enough force to slide a box across the floor, but it is much more difficult to push a car down the road. The motion of an object is determined by the strength of the force applied to move it, the weight of the object, and any other forces that might be acting to move the object in a different direction. Consider a game of tug-of-war (see Figure L8.1): if both people pull with equal strength, then the rope doesn't move, but if one person pulls harder, the rope moves in that direction.

FIGURE L8.1

In a game of tug-of-war, the overall movement of the rope is based on the strength of the pull in both directions.

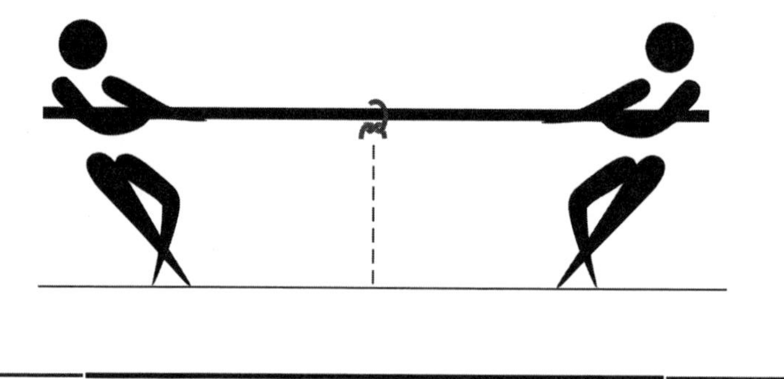

$\vec{F}_1$ ⟵ ⟶ $\vec{F}_2$

Isaac Newton (1642–1727) was a physicist who studied the motion of objects. He is perhaps most well known for the laws of motion he developed after extensive observation of the planets in our solar system. Newton described that (1) a stationary object will remain stationary unless an external force acts on it, (2) the change in an object's motion is proportional to the force acting on it, and (3) every force has an equal and opposite force. The motion of an object is the result of all the different forces that are pushing or pulling on that object. When all the forces are acting in the same direction, the object will move in that direction. If all the forces acting on an object are balanced, then the object either will not move or will move with a constant speed. When there are forces acting in different directions on the same object but they are not the same strength, then the forces are unbalanced; the object will move, but how it moves (e.g., fast, slow, constant speed, speeds up, or slows down) depends on the relationship of all the forces acting on the object.

Your Task

Use what you know about forces, systems, and stability and change to design and carry out an investigation that will allow you to predict how different pulling forces (hanging weights) influence the motion of a cart (e.g., does it speed up, slow down, or travel at a constant speed).

The guiding question of this investigation is, **How do changes in pulling force affect the motion of an object?**

Materials

You may use any of the following materials during your investigation:

- Pull cart
- Pull car track or flat table
- Pulley
- Pulley clamp
- String
- Hanging weights
- Meterstick
- Electronic or triple beam balance
- Motion sensor with interface
- Safety glasses or goggles

Safety Precautions

Follow all normal lab safety rules. In addition, take the following safety precautions:

1. Wear sanitized safety glasses or goggles during lab setup, hands-on activity, and takedown.
2. Use caution when working with the moving pulleys, cart, and weights.
3. Keep your fingers and toes out of the way of the moving objects.
4. Wash hands with soap and water after completing the lab activity.

Investigation Proposal Required? ☐ Yes ☐ No

FIGURE L8.2

Setting up the motion sensor, cart, and pulley

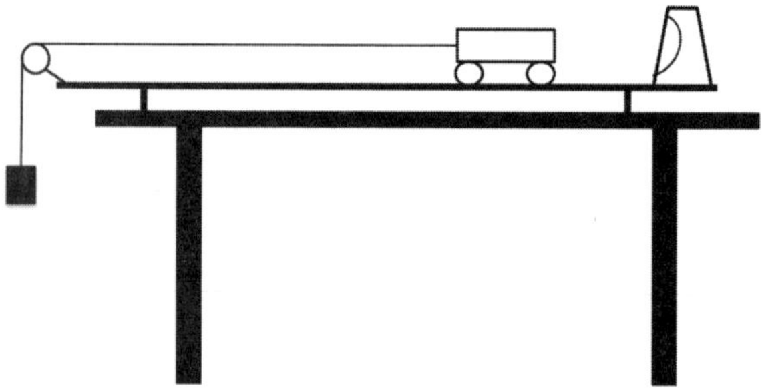

Getting Started

To answer the guiding question, you will need to plan an investigation to measure the motion of a cart as it is pulled across the tabletop. Figure L8.2 shows how you can set up the cart and motion sensor to collect your data; however, to accomplish this task, you must determine what type of data you need to collect, how you will collect it, and how you will analyze it.

To determine *what type of data you need to collect,* think about the following questions:

- What information do you need to describe the motion of the cart?
- What information or measurements do you need to calculate the speed of the cart?

To determine *how you will collect your data,* think about the following questions:

- What equipment will you need to collect the data you need?
- How will you make sure that your data are of high quality (i.e., how will you reduce error)?
- How will you keep track of the data you collect?
- How will you organize your data?

To determine *how you will analyze your data,* think about the following questions:

- What type of calculations will you need to make?
- What type of table or graph could you create to help make sense of your data?
- How will you determine the effect of different pulling forces on the cart's motion?

Connections to Crosscutting Concepts, the Nature of Science, and the Nature of Scientific Inquiry

As you work through your investigation, be sure to think about

- how scientists often need to understand and define systems under study and that making a model is a tool for developing a better understanding of natural phenomena in science,
- the importance of understanding what makes a system stable or unstable and what controls rates of change within a system,
- how scientists use different types of methods to answer different types of questions, and
- the nature and role of experiments within science.

Initial Argument

Once your group has finished collecting and analyzing your data, your group will need to develop an initial argument. Your initial argument needs to include a *claim, evidence* to support your claim, and a *justification* of the evidence. The claim is your group's answer to the guiding question. The evidence is an analysis and interpretation of your data. Finally, the justification of the evidence is why your group thinks the evidence matters. The justification of the evidence is important because scientists can use different kinds of evidence to support their claims. Your group will create your initial argument on a whiteboard. Your whiteboard should include all the information shown in Figure L8.3.

FIGURE L8.3

Argument presentation on a whiteboard

The Guiding Question:	
Our Claim:	
Our Evidence:	Our Justification of the Evidence:

Argumentation Session

The argumentation session allows all of the groups to share their arguments. One member of each group will stay at the lab station to share that group's argument, while the other members of the group go to the other lab stations to listen to and critique the arguments developed by their classmates. This is similar to how scientists present their arguments to other scientists at conferences. If you are responsible for critiquing your classmates' arguments, your goal is to look for mistakes so these mistakes can be fixed and they can make their argument better. The argumentation session is also a good time to think about ways you can make your initial argument better. Scientists must share and critique arguments like this to develop new ideas.

To critique an argument, you might need more information than what is included on the whiteboard. You will therefore need to ask the presenter lots of questions. Here are some good questions to ask:

- How did you collect your data? Why did you use that method? Why did you collect those data?
- What did you do to make sure the data you collected are reliable? What did you do to decrease measurement error?
- How did your group analyze the data? Why did you decide to do it that way? Did you check your calculations?
- Is that the only way to interpret the results of your analysis? How do you know that your interpretation of your analysis is appropriate?
- Why did your group decide to present your evidence in that way?
- What other claims did your group discuss before you decided on that one? Why did your group abandon those alternative ideas?

- How confident are you that your claim is valid? What could you do to increase your confidence?

Once the argumentation session is complete, you will have a chance to meet with your group and revise your initial argument. Your group might need to gather more data or design a way to test one or more alternative claims as part of this process. Remember, your goal at this stage of the investigation is to develop the most acceptable and valid answer to the research question!

Report

Once you have completed your research, you will need to prepare an *investigation report* that consists of three sections. Each section should provide an answer to the following questions:

1. What question were you trying to answer and why?
2. What did you do to answer your question and why?
3. What is your argument?

Your report should answer these questions in two pages or less. This report must be typed, and any diagrams, figures, or tables should be embedded into the document. Be sure to write in a persuasive style; you are trying to convince others that your claim is acceptable and valid!

Checkout Questions

Lab 8. Force and Motion

How Do Changes in Pulling Force Affect the Motion of an Object?

1. Describe a general rule for predicting the motion of an object that is being pushed or pulled by unbalanced forces.

2. Below is a position versus time graph for a car accelerating away from a stoplight.

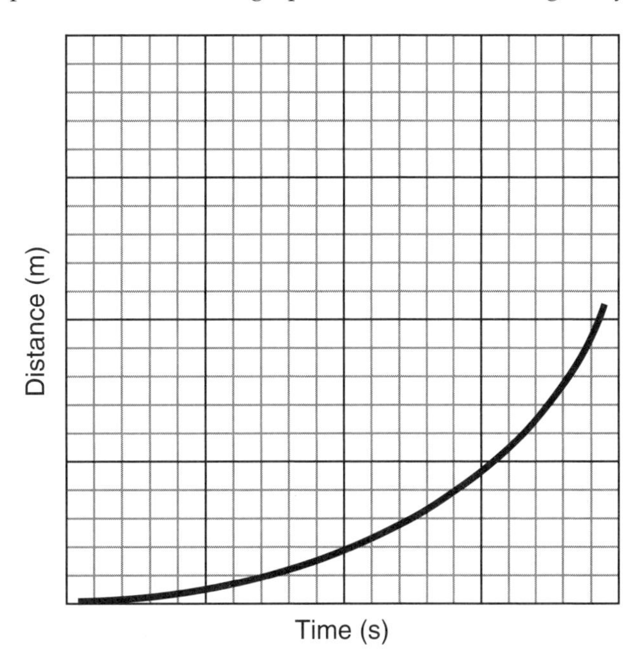

Draw a velocity versus time graph for the same car.

Explain your answer. Why did you draw your graph that way?

3. Experiments are the best way to get answers during a scientific investigation.

    a. I agree with this statement.
    b. I disagree with this statement.

    Explain your answer, using an example from your investigation about force and motion.

4. The scientific method guides scientists when they do their work.

    a. I agree with this statement.
    b. I disagree with this statement.

    Explain your answer, using an example from your investigation about force and motion.

5. Scientists sometimes study systems that are very large or very small, and sometimes scientists study systems that have lots of components. It is useful to make models of complex systems to better understand what is going on. Explain how models can be useful for understanding systems, using an example from your investigation about force and motion.

6. It is often important to understand the relationships between components of a system. Explain why it is important to identify factors that cause a system to become unstable, using an example from your investigation about force and motion.

Teacher Notes

Lab 9. Mass and Motion:

How Do Changes in the Mass of an Object Affect Its Motion?

Purpose

The purpose of this lab is to *introduce* students to the relationship between the force acting on an object, the object's mass, and the resulting motion of that object. This lab gives students an opportunity to investigate what makes a system stable and what causes changes within a system, as well as draw conclusions from patterns they observe during an investigation. Students will also learn about the difference between observations and inferences and the difference between data and evidence.

The Content

This lab is intended to scaffold students' introduction to Newton's second law of motion, which describes that the acceleration of an object is directly proportional to the sum of the forces acting on that object and inversely proportional to the mass of the object. Newton's second law of motion is often written as follows:

$\Sigma F = ma$, where

- ΣF is the sum of the forces (F) or net force on the object,
- measured in newtons ($1 \text{ N} = 1 \text{ kg*m/s}^2$);
- m is the mass of the object, measured in kilograms (kg); and,
- a is the acceleration of the object, or the change in velocity for the object, measured in meters per second squared (m/s^2)

This equation describes that the sum of the forces acting on an object (ΣF) is equal to the product of the mass of the object (m) and the acceleration of the object (a). For an object with a constant mass, the acceleration of the object is directly related to the net force acting on it. If the net force on the object increases, then the acceleration of the object will increase. Likewise, if the net force on the object decreases, then the acceleration of the object will decrease. If the force acting on an object is held constant but the mass of the object changes, the change in acceleration is inversely proportional to the change in mass. If force is constant and the mass increases, then the acceleration of the object will decrease; similarly, if force is constant and the mass of the object decreases, then the acceleration will increase.

It is important to remember that Newton's second law of motion describes the motion of an object when taking into account the sum of all the forces acting on that object. Therefore, it is important to remember two conditions related to the motion of an object; whether the forces acting on that object are *balanced* or *unbalanced.* When the forces acting on an object are balanced, that means that for every push or pull, there is a push or pull in the other direction that is equal strength. Consider two teams playing tug-of-war. If the teams pull with equal strength, the forces are balanced and there is no acceleration in either direction. A common misconception is that when the forces acting on an object are balanced, then there is no movement. This is not correct. When the forces acting on an object are balanced the object experiences zero acceleration, which simply means its velocity is not changing. Therefore, an object under the influence of balanced forces may have a speed of 0 m/s (i.e., sitting still with no movement) or it may have a constant speed of some value (e.g., 10 m/s or 50 mph). A car traveling down the road at a constant 50 mph is under the influence of balanced forces. Air resistance and friction from the tires and road balance the force from the engine; the car is moving but there is no acceleration. If the driver presses the gas pedal, which changes the strength of the force from the engine, the car will accelerate to a new speed; this represents unbalanced forces. When the forces (i.e., all the pushes and pulls) acting on an object do not add up to zero, the forces are unbalanced, which means that there is a net force in some direction. In this case the object will accelerate in the direction of that net force.

When physicists study the motion of objects, they often reduce the scenario to its most basic components and draw a diagram. These diagrams are called free-body diagrams and include a representation of the object under study (usually drawn as a dot or a box) and arrows indicating the different forces acting on the object, which are drawn in the direction that the force is acting on the object. Consider the scenario in the free-body diagram in Figure 9.1.

FIGURE 9.1

Sample free-body diagram

A 5 kg box sitting on the floor is being pulled to the right with a force of 12 N.
There is a friction force between the box and the floor of 2 N.
What is the acceleration of the box?

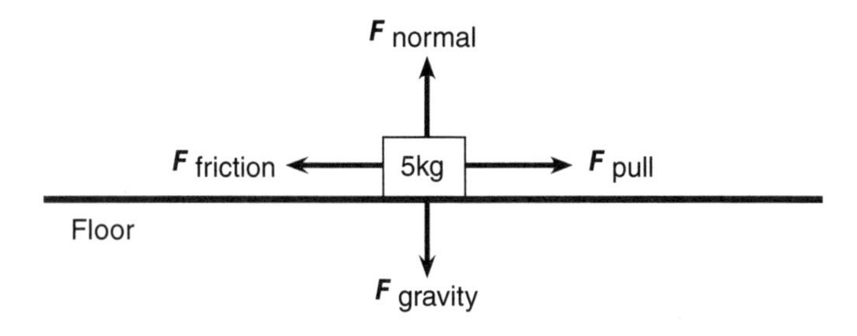

In this scenario, there are four forces displayed: F_{pull}, the force of the pull on the box (12 N); $F_{friction}$, the force of friction for the box sliding across the floor (2 N); $F_{gravity}$, the force of gravity pulling the box down; and F_{normal}, the force of the floor pushing up on the box. To determine the acceleration of the box, it is necessary to know the sum of all the forces acting on the box. The pulling force and the force due to friction are both given in the scenario. The force due to gravity is determined by multiplying the mass of the box (5 kg) by the acceleration the box experiences due to the force of gravity, which is a constant on Earth of 9.8 m/s^2. As a result, the force of gravity acting on the box is 49 N [$F = ma$, where $F = (5\text{ kg})*(9.8\text{ m/s}^2)$]. In this example, the normal force is equal in strength to the force of gravity but in the opposite direction. Therefore, these two forces are considered balanced and cancel each other out, and there is no acceleration up or down due to either of these forces. In this scenario, the pulling force is greater than the force of friction, meaning those forces are unbalanced, so intuitively we can predict that the box will accelerate to the right in the direction of the pull. For this example, we can simply subtract the force due to friction from the force of the pull to determine that the net force is 10 N [$\Sigma F = F_{pull} - F_{friction} = 12\text{ N} - 2\text{ N} = 10\text{ N}$].

Knowing the net force applied to the box and the mass of the box allows us to solve for the acceleration of the box using Newton's second law of motion as follows:

$$\Sigma F = ma, \text{ where } \Sigma F = 10\text{ N and } m = 5\text{ kg}$$

$$\Sigma F = ma$$

$$10\text{ N} = (5\text{ kg})*a$$

$$10\text{ N}/5\text{ kg} = a$$

$$a = 2\text{ m/s}^2$$

If all other factors remain the same, changing the pulling force will result in a greater acceleration. Also, in many contexts the forces due to friction and air resistance are very small and therefore are not taken into account in calculations. This can result in small errors between predicted and observed values. *Note:* Ignoring friction in the above example results in an acceleration of 2.4 m/s^2.

To illustrate how changes in mass affect the acceleration of an object, consider the simplified scenarios in Figure 9.2. Both scenarios ignore friction.

In the scenarios in Figure 9.2, the pulling force applied to each box is the same, 15.0 N. In both cases, the force due to gravity and the normal force acting on the box are balanced, therefore they do not cause any acceleration either up or down. The only force affecting the motion of the boxes is the pulling force to the right. The calculations in Figure 9.2 show that the acceleration for the 3.0 kg box (a = 5.0 m/s^2) is twice that of the 6.0 kg box (a = 2.5 m/s^2). These results demonstrate the inverse relationship between the mass of an object and the acceleration of that object if the net force acting on the objects is the same.

FIGURE 9.2

Sample scenarios for how changes in mass affect acceleration

A 3.0 kg box is being pulled by a force of 15.0 N. What is the acceleration of the box?

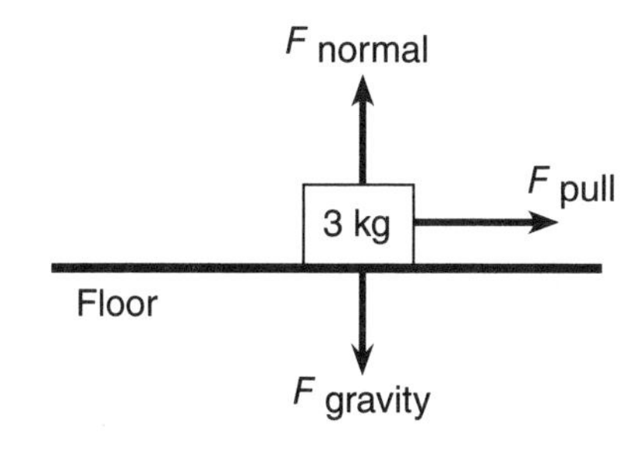

$\Sigma F = ma$, where $\Sigma F = 15.0$ N and $m = 3.0$ kg

$$\Sigma F = ma$$
$$15.0 \text{ N} = (3.0 \text{ kg})*a$$
$$15.0 \text{ N}/3.0 \text{ kg} = a$$
$$a = 5.0 \text{ m/s}^2$$

A 6.0 kg box is being pulled by a force of 15.0 N. What is the acceleration of the box?

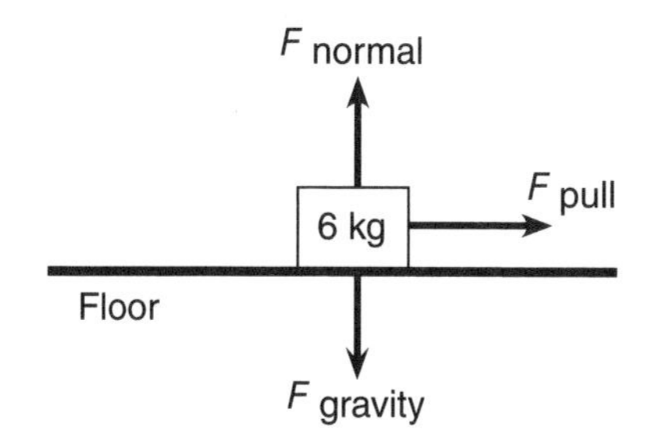

$\Sigma F = ma$, where $\Sigma F = 15.0$ N and $m = 6.0$ kg

$$\Sigma F = ma$$
$$15.0 \text{ N} = (6.0 \text{ kg})*a$$
$$15.0 \text{ N}/6.0 \text{ kg} = a$$
$$a = 2.5 \text{ m/s}^2$$

Timeline

The instructional time needed to complete this lab investigation is 200–280 minutes. Appendix 2 (p. 411) provides options for implementing this lab investigation over several class periods. Option A (280 minutes) should be used if students are unfamiliar with scientific writing, because this option provides extra instructional time for scaffolding the writing process. You can scaffold the writing process by modeling, providing examples, and providing hints as students write each section of the report. Option F (200 minutes) should be used if students are familiar with scientific writing and have developed the skills needed to write an investigation report on their own. In option F, students complete stage 6 (writing the investigation report) and stage 8 (revising the investigation report) as homework.

Materials and Preparation

The materials needed to implement this investigation are listed in Table 9.1 (p. 168).

TABLE 9.1

Materials list for Lab 9

Item	Quantity
Safety glasses or goggles	1 per student
Pull cart	1 per group
Pull cart track (or flat table)	1 per group
Pulley	1 per group
Pulley clamp	1 per group
Cart masses (500 g)	3 per group
String (approximately 1 m long)	1 per group
Hanging weight set	1 per group
Meterstick	1 per group
Electronic or triple beam balance	1 per group
Motion sensor with interface	1 per group
Investigation Proposal A (optional)	1 per group
Whiteboard, 2' × 3'*	1 per group
Lab Handout	1 per student
Peer-review guide	1 per student
Checkout Questions	1 per student

* As an alternative, students can use computer and presentation software, such as Microsoft PowerPoint or Apple Keynote, to create their arguments.

If you choose to use motion sensors for this investigation, that equipment, along with carts and accompanying tracks, can be purchased from vendors such as Pasco or Vernier. A sample equipment setup for using motion sensors is shown in Figure 9.3. We recommend using motion sensors for this lab because students will have the opportunity to collect data related to the cart's position, velocity, and acceleration simultaneously, which provides a variety of data that can be incorporated into their scientific arguments. Suitable motion sensor and cart equipment from these vendors include the following (item numbers are accurate at the time of writing this book but may change in the future):

- 1.2 m PAScar Dynamics Systems from Pasco (item ME-6955A)
- PASPORT Motion Sensor from Pasco (item PS-2103A)
- Dynamics Cart and Track System from Vernier (item DTS)

- Go!Motion motion detector from Vernier (item GO-MOT)

FIGURE 9.3

Setting up the motion sensor, cart, pulley, and masses

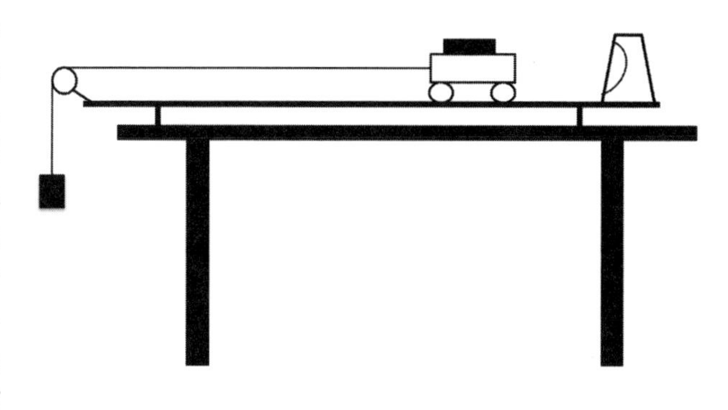

Alternatively, basic tabletop pull carts, table clamps, and pulleys can be purchased from a science supply company such as Carolina, Flinn Scientific, or Ward's Science. Tabletop pull carts can be coupled with a motion sensor without a track system or can be use with a stand-alone timing device such as a spark timer. The data generated from a spark timer will require additional measurements and calculations to determine the velocity and/or acceleration of the cart. However, the marks on several strips of spark timer paper can also be compared side by side, in a qualitative manner, to generate evidence related to the cart's motion under different scenarios. Spark timers can be purchased from a variety of vendors.

We recommend that you use a set routine for distributing and collecting the materials during the lab investigation. For example, equipment for each group can be set up at each group's lab station before class begins, or one member from each group can collect them from a table or a cart when needed during class.

Safety Precautions and Laboratory Waste Disposal

Remind students to follow all normal lab safety rules. In addition, tell students to take the following safety precautions.

1. Wear sanitized safety glasses or goggles during lab setup, hands-on activity, and takedown.
2. Use caution when working with the moving pulleys, cart, masses, and weights.
3. Keep their fingers and toes out of the way of the moving objects.
4. Wash hands with soap and water after completing the lab activity.

There is no laboratory waste associated with this activity.

Topics for the Explicit and Reflective Discussion

Concepts That Can Be Used to Justify the Evidence

To provide an adequate justification of their evidence, students must explain why they included the evidence in their arguments and make the assumptions underlying their analysis and interpretation of the data explicit. In this investigation, students can use the following concepts to help justify their evidence:

- Velocity
- Acceleration
- Mass versus weight
- Balanced and unbalanced forces
- Newton's second law of motion

We recommend that you review these concepts during the explicit and reflective discussion to help students make this connection.

How to Design Better Investigations

It is important for students to reflect on the strengths and weaknesses of the investigation they designed during the explicit and reflective discussion. Students should therefore be encouraged to discuss ways to eliminate potential flaws, measurement errors, or sources of bias in their investigations. To help students be more reflective about the design of their investigation, you can ask the following questions:

- What were some of the strengths of your investigation? What made it scientific?
- What were some of the weaknesses of your investigation? What made it less scientific?
- If you were to do this investigation again, what would you do to address the weaknesses in your investigation? What could you do to make it more scientific?

Crosscutting Concepts

This investigation is well aligned with two crosscutting concepts found in *A Framework for K–12 Science Education,* and you should review these concepts during the explicit and reflective discussion.

- *Patterns:* Scientists look for patterns in nature and attempt to understand the underlying cause of these patterns. In this lab students will identify patterns related to an object's mass and its motion.
- *Stability and change:* It is critical for scientists to understand what makes a system stable or unstable and what controls rates of change in a system. In this lab students will investigate how balanced or unbalanced forces influence the stability of a system.

The Nature of Science and the Nature of Scientific Inquiry

This investigation is well aligned with two important concepts related to the *nature of science* (NOS) and the *nature of scientific inquiry* (NOSI), and you should review these concepts during the explicit and reflective discussion.

- *The difference between observations and inferences in science:* An observation is a descriptive statement about a natural phenomenon, whereas an inference is an interpretation of an observation. Students should also understand that current scientific knowledge and the perspectives of individual scientists guide both observations and inferences. Thus, different scientists can have different but equally valid interpretations of the same observations due to differences in their perspectives and background knowledge.
- *The difference between data and evidence in science:* Data are measurements, observations, and findings from other studies that are collected as part of an investigation. Evidence, in contrast, is analyzed data and an interpretation of the analysis.

Hints for Implementing the Lab

- Learn how to use the motion sensors and interface software before the lab begins. It is important for you to know how to use the equipment so you can help students when they get stuck or confused or if technical issues arise.
- Allowing students to design their own procedures for collecting data gives students an opportunity to try, to fail, and to learn from their mistakes. However, you can scaffold students as they develop their procedure by having them fill out an investigation proposal. These proposals provide a way for you to offer students hints and suggestions without telling them how to do it. You can also check the proposals quickly during a class period. For this lab we suggest using Investigation Proposal A.
- Allow the students to become familiar with the motion sensor and interface as part of the tool talk before they begin to design their investigation. This gives students a chance to see what they can and cannot do with the equipment.
- Be sure that students record actual values (e.g., mass, velocity, acceleration, or save/print graphs) and do not just attempt to hand draw what they see on the computer screen.
- Depending on how you set up the equipment for this lab, the hanging weights can sometimes fall to the floor during trials. It can be helpful if you provide each group with a cushioned landing spot for their hanging weights. A cardboard box lined with a towel or paper towels works well for this.

Topic Connections

Table 9.2 (p. 172) provides an overview of the scientific practices, crosscutting concepts, disciplinary core ideas, and supporting ideas at the heart of this lab investigation. In addition, it lists NOS and NOSI concepts for the explicit and reflective discussion. Finally, it lists literacy and mathematics skills (*CCSS ELA* and *CCSS Mathematics*) that are addressed during the investigation.

TABLE 9.2

Lab 9 alignment with standards

Scientific practices	• Asking questions and defining problems • Developing and using models • Planning and carrying out investigations • Analyzing and interpreting data • Using mathematics and computational thinking • Constructing explanations and designing solutions • Engaging in argument from evidence • Obtaining, evaluating, and communicating information
Crosscutting concepts	• Patterns • Stability and change
Core idea	• PS2.A: Forces and motion
Supporting ideas	• Velocity • Acceleration • Mass versus weight • Balanced and unbalanced forces • Newton's second law of motion
NOS and NOSI concepts	• Observations and inferences • Difference between data and evidence
Literacy connections (*CCSS ELA*)	• *Reading*: Key ideas and details, craft and structure, integration of knowledge and ideas • *Writing*: Text types and purposes, production and distribution of writing, research to build and present knowledge, range of writing • *Speaking and listening*: Comprehension and collaboration, presentation of knowledge and ideas
Mathematics connections (*CCSS Mathematics*)	• Reason abstractly and quantitatively • Construct viable arguments and critique the reasoning of others • Attend to precision • Look for and make use of structure

Lab Handout

Lab 9. Mass and Motion

How Do Changes in the Mass of an Object Affect Its Motion?

Introduction

The motion of an object depends on all the different forces that are acting on the object, how strong those different forces are, and how much mass the object has. Changing the number of forces acting on an object, the direction of those forces, or the strength of the forces will have an effect on the object's motion. But what happens if you apply all the same forces to two different objects, one with a small mass and one with a larger mass?

The amount of mass of an object is a measure of the amount of matter in that object. Objects with more mass have more weight (a measure of the force of gravity) because there is a stronger attraction between the object and Earth due to gravity. Motorcycles are lightweight vehicles, about 250 kg, with strong engines that help them travel at high speeds (see Figure L9.1). How might the speed of a motorcycle change if it is carrying one rider or two riders? In this example the strength of the engine doesn't change, but the total mass of the motorcycle and riders increases. It is important for scientists and engineers to understand the relationship between the forces applied to an object and the mass of that object so that they can predict the object's motion.

FIGURE L9.1

Motorcycles are lightweight, but their engines generate a lot of force.

Your Task

Use what you know about forces, stability and change, and patterns to design and conduct an investigation that will allow you to describe the motion of a cart (e.g., does it speed up, slow down, or travel at a constant speed) and how its motion is affected by changing the mass of the cart while keeping the pulling force the same.

The guiding question of this investigation is, **How do changes in the mass of an object affect its motion?**

LAB 9

Materials

You may use any of the following materials during your investigation:

- Pull cart
- Pull car track or flat table
- Pulley
- Pulley clamp
- Cart masses
- String
- Hanging weights
- Meterstick
- Electronic or triple beam balance
- Motion sensor with interface
- Safety glasses or goggles

Safety Precautions

Follow all normal lab safety rules. In addition, take the following safety precautions:

1. Wear sanitized safety glasses or goggles during lab setup, hands-on activity, and takedown.
2. Use caution when working with the moving pulleys, cart, masses, and weights.
3. Keep your fingers and toes out of the way of the moving objects.
4. Wash hands with soap and water after completing the lab activity.

Investigation Proposal Required? ☐ Yes ☐ No

FIGURE L9.2

Setting up the motion sensor, cart, pulley, and masses

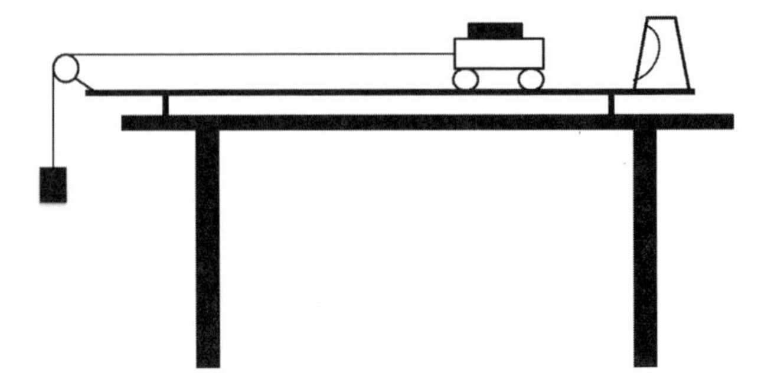

Getting Started

To answer the guiding question, you will need to plan an investigation to measure the motion of a cart as it is pulled across the tabletop. Figure L9.2 shows how you can set up the cart and motion sensor to collect your data; however, to accomplish this task, you must determine what type of data you need to collect, how you will collect it, and how you will analyze it.

To determine *what type of data you need to collect,* think about the following questions:

- What information do you need to describe the motion of the cart?
- What information or measurements do you need to calculate the speed of the cart?

To determine *how you will collect your data,* think about the following questions:

- What equipment will you need to collect the data you need?

- How will you make sure that your data are of high quality (i.e., how will you reduce error)?
- How will you keep track of the data you collect?
- How will you organize your data?

To determine *how you will analyze your data,* think about the following questions:

- What type of calculations will you need to make?
- What type of table or graph could you create to help make sense of your data?
- How will you determine the effect of different pulling forces on the cart's motion?

Connections to Crosscutting Concepts, the Nature of Science, and the Nature of Scientific Inquiry

As you work through your investigation, be sure to think about

- how scientists often look for patterns in nature and attempt to understand the underlying cause of these patterns,
- the importance of understanding what makes a system stable or unstable and what controls rates of change within a system,
- the differences between observations and inferences in science, and
- the difference between data and evidence in science.

Initial Argument

Once your group has finished collecting and analyzing your data, your group will need to develop an initial argument. Your initial argument needs to include a *claim, evidence* to support your claim, and a *justification* of the evidence. The claim is your group's answer to the guiding question. The evidence is an analysis and interpretation of your data. Finally, the justification of the evidence is why your group thinks the evidence matters. The justification of the evidence is important because scientists can use different kinds of evidence to support their claims. Your group will create your initial argument on a whiteboard. Your whiteboard should include all the information shown in Figure L9.3.

FIGURE L9.3

Argument presentation on a whiteboard

The Guiding Question:	
Our Claim:	
Our Evidence:	Our Justification of the Evidence:

Argumentation Session

The argumentation session allows all of the groups to share their arguments. One member of each group will stay at the lab station to share that group's argument, while the other members of the group go to the other lab stations to listen to and critique the arguments developed by their classmates. This is similar to how scientists present their arguments to other scientists at conferences. If you are responsible for critiquing your classmates' arguments, your goal is to look for mistakes so these mistakes can be fixed and they can make their argument better. The argumentation session is also a good time to think about ways you can make your initial argument better. Scientists must share and critique arguments like this to develop new ideas.

To critique an argument, you might need more information than what is included on the whiteboard. You will therefore need to ask the presenter lots of questions. Here are some good questions to ask:

- How did you collect your data? Why did you use that method? Why did you collect those data?
- What did you do to make sure the data you collected are reliable? What did you do to decrease measurement error?
- How did your group analyze the data? Why did you decide to do it that way? Did you check your calculations?
- Is that the only way to interpret the results of your analysis? How do you know that your interpretation of your analysis is appropriate?
- Why did your group decide to present your evidence in that way?
- What other claims did your group discuss before you decided on that one? Why did your group abandon those alternative ideas?
- How confident are you that your claim is valid? What could you do to increase your confidence?

Once the argumentation session is complete, you will have a chance to meet with your group and revise your initial argument. Your group might need to gather more data or design a way to test one or more alternative claims as part of this process. Remember, your goal at this stage of the investigation is to develop the most acceptable and valid answer to the research question!

Report

Once you have completed your research, you will need to prepare an *investigation report* that consists of three sections. Each section should provide an answer to the following questions:

1. What question were you trying to answer and why?
2. What did you do to answer your question and why?
3. What is your argument?

Your report should answer these questions in two pages or less. This report must be typed, and any diagrams, figures, or tables should be embedded into the document. Be sure to write in a persuasive style; you are trying to convince others that your claim is acceptable and valid!

Checkout Questions

Lab 9. Mass and Motion

How Do Changes in the Mass of an Object Affect Its Motion?

1. Describe a general relationship between the acceleration of two objects that are being pushed by the same-strength force, but one object is twice as heavy as the other.

2. Ashley is a race car driver and she wants her car to go faster. She is trying to decide between two plans to increase the acceleration of her car. Her car has a mass of 900 kg, and when she races it can accelerate at 20 m/s^2. Ashley would like her car to accelerate at 25 m/s^2. Her first plan is to make her car lighter by using some new materials; if she does that her car will have a mass of 675 kg. Her second plan is to get a stronger engine; the new engine would be 25% stronger than her current engine. But the new engine will make the car weigh 1,100 kg. Ashley only has enough money for one option. Which would you recommend?

   Explain your answer. Why did you make that recommendation?

3. In science, observations and inferences are the same thing.

   a. I agree with this statement.
   b. I disagree with this statement.

   Explain your answer, using an example from your investigation about mass and the motion of objects.

4. When discussing the results of an investigation, there is no need to differentiate between data and evidence—they are really the same thing.

   a. I agree with this statement.
   b. I disagree with this statement.

   Explain your answer, using an example from your investigation about mass and the motion of objects.

5. Identifying patterns in nature is important to the work of many scientists. Explain how understanding patterns and their causes is helpful to scientists. Use an example from your investigation about mass and the motion of objects to help in your explanation.

6. It is important for scientist to make predictions about natural systems. Use an example from your investigation about mass and the motion of objects to explain why it is important to identify factors that cause changes in a system or cause the system to become unstable.

Teacher Notes

Lab 10. Magnetic Force

How Is the Strength of an Electromagnet Affected by the Number of Turns of Wire in a Coil?

Purpose

The purpose of this lab is to *introduce* students to the relationship between electricity and magnetism and the factors that influence the strength of electromagnets. In addition, this lab gives students an opportunity to investigate the relationship between the structure of an electromagnet and its function, as well as how the flow of energy affects the function of an electromagnet. Students will also learn about the difference between laws and theories and the difference between data and evidence.

The Content

The relationship between electricity and magnetism is important within science as well as within broader society. Scientists and engineers take advantage of this relationship in relatively simply ways, such as making a doorbell ring with a solenoid, and in much more complex tasks, such as using magnetic resonance to take images of the body for medical purposes. Understanding the connection between electricity and magnetism requires taking a detailed view of atoms.

FIGURE 10.1

Model iron atom showing electrons in orbitals around the nucleus

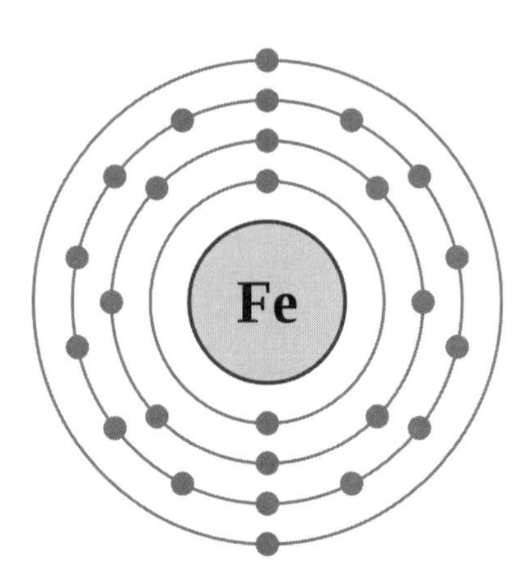

All matter is composed of atoms, and all atoms are composed of a nucleus of protons and neutrons, and electrons outside that nucleus. The electrons within an atom are responsible for the properties that allow electricity to flow through conductors such as copper wire. A simplified model of an atom, as shown in Figure 10.1, includes a core nucleus made of protons and neutrons with electrons orbiting that core. The electrons in an atom are also responsible for the magnetic properties of some elements. Naturally occurring elements, such as nickel or iron, can act like magnets because of the arrangement of the electrons and atoms within those substances. These substances are called ferromagnetic materials.

The electrons within an atom are constantly spinning as they orbit the nucleus of the atom. The direction of that spin influences the magnetic properties of the substances. When the atoms are arranged in such a way that the spins of the electrons are working together, the metal becomes a magnet. Ferromagnetic materials have properties that make it easy for them to become magnetized. When ferromagnetic materials become magnetized, they generate a *magnetic field* and are able to apply a force on other metal substances containing nickel or iron.

It is sometimes helpful to imagine an individual iron atom as a very small magnet with a north pole and a south pole. In a ferromagnetic material that has been magnetized, essentially all the individual atoms are arranged so that their north poles point in one direction and their south poles point in the opposite direction. When the individual atoms are shifted into this arrangement, the substance becomes a solid magnet.

Figure 10.2 shows how atoms in a ferromagnetic material are arranged when they demonstrate nonmagnetic and magnetic properties. North and south poles of magnets are attracted to each other because of the arrangement and orientation of the atoms within the magnet. As seen in Figure 10.2(b), when atoms are aligned in a magnet their poles point in opposite directions, and this is related to the spin of the electrons within individual atoms; the north end of one magnet is repelled from the north end of another magnet (the same is true for south end vs. south end) because they have similar arrangements of atoms and spinning electrons within those atoms. In contrast, north and south poles of magnets are attracted to each other because they have opposite arrangements of atoms and spinning electrons compared with one another.

FIGURE 10.2

Substance with nonmagnetic (a) and magnetic (b) characteristics

(a) In nonmagnetic substances, the atoms have no special arrangement and the poles point in a variety of directions.

(b) When a substance is magnetized, the atoms align so that like poles point in the same direction.

Electrons are also responsible for the magnetic fields that are produced due to a moving *electric current*. When electricity flows through a wire, a magnetic field is produced around that wire, meaning that an electric current can influence magnetic materials. The magnetic properties of a solid magnet are determined by the alignment of atoms and electrons. Similarly, when an electric current flows through a wire, electrons are generally moving in the same direction within the wire. This movement of electrons has similar effects as the alignment of electrons within a solid magnet. The magnetic field that is created from a moving electric current circles the wire that carries the current, and by coiling the wire into

FIGURE 10.3

Magnetic field for a wire (a) and a solenoid (b)

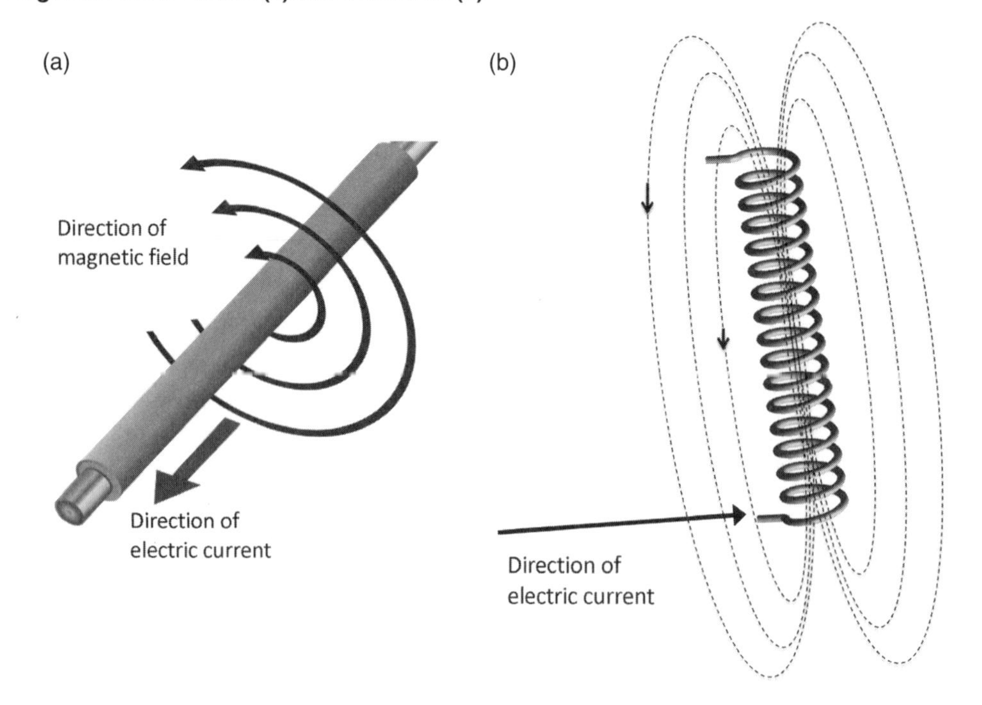

a cylinder shape—called a solenoid—the magnetic field can be concentrated inside the coil. Figure 10.3 shows the magnetic field around a single piece of wire and within a solenoid.

Given the relationship between electricity and magnetism, each phenomenon can be used to create the other. A moving magnet can create an electric current, which is the process by which electricity is generated within generators in power plants. Also, an electric current can create a magnet, which is the case with electromagnets used for a variety of purposes. When a solid magnet spins inside a coil of wire (such as a solenoid), the changing magnetic field causes an electric current to form in the wire as the electrons begin to move in response to the magnetic field. Likewise, when a coil of wire that is carrying an electric current surrounds a ferromagnetic material, the ferromagnetic material becomes magnetized as the individual atoms become aligned in response to the moving electrons in the wire.

With regard to creating an electromagnet, there are several factors that contribute to its overall strength. In general, the more atoms that come into alignment within the ferromagnetic material, the stronger the magnet. There are several ways to cause the atoms to change their alignment. The stronger the electric current in the wire of the electromagnet, the stronger the electric field that influences the alignment of the atoms in the ferromagnetic material; therefore, the alignment of more atoms is changed, which results in greater magnetic strength. Another way to increase the strength of the electromagnet is to use more turns of wire in the coil. Adding more turns of wire in the coil does not change the strength of the electric current, but it does provide a larger area over which the electric field can work. With more turns of the wire surrounding the ferromagnetic material, more of the atoms are exposed to the electric field and therefore change their alignment. A third option is to use a bigger ferromagnetic core in the electromagnet. A bigger ferromagnetic core will contain more atoms, each of which has the potential to become realigned and contribute to the strength of the electromagnet.

Timeline

The instructional time needed to complete this lab investigation is 170–230 minutes. Appendix 2 (p. 411) provides options for implementing this lab investigation over several class periods. Option C (230 minutes) should be used if students are unfamiliar with scientific writing, because this option provides extra instructional time for scaffolding the writing process. You can scaffold the writing process by modeling, providing examples, and providing hints as students write each section of the report. Option D (170 minutes) should be used if students are familiar with scientific writing and have developed the skills needed to write an investigation report on their own. In option D, students complete stage 6 (writing the investigation report) and stage 8 (revising the investigation report) as homework.

Materials and Preparation

The materials needed to implement this investigation are listed in Table 10.1 (p. 184). Most of the consumables and equipment can be purchased from a science supply company such as Carolina, Flinn Scientific, or Ward's Science. Batteries and paper clips can be purchased at office supply or general retail stores.

To prepare for this lab it is useful to have sections of copper wire cut and ends stripped before class. Although students are capable of doing this work, having wires prepared before class will save time and ensure that each group has an equal length of wire to work with. Also, we suggest using a battery holder with Fahnestock clips on the ends so that the wire can be easily attached to and removed from the battery. Batteries are listed as a consumable item for this lab; they will not need to be replaced for each class period conducting the lab, but it may be a good idea to replace the batteries each year.

We recommend that you use a set routine for distributing and collecting the materials during the lab investigation. For example, the consumables and equipment for each group can be set up at each group's lab station before class begins, or one member from each group can collect them from a table or a cart when needed during class.

TABLE 10.1

Materials list for Lab 10

Item	Quantity
Consumable	
Size D battery	1 per group
Equipment and other materials	
Safety glasses or goggles	1 per student
Battery holder	1 per group
Copper wire	1 per group
Iron nail	1 per group
Paper clips	100 per group
Gauss meter (optional)	1 per group
Electronic or triple beam balance	1 per group
Ruler	1 per group
Investigation Proposal C (optional)	1 per group
Whiteboard, 2' × 3'*	1 per group
Lab Handout	1 per student
Peer-review guide	1 per student
Checkout Questions	1 per student

* As an alternative, students can use computer and presentation software, such as Microsoft PowerPoint or Apple Keynote, to create their arguments.

Safety Precautions and Laboratory Waste Disposal

Remind students to follow all normal lab safety rules. In addition, tell students to take the following safety precautions:

1. Wear sanitized safety glasses or goggles during lab setup, hands-on activity, and takedown.
2. Given that the electromagnet setup will generate heat, disconnect the wire from the battery when not actively collecting data, to prevent skin burn.
3. Use caution when wrapping the copper wire around the nail. It is possible that the loose end will swing around during the wrapping process and present a hazard.

4. Use caution in handling the wire end or nails. They are sharp and can cut or puncture skin.
5. Wash hands with soap and water after completing the lab activity.
6. Batteries and wire may be stored for future use. When batteries need replacing, dispose of old batteries according to manufacturer's recommendations.

Topics for the Explicit and Reflective Discussion

Concepts That Can Be Used to Justify the Evidence

To provide an adequate justification of their evidence, students must explain why they included the evidence in their arguments and make the assumptions underlying their analysis and interpretation of the data explicit. In this investigation, students can use the following concepts to help justify their evidence:

- Electric current
- Electric fields
- Magnetic fields

We recommend that you review these concepts during the explicit and reflective discussion to help students make this connection.

How to Design Better Investigations

It is important for students to reflect on the strengths and weaknesses of the investigation they designed during the explicit and reflective discussion. Students should therefore be encouraged to discuss ways to eliminate potential flaws, measurement errors, or sources of bias in their investigations. To help students be more reflective about the design of their investigation, you can ask the following questions:

1. What were some of the strengths of your investigation? What made it scientific?
2. What were some of the weaknesses of your investigation? What made it less scientific?
3. If you were to do this investigation again, what would you do to address the weaknesses in your investigation? What could you do to make it more scientific?

Crosscutting Concepts

This investigation is well aligned with two crosscutting concepts found in *A Framework for K–12 Science Education,* and you should review these concepts during the explicit and reflective discussion.

- *Energy and matter: Flows, cycles, and conservation:* In science it is important to track how energy and matter move into, out of, and within systems. In this lab students will investigate how electrical energy can be used to generate magnetic fields.
- *Structure and function:* The way an object is shaped or structured determines many of its properties and functions. In this lab students will investigate how the structure and features of an electromagnet influence its strength.

The Nature of Science and the Nature of Scientific Inquiry

This investigation is well aligned with two important concepts related to the *nature of science* (NOS) and the *nature of scientific inquiry* (NOSI), and you should review these concepts during the explicit and reflective discussion.

- The *difference between laws and theories in science:* A scientific law describes the behavior of a natural phenomenon or a generalized relationship under certain conditions; a scientific theory is a well-substantiated explanation of some aspect of the natural world. Theories do not become laws even with additional evidence; they explain laws. However, not all scientific laws have an accompanying explanatory theory. It is also important for students to understand that scientists do not discover laws or theories; the scientific community develops them over time.
- *The difference between data and evidence in science:* Data are measurements, observations, and findings from other studies that are collected as part of an investigation. Evidence, in contrast, is analyzed data and an interpretation of the analysis.

Hints for Implementing the Lab

- Allowing students to design their own procedures for collecting data gives students an opportunity to try, to fail, and to learn from their mistakes. However, you can scaffold students as they develop their procedure by having them fill out an investigation proposal. These proposals provide a way for you to offer students hints and suggestions without telling them how to do it. You can also check the proposals quickly during a class period. We suggest using Investigation Proposal C for this lab.
- There are a variety of ways for students to collect data for this lab. One way is to simply count the number of paper clips that "stick" to the electromagnet. A second option is to pick up a sample of paper clips and then determine the mass that was picked up. Knowing the mass allows students to then calculate a weight for the items picked up, which could serve as a proxy for the force (or strength) of the electromagnet.
- Allow the students to collect their data as they see fit, but be mindful that the tighter the coils of wire are wrapped around the nail the better. Also, it is usually

easier for students to pick up paper clips using the head end of the nail rather than the point.

- Students will need to be able to wrap the copper wire around the nail for nearly the whole length of the nail to test how the number of wraps affects the strength of their electromagnet. Be sure to provide each group a single piece of wire that is long enough for this task; using more than one piece of wire will result in poor data.

Topic Connections

Table 10.2 provides an overview of the scientific practices, crosscutting concepts, disciplinary core ideas, and supporting ideas at the heart of this lab investigation. In addition, it lists NOS and NOSI concepts for the explicit and reflective discussion. Finally, it lists literacy and mathematics skills (*CCSS ELA* and *CCSS Mathematics*) that are addressed during the investigation.

TABLE 10.2

Lab 10 alignment with standards

Scientific practices	• Asking questions and defining problems • Planning and carrying out investigations • Analyzing and interpreting data • Using mathematics and computational thinking • Constructing explanations and designing solutions • Engaging in argument from evidence • Obtaining, evaluating, and communicating information
Crosscutting concepts	• Energy and matter: Flows, cycles, and conservation • Structure and function
Core idea	• PS2.B: Types of interactions
Supporting ideas	• Electric current • Electric fields • Magnetic fields
NOS and NOSI concepts	• Scientific laws and theories • Difference between data and evidence
Literacy connections (*CCSS ELA*)	• *Reading*: Key ideas and details, craft and structure, integration of knowledge and ideas • *Writing*: Text types and purposes, production and distribution of writing, research to build and present knowledge, range of writing • *Speaking and listening*: Comprehension and collaboration, presentation of knowledge and ideas
Mathematics connections (*CCSS Mathematics*)	• Reason abstractly and quantitatively • Construct viable arguments and critique the reasoning of others

Lab Handout

Lab 10. Magnetic Force

How Is the Strength of an Electromagnet Affected by the Number of Turns of Wire in a Coil?

Introduction

Magnets and magnetic fields are useful for many applications. For example, small permanent magnets and electromagnets are used in speakers that are found in cell phones or headphones used to listen to music. In a speaker, the changes in the magnetic field of the electromagnet cause parts of the speaker to vibrate, which produces the sounds we hear when we listen to music. The electromagnets in headphone speakers are small and fairly weak, but other electromagnets can be much larger and stronger, such as those used in junkyards to pick up and move old cars. Electromagnets are also used in the medical field in devices such as MRI (magnetic resonance imaging) machines. The powerful electromagnets in MRI machines influence the atoms in our bodies and allow doctors to create images that are useful in diagnosing injuries.

Permanent magnets, such as refrigerator magnets or those made from combinations of metals such as iron (Fe), nickel (Ni), or neodymium (Nd), always demonstrate magnetic properties. Permanent magnets are surrounded by a magnetic field. This magnetic field can influence other magnets or some materials (like some metals) and cause the objects to be pulled toward the magnet or pushed away from the magnet.

Magnetic fields can also be created when electricity passes through a wire. The electric current (moving electrical charges) in the wire creates a magnetic field surrounding the wire (see Figure L10.1). The magnetic field surrounding the wire is usually weak, but it can still have an effect on other magnets or materials. Coiling the wire will help concentrate the magnetic field on the inside of the coil (see Figure L10.2).

Turning a coil of wire into an electromagnet is as simple as wrapping the coil of wire around a piece of metal, such as an iron nail (see Figure L10.3, p. 190). When a wire is coiled around the nail, the magnetic field from the wire that is concentrated inside the coils magnetizes the iron nail and produces the electromagnet. Individual iron atoms can act like very small magnets, but inside a nail, the iron atoms point in random directions; therefore, the nail on its own does not act like a magnet. But when the iron atoms inside the nail are influenced by the magnetic field from the coil of wire, they change their alignment and point in similar directions. Only iron atoms inside the coil of wire will change their alignment, and the more atoms that point in the same direction, the greater the magnetic strength of the nail. The nail will only act like a magnet when the electric current is flowing through the wire; when the electric current stops, the iron atoms return to their original and random alignment and no longer act like a magnet.

FIGURE L10.1
Magnetic field surrounding a wire

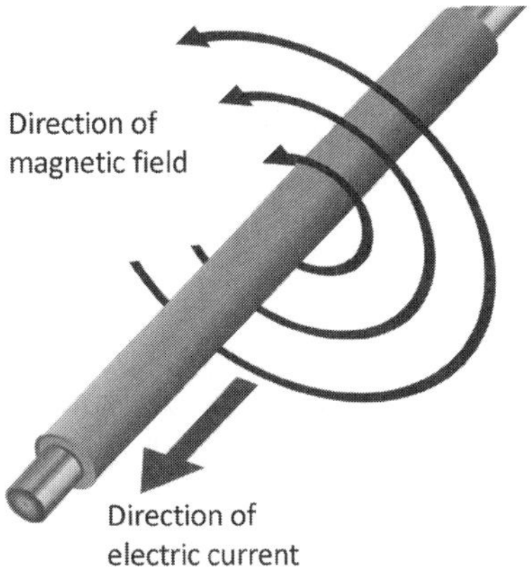

Direction of magnetic field

Direction of electric current

FIGURE L10.2
Concentrated magnetic field in a coil of wire

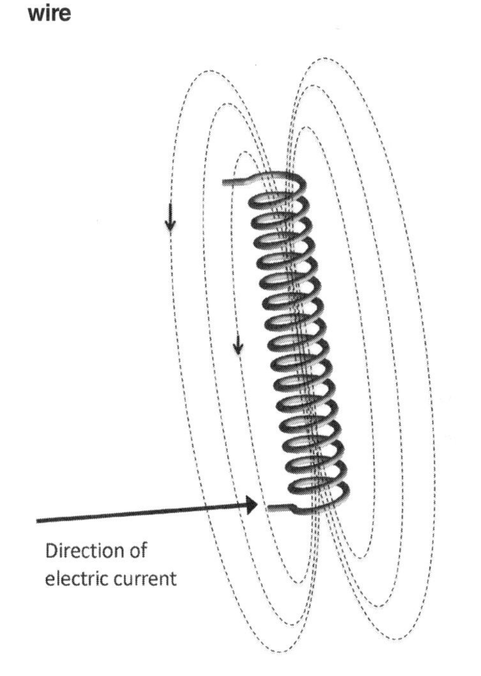

Your Task

Build an electromagnet using a battery, some wire, and a nail. Then, use what you know about magnetic fields, tracking energy in a system, and structure and function to design and carry out an investigation that will allow you to determine how the number of turns of wire wrapped around the nail affects the strength of the electromagnet.

The guiding question of this investigation is, **How is the strength of an electromagnet affected by the number of turns of wire in a coil?**

Materials

You may use any of the following materials during your investigation:

- Size D battery
- Battery holder
- Copper wire
- Iron nail
- Paper clips
- Gauss meter (optional)
- Electronic or triple beam balance
- Ruler
- Safety glasses or goggles

Safety Precautions

Follow all normal lab safety rules. In addition, take the following safety precautions:

1. Wear sanitized safety glasses or goggles during lab setup, hands-on activity, and takedown.

2. Given that the electromagnet setup will generate heat, disconnect the wire from the battery when not actively collecting data, to prevent skin burn.
3. Use caution when wrapping the copper wire around the nail. It is possible that the loose end will swing around during the wrapping process and present a hazard.
4. Use caution in handling the wire end or nails. They are sharp and can cut or puncture skin.
5. Wash hands with soap and water after completing the lab activity.

Investigation Proposal Required? ☐ Yes ☐ No

Getting Started

To answer the guiding question, you will need to design and conduct an investigation to measure the strength of your electromagnet. To accomplish this task, you must determine what type of data you need to collect, how you will collect it, and how you will analyze it. Figure L10.3 shows how to construct a simple electromagnet from a battery, a copper wire, and a nail.

To determine *what type of data you need to collect,* think about the following questions:

- How will you determine the strength of the electromagnet?
- What information or measurements will you need to record?
- What parts of the electromagnet will you change and what parts will you keep consistent?

FIGURE L10.3

Electromagnet made from a D-cell battery, a copper wire, and an iron nail

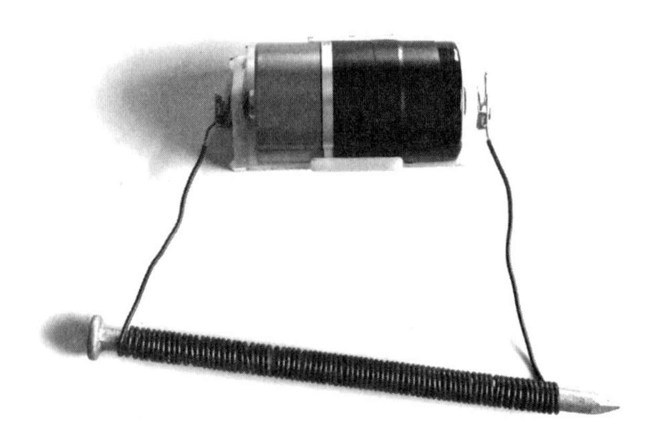

To determine *how you will collect your data,* think about the following questions:

- What equipment will you need to collect the data you need?
- How will you make sure that your data are of high quality (i.e., how will you reduce error)?
- Are there different ways you can measure the amount of coils you used?
- How will you keep track of the data you collect?
- How will you organize your data?

To determine *how you will analyze your data,* think about the following questions:

- How will you determine if the number of coils affects the strength of the electromagnet?
- What type of table or graph could you create to help make sense of your data?

Connections to Crosscutting Concepts, the Nature of Science, and the Nature of Scientific Inquiry

As you work through your investigation, be sure to think about

- why it is important to track how energy and matter move into, out of, and within systems;
- how the structure or shape of something can influence how it functions and places limits on what it can and cannot do;
- the difference between theories and laws in science; and
- the difference between data and evidence in science.

Initial Argument

Once your group has finished collecting and analyzing your data, your group will need to develop an initial argument. Your initial argument needs to include a *claim, evidence* to support your claim, and a *justification* of the evidence. The claim is your group's answer to the guiding question. The evidence is an analysis and interpretation of your data. Finally, the justification of the evidence is why your group thinks the evidence matters. The justification of the evidence is important because scientists can use different kinds of evidence to support their claims. Your group will create your initial argument on a whiteboard. Your whiteboard should include all the information shown in Figure L10.4.

FIGURE L10.4

Argument presentation on a whiteboard

The Guiding Question:	
Our Claim:	
Our Evidence:	Our Justification of the Evidence:

Argumentation Session

The argumentation session allows all of the groups to share their arguments. One member of each group will stay at the lab station to share that group's argument, while the other members of the group go to the other lab stations to listen to and critique the arguments developed by their classmates. This is similar to how scientists present their arguments to other scientists at conferences. If you are responsible for critiquing your classmates' arguments, your goal is to look for mistakes so these mistakes can be fixed and they can make their argument better. The argumentation session is also a good time to think about ways you can make your initial argument better. Scientists must share and critique arguments like this to develop new ideas.

To critique an argument, you might need more information than what is included on the whiteboard. You will therefore need to ask the presenter lots of questions. Here are some good questions to ask:

- How did you collect your data? Why did you use that method? Why did you collect those data?
- What did you do to make sure the data you collected are reliable? What did you do to decrease measurement error?
- How did your group analyze the data? Why did you decide to do it that way? Did you check your calculations?
- Is that the only way to interpret the results of your analysis? How do you know that your interpretation of your analysis is appropriate?
- Why did your group decide to present your evidence in that way?
- What other claims did your group discuss before you decided on that one? Why did your group abandon those alternative ideas?
- How confident are you that your claim is valid? What could you do to increase your confidence?

Once the argumentation session is complete, you will have a chance to meet with your group and revise your initial argument. Your group might need to gather more data or design a way to test one or more alternative claims as part of this process. Remember, your goal at this stage of the investigation is to develop the most acceptable and valid answer to the research question!

Report

Once you have completed your research, you will need to prepare an *investigation report* that consists of three sections. Each section should provide an answer to the following questions:

1. What question were you trying to answer and why?
2. What did you do to answer your question and why?
3. What is your argument?

Your report should answer these questions in two pages or less. This report must be typed, and any diagrams, figures, or tables should be embedded into the document. Be sure to write in a persuasive style; you are trying to convince others that your claim is acceptable and valid!

Checkout Questions

Lab 10. Magnetic Force

How Is the Strength of an Electromagnet Affected by the Number of Turns of Wire in a Coil?

1. Malik and Jason are both making electromagnets. Malik wants to use two batteries to make his electromagnet the strongest. Jason plans on using only one battery but wrapping his wire twice as much as Malik. Use what you know about electromagnets to explain why both students have good strategies for making strong electromagnets.

2. Below are two samples of different materials. The samples show the general alignment of atoms within that material. Each individual atom acts like a miniature magnet, with the light gray side representing the south end of the magnet and the dark gray side representing the north end of the magnet. Which sample as a whole would be better for making a large magnet?

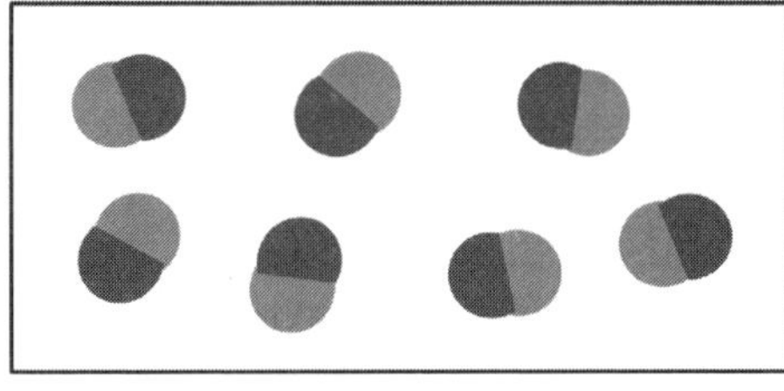

Sample A

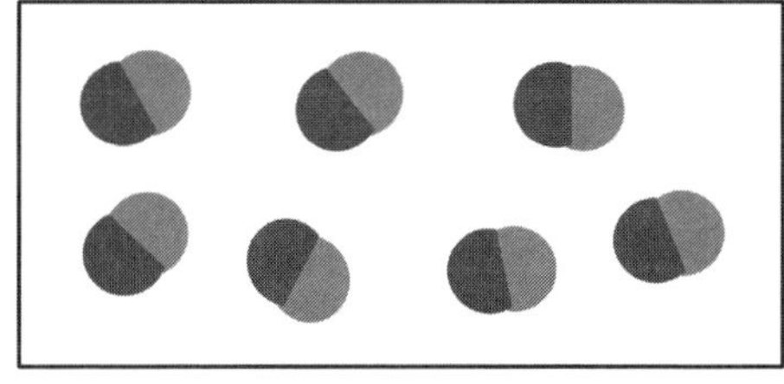

Sample B

Explain your answer. Why did you make that decision?

3. In science, laws are more important than theories because laws are true.

    a. I agree with this statement.
    b. I disagree with this statement.

    Explain your answer, using an example from your investigation about magnetic force.

4. In science, evidence for our claims comes from the data we collect.

    a. I agree with this statement.
    b. I disagree with this statement.

    Explain your answer, using an example from your investigation about magnetic force.

5. The amount of matter and energy in the universe is constant. Explain how understanding the movement of energy and matter within and between systems is helpful to scientists. Use an example from your investigation about magnetic force to help in your explanation.

6. The structure of an object is usually related to the function of that object. Use an example from your investigation about magnetic force to explain why it is important to understand the relationship between structure and function within science.

Application Labs

Teacher Notes

Lab 11. Design Challenge

Which Electromagnet Design Is Best for Picking Up 50 Paper Clips?

Purpose

The purpose of this lab is to allow students to *apply* what they know about the relationship between electricity and magnetism and the factors that influence the strength of electromagnets to design an electromagnet to solve a specific problem. In addition, this lab gives students an opportunity to investigate the cause-and-effect relationships when changing various features of an electromagnet. Students will also learn about the different methods used in scientific investigations and about the role of imagination and creativity in science.

The Content

The relationship between electricity and magnetism is important within science as well as within broader society. Scientists and engineers take advantage of this relationship in relatively simply ways, such as making a doorbell ring with a solenoid, and in much more complex tasks, such as using magnetic resonance to take images of the body for medical purposes. Understanding the connection between electricity and magnetism requires taking a detailed view of atoms.

FIGURE 11.1

Model iron atom showing electrons in orbitals around the nucleus

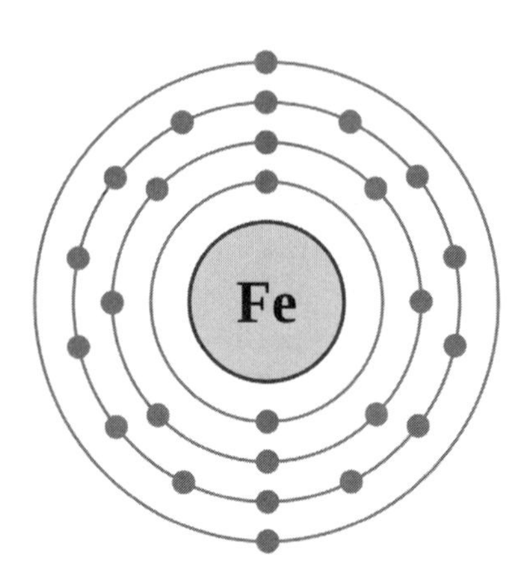

All matter is composed of atoms, and all atoms are composed of a nucleus of protons and neutrons, and electrons outside that nucleus. The electrons within an atom are responsible for the properties that allow electricity to flow through conductors such as copper wire. A simplified model of an atom, as shown in Figure 11.1, includes a core nucleus made of protons and neutrons with electrons orbiting that core. The electrons in an atom are also responsible for the magnetic properties of some elements. Naturally occurring elements, such as nickel or iron, can act like magnets because of the arrangement of the electrons and atoms within those substances. These substances are called *ferromagnetic materials.*

The electrons within an atom are constantly spinning as they orbit the nucleus of the atom. The direction of that spin influences the magnetic properties of the substances. When the atoms are arranged in such a way that the spins of the electrons are working together, the metal becomes a magnet. Ferromagnetic materials have properties that make it easy for them to become magnetized. When ferromagnetic materials become magnetized, they generate a *magnetic field* and are able to apply a force on other metal substances containing nickel or iron.

It is sometimes helpful to imagine an individual iron atom as a very small magnet with a north pole and a south pole. In a ferromagnetic material that has been magnetized, essentially all the individual atoms are arranged so that their north poles point in one direction and their south poles point in the opposite direction. When the individual atoms are shifted into this arrangement, the substance becomes a solid magnet.

Figure 11.2 shows how atoms in a ferromagnetic material are arranged when they demonstrate nonmagnetic and magnetic properties. North and south poles of magnets are attracted to each other because of the arrangement and orientation of the atoms within the magnet. As seen in Figure 11.2(b), when atoms are aligned in a magnet, their poles point in opposite directions, and this is related to the spin of the electrons within individual atoms; the north end of one magnet is repelled from the north end of another magnet (the same is true for south end vs. south end) because they have similar arrangements of atoms and spinning electrons within those atoms. In contrast, north and south poles of magnets are attracted to each other because they have opposite arrangements of atoms and spinning electrons compared with one another.

FIGURE 11.2

Substance with nonmagnetic (a) and magnetic (b) characteristics

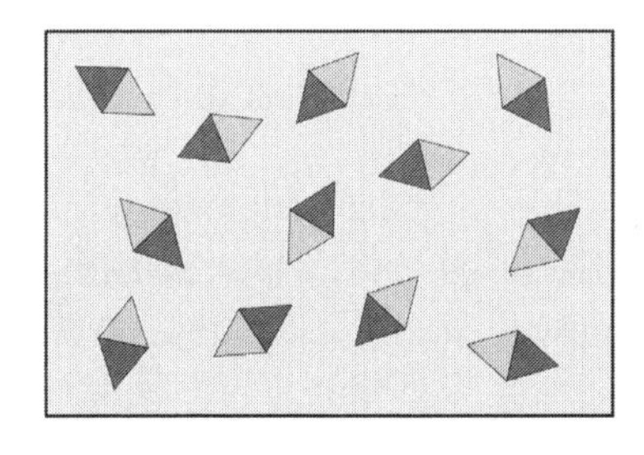

(a) In nonmagnetic substances, the atoms have no special arrangement and the poles point in a variety of directions.

(b) When a substance is magnetized, the atoms align so that like poles point in the same direction.

Electrons are also responsible for the magnetic fields that are produced due to a moving *electric current.* When electricity flows through a wire, a magnetic field is produced around that wire, meaning that an electric current can influence magnetic materials. The magnetic properties of a solid magnet are determined by the alignment of atoms and electrons. Similarly, when an electric current flows through a wire, electrons are generally moving in the same direction within the wire. This movement of electrons has similar effects to the alignment of electrons within a solid magnet. The magnetic field that is created from a moving electric current circles the wire that carries the current, and by coiling the wire into a cylinder shape—called a solenoid—the magnetic field can be concentrated inside the coil. Figure 11.3 (p. 198) shows the magnetic field around a single piece of wire and within a solenoid.

FIGURE 11.3

Magnetic field for a wire (a) and a solenoid (b)

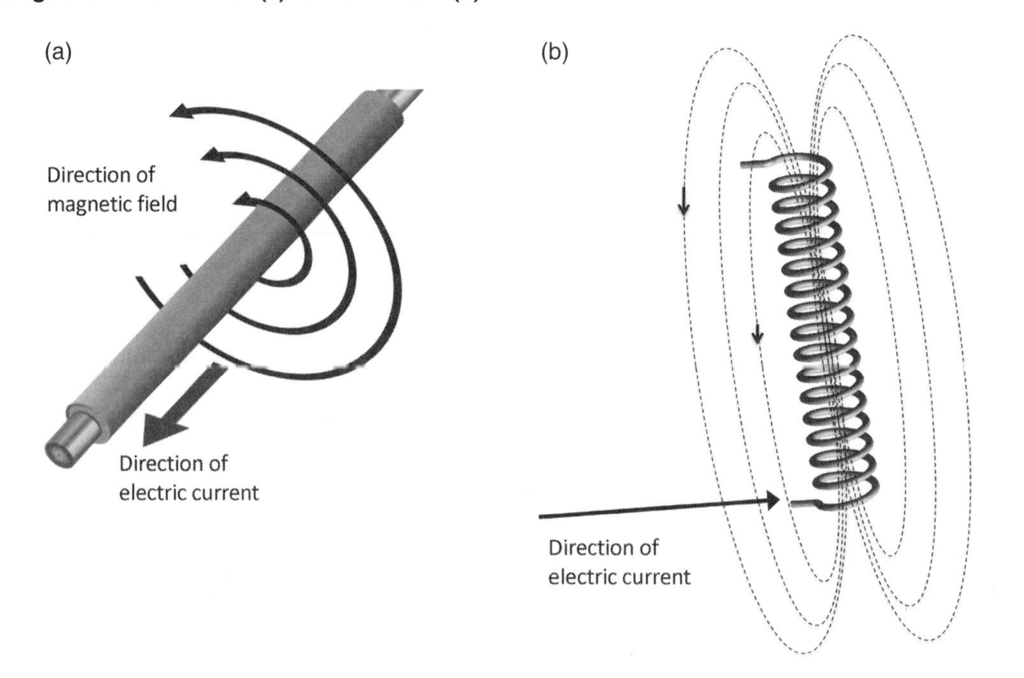

Given the relationship between electricity and magnetism, each phenomenon can be used to create the other. A moving magnet can create an electric current, which is the process by which electricity is generated within generators in power plants. Also, an electric current can create a magnet, which is the case with electromagnets used for a variety of purposes. When a solid magnet spins inside a coil of wire (such as a solenoid), the changing magnetic field causes an electric current to form in the wire as the electrons begin to move in response to the magnetic field. Likewise, when a coil of wire that is carrying an electric current surrounds a ferromagnetic material, the ferromagnetic material becomes magnetized as the individual atoms become aligned in response to the moving electrons in the wire.

With regard to creating an electromagnet, there are several factors that contribute to its overall strength. In general, the more atoms that come into alignment within the ferromagnetic material, the stronger the magnet. There are several ways to cause the atoms to change their alignment. The stronger the electric current in the wire of the electromagnet, the stronger the electric field that influences the alignment of the atoms in the ferromagnetic material; therefore, the alignment of more atoms is changed, which results in greater magnetic strength. Another way to increase the strength of the electromagnet is to use more turns of wire in the coil. Adding more turns of wire in the coil does not change the strength of the electric current, but it does provide a larger area over which the electric field can work. With more turns of wire surrounding the ferromagnetic material, more of the atoms are exposed to the electric field and therefore change their alignment. A third option is to use a bigger ferromagnetic core in the electromagnet. A bigger ferromagnetic core will contain more atoms, each of which has the potential to become realigned and contribute to the strength of the electromagnet.

Engineering Connection

This investigation engages students in a design challenge to construct an electromagnet with a specific strength, which is inferred by the ability to pick up a specific amount of paper clips or mass. Unlike a typical scientific investigation, in this lab students are not trying to explain a natural phenomenon; rather, they are determining the best solution to a problem. To determine the best solution, students will need to develop multiple solutions and test them using an iterative process of design, test, refine, and optimize. During this iterative cycle of design, students' solutions are tested and refined based on data related to how well the design helps solve the given problem. The optimization stage of the design cycle is where the final design is improved based on trading off more or less important features of the design given the constraints of the task. Design constraints may include the size of the device, cost of the materials, the ratio of cost to outcome, or acceptable margin of error for a specified outcome. We suggest that you and your students work collaboratively to determine what the constraints will be for this design challenge.

The outcome for this investigation will be a design solution that meets the constraints identified for the given problem. At the conclusion of the design cycle, student will still generate an argument that includes a claim, evidence, and justification; however, the arguments will be slightly modified with respect to a more typical argument-driven inquiry investigation. Students may make a general claim about the features of a successful design, or their claim may be the specific design developed by their group. The evidence portion of the argument will include data that have been analyzed and interpreted to support the success of their design. Finally, the justification of their evidence will include a connection to scientific ideas like a typical scientific argument, with the addition of how their design also addresses the constraints of the problem they are solving.

Timeline

The instructional time needed to complete this lab investigation is 170–230 minutes. Appendix 2 (p. 411) provides options for implementing this lab investigation over several class periods. Option C (230 minutes) should be used if students are unfamiliar with scientific writing, because this option provides extra instructional time for scaffolding the writing process. You can scaffold the writing process by modeling, providing examples, and providing hints as students write each section of the report. Option D (170 minutes) should be used if students are familiar with scientific writing and have developed the skills needed to write an investigation report on their own. In option D, students complete stage 6 (writing the investigation report) and stage 8 (revising the investigation report) as homework.

Materials and Preparation

The materials needed to implement this investigation are listed in Table 11.1. Most of the consumables and equipment can be purchased from a science supply company such as Carolina, Flinn Scientific, or Ward's Science. Batteries and paper clips can be purchased at office supply or general retail stores.

TABLE 11.1

Materials list for Lab 11

Item	Quantity
Consumables	
Size D battery	As needed
Size C battery	As needed
Size AA battery	As needed
Equipment and other materials	
Safety glasses or goggles	1 per student
Battery holder	1 per group
Copper wire	1 per group
Iron nails (assorted sizes)	As needed
Paper clips	100 per group
Gauss meter (optional)	1 per group
Ruler	1 per group
Investigation Proposal C (optional)	1 per group
Whiteboard, 2' × 3'*	1 per group
Lab Handout	1 per student
Peer-review guide	1 per student
Checkout Questions	1 per student

* As an alternative, students can use computer and presentation software, such as Microsoft PowerPoint or Apple Keynote, to create their arguments.

To prepare for this lab, it is useful to have sections of copper wire cut and ends stripped before class. Although students are capable of doing this work, having wires prepared before class will save time and ensure that each group has an equal length of wire to work with. Also, we suggest using a battery holder with Fahnestock clips on the ends so that the wire can be easily attached to and removed from the battery. Batteries are listed as a consumable item for this lab; they will not need to be replaced for each class period conducting the lab, but it may be a good idea to replace the batteries each year.

We recommend that you use a set routine for distributing and collecting the materials during the lab investigation. For example, the consumables and equipment for each group can be set up at each group's lab station before class begins, or one member from each group can collect them from a table or a cart when needed during class.

Safety Precautions and Laboratory Waste Disposal

Remind students to follow all normal lab safety rules. In addition, tell students to take the following safety precautions:

1. Wear sanitized safety glasses or goggles during lab setup, hands-on activity, and takedown.
2. Given that the electromagnet setup will generate heat, disconnect the wire from the battery when not actively collecting data to prevent skin burn.
3. Use caution when wrapping the copper wire around the nail. It is possible that the loose end will swing around during the wrapping process and present a hazard.
4. Use caution in handling the wire end or nails. They are sharp and can cut or puncture skin.
5. Wash hands with soap and water after completing the lab activity.

Batteries and wire may be stored for future use. When batteries need replacing, dispose of old batteries according to manufacturer's recommendations.

Topics for the Explicit and Reflective Discussion

Concepts That Can Be Used to Justify the Evidence

To provide an adequate justification of their evidence, students must explain why they included the evidence in their arguments and make the assumptions underlying their analysis and interpretation of the data explicit. In this investigation, students can use the following concepts to help justify their evidence:

- Electric current
- Electric fields
- Magnetic fields

We recommend that you review these concepts during the explicit and reflective discussion to help students make this connection.

How to Design Better Investigations

It is important for students to reflect on the strengths and weaknesses of the investigation they designed during the explicit and reflective discussion. Students should therefore be encouraged to discuss ways to eliminate potential flaws, measurement errors, or sources of bias in their investigations. To help students be more reflective about the design of their investigation, you can ask the following questions:

1. What were some of the strengths of your investigation? What made it scientific?
2. What were some of the weaknesses of your investigation? What made it less scientific?
3. If you were to do this investigation again, what would you do to address the weaknesses in your investigation? What could you do to make it more scientific?
4. Did you meet the goal of the design challenge?
5. Did you ensure that your solution is consistent with the design parameters?

Crosscutting Concepts

This investigation is well aligned with two crosscutting concepts found in *A Framework for K–12 Science Education,* and you should review these concepts during the explicit and reflective discussion.

- *Cause and effect: Mechanism and explanation:* One of the main objectives of science is to identify and establish relationships between a cause and an effect. In this lab students will investigate how changes in one aspect of their design affect the overall performance of their electromagnet.
- *Systems and system models:* It is critical for scientists to be able to define the system under study and then make a model of it to understand it. Students can generate a variety of physical models as well as mathematical models to help them design an electromagnet to solve the problem provided in this lab.

The Nature of Science and the Nature of Scientific Inquiry

This investigation is well aligned with two important concepts related to the *nature of science* (NOS) and the *nature of scientific inquiry* (NOSI), and you should review these concepts during the explicit and reflective discussion.

- *Methods used in scientific investigations:* Examples of methods include experiments, systematic observations of a phenomenon, literature reviews, and analysis of existing data sets; the choice of method depends on the objectives of the research. There is no universal step-by-step scientific method that all scientists follow; rather, different scientific disciplines (e.g., chemistry vs. physics) and fields within a discipline (e.g., organic vs. physical chemistry) use different types of methods, use different core theories, and rely on different standards to develop scientific knowledge.

- *The importance of imagination and creativity in science:* Students should learn that developing explanations for or models of natural phenomena and then figuring out how they can be put to the test of reality is as creative as writing poetry, composing music, or designing video games. Scientists must also use their imagination and creativity to figure out new ways to test ideas and collect or analyze data.

Hints for Implementing the Lab

- Allowing students to design their own procedures for collecting data gives students an opportunity to try, to fail, and to learn from their mistakes. However, you can scaffold students as they develop their procedure by having them fill out an investigation proposal. These proposals provide a way for you to offer students hints and suggestions without telling them how to do it. You can also check the proposals quickly during a class period. We suggest using Investigation Proposal C for this lab. Alternatively, you can have students generate mathematical models based on preliminary data to help them narrow down the characteristics for their electromagnet design.
- Allow the students to collect their data as they see fit, but be mindful that the tighter the turns of wire are wrapped around the nail, the better. Also, it is usually easier for students to pick up paper clips using the head end of the nail rather than the point.
- Students will need to be able to wrap the copper wire around the nail for nearly the whole length of the nail to test how the number of turns of wire affects the strength of their electromagnet. Be sure to provide each group a single piece of wire that is long enough for this task; using more than one piece of wire will result in poor data.
- It is important for you to encourage students to test several variables and designs. There are many options for how to build an electromagnet to accomplish the task in this lab, such as mixing and matching different-size batteries, different-size nails, and different numbers of turns of wire around a nail. Other advanced options include using a combination of batteries connected in series or in parallel to alter the strength of the electric current.
- When students are designing their electromagnets, make sure you remind them of the cost constraint—the goal is to build the cheapest electromagnet that will still accomplish the assigned task.

Topic Connections

Table 11.2 provides an overview of the scientific practices, crosscutting concepts, disciplinary core ideas, and supporting ideas at the heart of this lab investigation. In addition, it lists NOS and NOSI concepts for the explicit and reflective discussion. Finally, it lists literacy and mathematics skills (*CCSS ELA* and *CCSS Mathematics*) that are addressed during the investigation.

TABLE 11.2

Lab 11 alignment with standards

Scientific practices	• Asking questions and defining problems • Developing and using models • Planning and carrying out investigations • Analyzing and interpreting data • Using mathematics and computational thinking • Constructing explanations and designing solutions • Engaging in argument from evidence • Obtaining, evaluating, and communicating information
Crosscutting concepts	• Cause and effect: Mechanism and explanation • Systems and system models
Core ideas	• PS2.B: Types of interactions • ETS1.A: Defining and delimiting an engineering problem • ETS1.B: Developing possible solutions
Supporting ideas	• Electric current • Electric fields • Magnetic fields
NOS and NOSI concepts	• Methods used in scientific investigations • Imagination and creativity in science
Literacy connections (*CCSS ELA*)	• *Reading*: Key ideas and details, craft and structure, integration of knowledge and ideas • *Writing*: Text types and purposes, production and distribution of writing, research to build and present knowledge, range of writing • *Speaking and listening*: Comprehension and collaboration, presentation of knowledge and ideas
Mathematics connections (*CCSS Mathematics*)	• Reason abstractly and quantitatively • Construct viable arguments and critique the reasoning of others • Model with mathematics

Lab Handout

Lab 11. Design Challenge

Which Electromagnet Design Is Best for Picking Up 50 Paper Clips?

Introduction

Electromagnets are formed when a wire is wrapped around a metal core and electricity flows through the wire. When electricity flows through the wire, a magnetic field is produced, which is then concentrated in the metal core. It is possible to make a simple electromagnet by wrapping copper wire around an iron rod or nail and then connecting the wire to a small battery (see Figure L11.1). When electricity from the battery flows through the wire, the magnetic field from the electric current causes the iron core to have magnetic properties.

FIGURE L11.1

Electromagnet made from a D-cell battery, a copper wire, and an iron nail

The iron core of an electromagnet is a key component for making the device work. The atoms of iron contain a large number of electrons. Within the iron core, the atoms and their electrons are moving about in random directions. However, when the atoms are exposed to the magnetic field from the electric current in the wire, the atoms and electrons change their orientations and movement to align with the magnetic field of the electric current. This rearrangement of the atoms amplifies the magnetic field from the wire and results in a stronger electromagnet. When the electric current stops (by turning it off or disconnecting the battery), the atoms in the iron core go back to their random motion and positions and no longer have magnetic properties.

There are three main parts required to build an electromagnet: the iron core, copper wire, and an electricity source. Changes in each of these pieces of the electromagnet will influence the overall strength of the magnet. The more powerful the source of electricity, the more electric current there will be flowing through the wire. The larger the iron core, the more atoms there are that can be aligned to help amplify the magnetic force. The number of times the wire is wrapped around the iron rod will also influence the strength of the electromagnet.

Electromagnets are used in many different contexts, ranging from small applications, such as the audio speakers in a cell phone, to large applications, such as picking up old cars in a junkyard. In each of these contexts, the electromagnet must be strong enough to accomplish its intended purpose but not so strong that it has unintended effects on nearby materials. In some cases, an electromagnet may be used to pick up a specific weight or even a specific amount of objects. If an electromagnet is designed for a specific task, it must have the right combination of properties, including the amount of electricity, the size of the iron core, and the number of times the wire is wrapped around the iron core so that it has a consistent and predictable performance.

LAB 11

Your Task

Use what you know about magnetic fields and cause-and-effect relationships to design a small electromagnet that can be used for a specific task. Your specific design challenge is as follows: The International Paperclip Company (IPC) needs an electromagnet that will pick up approximately 50 paper clips to help fill and package a single box of clips. For the IPC to keep their customers happy and the company's costs consistent, the electromagnet must reliably pick up between 47 and 53 paper clips. The IPC would also like to spend as little money as possible to construct their new electromagnet, while making sure that their new electromagnet will accomplish its task.

The guiding question of this investigation is, **Which electromagnet design is best for picking up 50 paper clips?**

Materials

You may use any of the following materials during your investigation:

- Batteries (different sizes)
- Copper wire
- Iron nails (different sizes)
- Paperclips
- Gauss meter (optional)
- Safety glasses or goggles

Safety Precautions

Follow all normal lab safety rules. In addition, take the following safety precautions:

1. Wear sanitized safety glasses or goggles during lab setup, hands-on activity, and takedown.
2. Given that the electromagnet setup will generate heat, disconnect the wire from the battery when not actively collecting data, to prevent skin burn.
3. Use caution when wrapping the copper wire around the nail. It is possible that the loose end will swing around during the wrapping process and present a hazard.
4. Use caution in handling the wire end or nails. They are sharp and can cut or puncture skin.
5. Wash hands with soap and water after completing the lab activity.

Investigation Proposal Required? ☐ Yes ☐ No

Getting Started

To answer the guiding question, you will need to design and test an electromagnet that is capable of picking up 50 paper clips. You must also consider the cost of the materials that are used in your electromagnet. You will need to systematically test how changes to each part of the electromagnet influence the strength of the magnet and its ability to accomplish its intended task. To complete this design challenge, you must determine what type of data you need to collect, how you will collect it, and how you will analyze it.

To determine *what type of data you need to collect,* think about the following questions:

- How will you determine the strength of the electromagnet?
- What information or measurements will you need to record?
- What parts of the electromagnet will you change and what parts will you keep consistent?
- What design factors will you consider when building your electromagnet?

To determine *how you will collect your data,* think about the following questions:

- What equipment will you need to collect the data you need?
- How will you make sure that your data are of high quality (i.e., how will you reduce error)?
- How will you keep track of the data you collect?
- How will you organize your data?

To determine *how you will analyze your data,* think about the following questions:

- How will you determine which parts of the electromagnet have the biggest impact on its strength?
- What type of table or graph could you create to help make sense of your data?
- How will you determine if one electromagnet design is better than another?

When you design your electromagnet for the IPC, it will be important to take into account the amount of materials you use, how much they cost, and how often the components will need to be replaced. Table L11.1 includes information about the different materials that might be used in making your electromagnet.

TABLE L11.1

Cost information for electromagnet materials

Item	Cost
Battery, 1.5 v, size D	$0.99/each
Battery, 1.5 v, size C	$0.89/each
Battery, 1.5 v, size AA	$0.45/each
Copper wire	$0.07/30 cm
Iron core	$0.015/cm

Connections to Crosscutting Concepts, the Nature of Science, and the Nature of Scientific Inquiry

As you work through your investigation, be sure to think about

- why it is important to understand cause-and-effect relationships,
- how defining a system under study and making a model of it is helpful in science,
- how scientists use different methods to answer different types of questions, and
- the role of imagination and creativity in science.

Initial Argument

Once your group has finished collecting and analyzing your data, your group will need to develop an initial argument. Your initial argument needs to include a *claim, evidence* to support your claim, and a *justification* of the evidence. The claim is your group's answer to the guiding question. The evidence is an analysis and interpretation of your data. Finally, the justification of the evidence is why your group thinks the evidence matters. The justification of the evidence is important because scientists can use different kinds of evidence to support their claims. Your group will create your initial argument on a whiteboard. Your whiteboard should include all the information shown in Figure L11.2.

FIGURE L11.2

Argument presentation on a whiteboard

The Guiding Question:	
Our Claim:	
Our Evidence:	Our Justification of the Evidence:

Argumentation Session

The argumentation session allows all of the groups to share their arguments. One member of each group will stay at the lab station to share that group's argument, while the other members of the group go to the other lab stations to listen to and critique the arguments developed by their classmates. This is similar to how scientists present their arguments to other scientists at conferences. If you are responsible for critiquing your classmates' arguments, your goal is to look for mistakes so these mistakes can be fixed and they can make their argument better. The argumentation session is also a good time to think about ways you can make your initial argument better. Scientists must share and critique arguments like this to develop new ideas.

To critique an argument, you might need more information than what is included on the whiteboard. You will therefore need to ask the presenter lots of questions. Here are some good questions to ask:

- How did you collect your data? Why did you use that method? Why did you collect those data?

- What did you do to make sure the data you collected are reliable? What did you do to decrease measurement error?
- How did you group analyze the data? Why did you decide to do it that way? Did you check your calculations?
- Is that the only way to interpret the results of your analysis? How do you know that your interpretation of your analysis is appropriate?
- Why did your group decide to present your evidence in that way?
- What other claims did your group discuss before you decided on that one? Why did your group abandon those alternative ideas?
- How confident are you that your claim is valid? What could you do to increase your confidence?

Once the argumentation session is complete, you will have a chance to meet with your group and revise your initial argument. Your group might need to gather more data or design a way to test one or more alternative claims as part of this process. Remember, your goal at this stage of the investigation is to develop the most acceptable and valid answer to the research question!

Report

Once you have completed your research, you will need to prepare an *investigation report* that consists of three sections. Each section should provide an answer to the following questions:

1. What question were you trying to answer and why?
2. What did you do to answer your question and why?
3. What is your argument?

Your report should answer these questions in two pages or less. This report must be typed, and any diagrams, figures, or tables should be embedded into the document. Be sure to write in a persuasive style; you are trying to convince others that your claim is acceptable and valid!

Checkout Questions

Lab 11. Design Challenge

Which Electromagnet Design Is Best for Picking Up 50 Paper Clips?

1. A compass needle is a small magnet that aligns with the magnetic field of Earth and is used to help people tell which direction is north when they are traveling. But whenever a compass gets close to a strong electric current, the compass needle points in a different direction—see the drawings below.

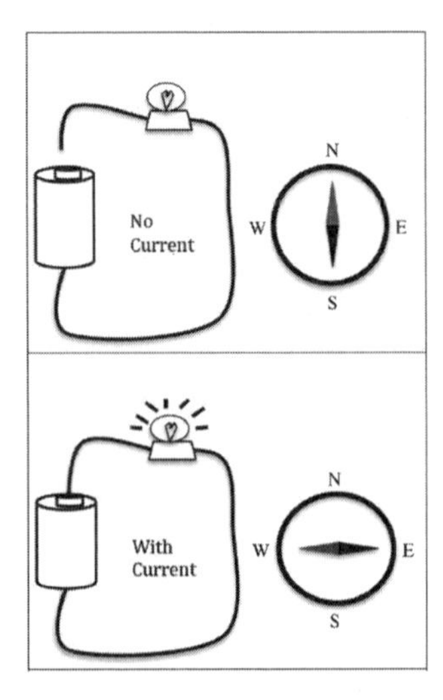

Use the images and what you know about magnets and electric currents to explain why the compass needle would move when there is an electric current present.

2. Look at the drawings of the electromagnets below. Which one will be the strongest—A, B, C, or D?

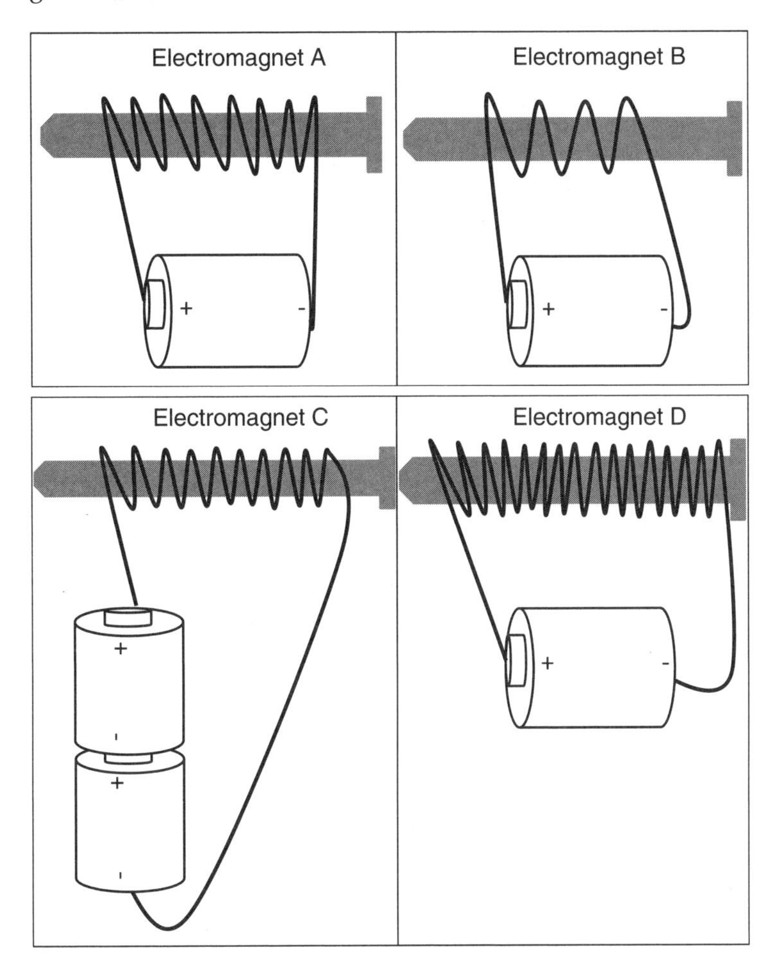

Explain your answer. Why did you make that decision?

3. When scientists use their imagination and creativity, it makes their work less scientific.

    a. I agree with this statement.
    b. I disagree with this statement.

    Explain your answer, using an example from your experience designing an electromagnet.

4. Scientists use the scientific method to design investigations to answer questions.

    a. I agree with this statement.
    b. I disagree with this statement.

    Explain your answer, using an example from your experience designing an electromagnet.

5. Changing one aspect of a system or one variable in an investigation can have an impact on many other things. Explain how understanding cause-and-effect relationships is helpful to scientists. Use an example from your experience designing an electromagnet to help in your explanation.

6. Before designing equipment, engineers usually make mathematical models or smaller versions of the equipment to test out how it will work. Using an example from your experience designing an electromagnet, explain why it is important for scientists and engineers to make models of systems.

Teacher Notes

Lab 12. Unbalanced Forces

How Does Surface Area Influence Friction and the Motion of an Object?

Purpose

The purpose of this lab is for students to *apply* what they know about balanced and unbalanced forces to investigate the impact of friction on the motion of an object. In addition, this lab gives students an opportunity to observe patterns and investigate their cause and to develop an understanding of stability and change within systems. Students will also learn about the differences between observations and inferences, as well as how scientists use different methods to investigate phenomena.

The Content

Balanced versus unbalanced forces is an important idea within physics. Understanding how forces affect an object allows scientists to predict and describe the motion of objects. When the forces acting on an object are *balanced,* that means that when all the forces—in their different directions—are added together, the sum is zero. In other words, there is *no* net force. When the forces acting on an object are *unbalanced,* that means that when all the forces—in their different directions—are added together, the sum is some value other than zero. In other words, there *is* a net force in some direction. The implications for balanced and unbalanced forces on an object's motion can be summarized simply. When the forces acting on an object are balanced (no net force), the motion of the object does not change. That is not to say that the object is stationary; rather, it means that there is no change in motion or no acceleration. When the forces acting on an object are balanced, the object will maintain a constant velocity, which may be 0 m/s (stationary) or any other value. When the forces acting on an object are unbalanced, the motion of the object does change, which means there is a change in speed or direction for the object. In other words, the object experiences some sort of acceleration (change in velocity).

The impact of balanced versus unbalanced forces is summed up in Newton's second law of motion, which describes that the acceleration of an object is directly proportional to the sum of the forces acting on that object and inversely proportional to the mass of the object. Also gleaned from Newton's second law of motion is the concept that the motion of an object will not change unless the object is acted on by an unbalanced force. Newton's second law of motion is often written as follows:

$\Sigma F = ma$, where

- ΣF is the sum of the forces (F) or net force on the object;
- ΣF is measured in newtons (1 N = 1 kg*m/s^2);
- m is the mass of the object, measured in kilograms (kg); and
- a is the acceleration of the object, or the change in velocity for the object, measured in meters per second squared (m/s^2).

This equation describes that the sum of the forces acting on an object (ΣF) is equal to the product of the mass of the object (m) and the acceleration of the object (a). For an object with a constant mass, the acceleration of the object is directly related to the net force acting on it. If the net force on the object increases, then the acceleration of the object will increase. Likewise, if the net force on the object decreases, then the acceleration of the object will decrease. If the force acting on an object is held constant but the mass of the object changes, the change in acceleration is inversely proportional to the change in mass. If force is constant and the mass increases, then the acceleration of the object will decrease; similarly, if force is constant and the mass of the object decreases, then the acceleration will increase.

It is usually easy to recognize some forces acting on objects, such as a person pushing on a box or pulling on a rope, or even the push of a car engine that propels a car forward. However, there are other forces that are not spotted as easily. One such force is friction. Friction occurs whenever two surfaces are in contact with each other. When a box slides across the floor, there is friction between the bottom surface of the box and the surface of the floor. Whenever one surface is sliding past another, the force of friction is always applied in the direction opposite of the movement of the object. Figure 12.1 shows a box sliding across the floor and the relationship between the two objects, their motion, and the force of friction. Friction is often referred to as a *resistive force* because friction always opposes motion or works against motion.

FIGURE 12.1

A box sliding across the floor experiences a friction force that opposes the motion of the box.

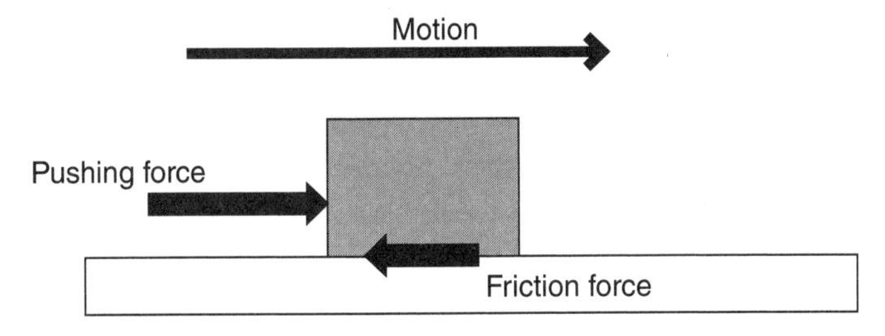

In the example in Figure 12.1, the pushing force is greater than the friction force; therefore, the forces acting on the box are unbalanced and the box will accelerate to the right. To understand the motion of objects more completely, it is important to understand the factors that influence friction. The amount of friction present is determined by two key factors. The first factor is the *coefficient of friction* for the two surfaces involved, which is a unique value for different combinations of surfaces and is usually determined experimentally. The second factor that determines the force due to friction is the *normal force* that is being applied to the moving object. The normal force is defined as the force applied perpendicular to the surface of contact. For the example in Figure 12.1, the normal force would be straight up, because that is the perpendicular to the surface of contact. For objects sitting on a flat surface parallel to the ground, the normal force is equal in magnitude to the weight of the object. When the object is not moving up or down, the normal force and the force due to gravity (weight) are balanced. When the normal force is equal to the weight of an object, the normal force can be calculated by $F_w = mg$, where m equals mass and g equals the acceleration due to gravity (9.8 m/s^2). The formula for determining the amount of friction between two objects is as follows:

$$F_{friction} = \mu_s F_n, \text{ where}$$

- $F_{friction}$ = force of friction, measured in newtons (N);
- μ_s = coefficient of static friction for the surfaces involved (no units); and
- F_n = normal force on the object, measured in newtons (N).

There are two types of friction, *static friction* and *kinetic friction*. Static friction refers to the amount of force needed to overcome the friction holding an object stationary. Kinetic friction is the amount of force of friction that is resisting the motion of an object as it slides along a surface. Static friction is always greater than kinetic friction, and each type of friction has its own coefficient of friction for different surface combinations. The formulas for calculating static and kinetic friction are the same, with the exception of using the correct coefficient of friction.

Determining the force of static friction using a spring scale or force sensor is relatively simple. With a spring scale or force sensor attached to an object such as a small block, begin pulling and make careful note of the force reading when the block begins to move. The pulling force at which the block begins to move is essentially equal to the static friction force that resists the motion of the block.

A common misconception related to static friction is the role of surface area. Many individuals think that the more surface area there is in contact between objects, the more friction there must be. This is incorrect; the coefficient of friction and the normal force determine the amount of friction. If the types of surfaces are the same and the total weight of the objects is the same, then the amount of friction will also be the same, regardless of the amount of surface area involved. As a result, the net force acting on the objects would be the same and the change in motion or acceleration for the objects would be the same. Figures 12.2 (p. 217) and 12.3 (p. 218) show two configurations for how a set of boxes could be pushed across the floor. Review these examples to see how the amount of surface area is not a factor for determining the amount of friction and the resulting movement of objects (assume a coefficient of friction of $\mu_s = 0.15$). From the examples presented in Figures 12.2 and 12.3, it is clear that surface area is not a factor when it comes to the influence of friction forces on the motion of an object.

FIGURE 12.2

Friction scenario for two side-by-side boxes of equal size

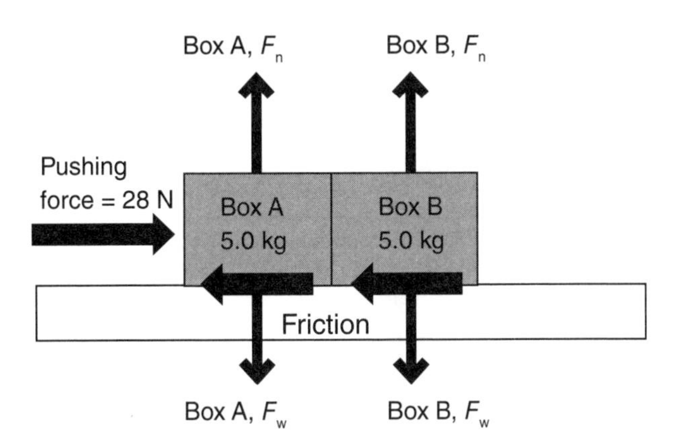

Friction for Box A	Friction for Box B	Motion for Boxes A and B
$F_A = \mu_s F_n$ and $F_n = F_w = mg$	$F_B = \mu_s F_n$ and $F_n = F_w = mg$	$\Sigma F = ma$, where m = mass of *Box A + Box B*
$F_A = \mu_s F_n = \mu_s(mg)$	$F_B = \mu_s F_n = \mu_s(mg)$	$\Sigma F = F_{push} - F_A - F_B = ma$
$F_A = (0.15)(mg)$	$F_B = (0.15)(mg)$	$\Sigma F = 28\text{ N} - 7.35\text{ N} - 7.35\text{ N} = ma$
$F_A = (0.15)(5.0\text{ kg}*9.8\text{ m/s}^2)$	$F_B = (0.15)(5.0\text{ kg}*9.8\text{ m/s}^2)$	$\Sigma F = 13.3\text{ N} = ma$
$F_A = (0.15)(49\text{ N})$	$F_B = (0.15)(49\text{ N})$	$13.2\text{ N} = ma$
$F_A = 7.35\text{ N}$	$F_B = 7.35\text{ N}$	$13.2\text{ N} = (5.0\text{ kg} + 5.0\text{ kg})*a$
		$13.2\text{ N} = (10\text{ kg})*a$
		$a = 13.2\text{ N} / 10\text{ kg}$
		$a = 1.32\text{ m/s}^2$

FIGURE 12.3

Friction scenario for two stacked boxes of equal size

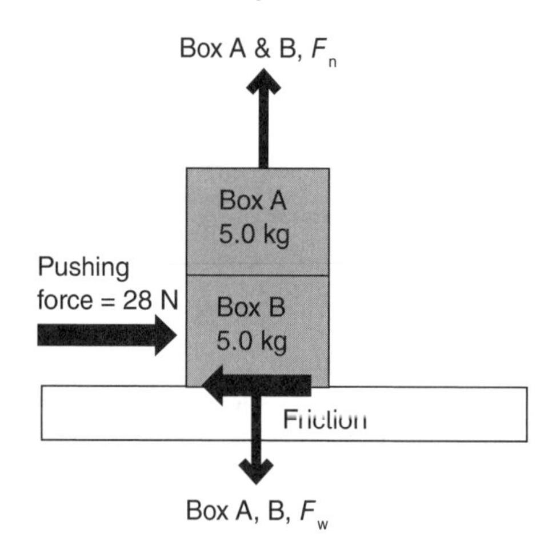

Friction for stacked Boxes A and B	Motion for Boxes A and B
$F_{A+B} = \mu_s F_n$, where $F_n = F_{wA} + F_{wB}$ and $F_w = mg$	$\Sigma F = ma$, where m = mass of *Box A* + *Box B*
$F_{A+B} = \mu_s F_n = \mu_s(F_{wA} + F_{wB})$	$\Sigma F = F_{push} - F_{A+B} = ma$
$F_{A+B} = (0.20)(F_{wA} + F_{wB})$	$\Sigma F = 28\ \text{N} - 14.7\ \text{N} = ma$
$F_{A+B} = (0.15)(m_A g + m_B g)$	$\Sigma F = 13.3\ \text{N} = ma$
$F_{A+B} = (0.15)[(5.0\ \text{kg}*9.8\ \text{m/s}^2) + (5.0\ \text{kg}*9.8\ \text{m/s}^2)]$	$13.2\ \text{N} = ma$
$F_{A+B} = (0.15)(49\ \text{N} + 49\ \text{N})$	$13.2\ \text{N} = (5.0\ \text{kg} + 5.0\ \text{kg})*a$
$F_{A+B} = (0.15)(98\ \text{N})$	$13.2\ \text{N} = (10\ \text{kg})*a$
$F_{A+B} = 14.7\ \text{N}$	$a = 13.2\ \text{N} / 10\ \text{kg}$
	$a = 1.32\ \text{m/s}^2$

Timeline

The instructional time needed to complete this lab investigation is 200–280 minutes. Appendix 2 (p. 411) provides options for implementing this lab investigation over several class periods. Option A (280 minutes) should be used if students are unfamiliar with scientific writing, because this option provides extra instructional time for scaffolding the writing process. You can scaffold the writing process by modeling, providing examples, and providing hints as students write each section of the report. Option F (200 minutes) should be used if students are familiar with scientific writing and have developed the skills needed to write an investigation report on their own. In option F, students complete stage 6 (writing the investigation report) and stage 8 (revising the investigation report) as homework.

Materials and Preparation

The materials needed to implement this investigation are listed in Table 12.1 (p. 220). The equipment and materials can be purchased from a science supply company such as Carolina, Flinn Scientific, or Ward's Science. Friction block sets are readily available, but it is also possible to make your own blocks to use during this lab. A 2 in. × 4 in. × 8 ft. piece of lumber from a hardware store can be cut into several blocks approximately 6 in. long. Cover one large face and one small face of the block with sandpaper and then add a small eyelet for attaching a spring scale for pulling. This basic friction block will allow students to test large versus small surface area for two different surfaces—sandpaper and plain wood. You can use this same approach to create a variety of friction blocks. Alternative surface coverings include felt, coarse-grained sandpaper, wax paper, aluminum foil, and carpet. Making your own friction blocks for this activity is generally more cost-effective than purchasing them from a vendor and allows for a greater variety of test surfaces.

We recommend that you use a set routine for distributing and collecting the materials during the lab investigation. For example, the consumables and equipment for each group can be set up at each group's lab station before class begins, or one member from each group can collect them from a table or a cart when needed during class.

TABLE 12.1
Materials list for Lab 12

Item	Quantity
Safety glasses or goggles	1 per student
Friction block set	1 per group
20 N spring scale	1 per group
10 N spring scale	1 per group
5 N spring scale	1 per group
Mass set	1 per group
Meterstick	1 per group
Stopwatch	1 per group
Investigation Proposal B (optional)	1 per group
Whiteboard, 2' × 3'*	1 per group
Lab Handout	1 per student
Peer-review guide	1 per student
Checkout Questions	1 per student

* As an alternative, students can use computer and presentation software, such as Microsoft PowerPoint or Apple Keynote, to create their arguments.

Safety Precautions and Laboratory Waste Disposal

Remind students to follow all normal lab safety rules. In addition, tell students to take the following safety precautions:

1. Wear sanitized safety glasses or goggles during lab setup, hands-on activity, and takedown.
2. Wash hands with soap and water after completing the lab activity.

There is no laboratory waste associated with this activity.

Topics for the Explicit and Reflective Discussion

Concepts That Can Be Used to Justify the Evidence

To provide an adequate justification of their evidence, students must explain why they included the evidence in their arguments and make the assumptions underlying their analysis and interpretation of the data explicit. In this investigation, students can use the following concepts to help justify their evidence:

- Balanced and unbalanced forces
- Static friction and kinetic friction
- Velocity and acceleration
- Newton's second law of motion

We recommend that you review these concepts during the explicit and reflective discussion to help students make this connection.

How to Design Better Investigations

It is important for students to reflect on the strengths and weaknesses of the investigation they designed during the explicit and reflective discussion. Students should therefore be encouraged to discuss ways to eliminate potential flaws, measurement errors, or sources of bias in their investigations. To help students be more reflective about the design of their investigation, you can ask the following questions:

1. What were some of the strengths of your investigation? What made it scientific?
2. What were some of the weaknesses of your investigation? What made it less scientific?
3. If you were to do this investigation again, what would you do to address the weaknesses in your investigation? What could you do to make it more scientific?

Crosscutting Concepts

This investigation is well aligned with two crosscutting concepts found in *A Framework for K–12 Science Education*, and you should review these concepts during the explicit and reflective discussion.

- *Patterns:* Scientists look for patterns in nature and attempt to understand the underlying cause of these patterns. In this lab students will look for patterns related to surface area, friction, and the motion of objects.
- *Stability and change:* It is critical for scientists to understand what makes a system stable or unstable and what controls rates of change in a system. In this lab students will investigate how balanced and unbalanced forces are related to the changes that occur within a system.

The Nature of Science and the Nature of Scientific Inquiry

This investigation is well aligned with two important concepts related to the *nature of science* (NOS) and the *nature of scientific inquiry* (NOSI), and you should review these concepts during the explicit and reflective discussion.

- *The difference between observations and inferences in science:* An observation is a descriptive statement about a natural phenomenon, whereas an inference is an interpretation of an observation. Students should also understand that current scientific knowledge and the perspectives of individual scientists guide both observations and inferences. Thus, different scientists can have different but equally valid interpretations of the same observations due to differences in their perspectives and background knowledge.
- *Methods used in scientific investigations:* Examples of methods include experiments, systematic observations of a phenomenon, literature reviews, and analysis of existing data sets; the choice of method depends on the objectives of the research. There is no universal step-by step scientific method that all scientists follow; rather, different scientific disciplines (e.g., chemistry vs. biology) and fields within a discipline (e.g., organic vs. physical chemistry) use different types of methods, use different core theories, and rely on different standards to develop scientific knowledge.

Hints for Implementing the Lab

- Allowing students to design their own procedures for collecting data gives students an opportunity to try, to fail, and to learn from their mistakes. However, you can scaffold students as they develop their procedure by having them fill out an investigation proposal. These proposals provide a way for you to offer students hints and suggestions without telling them how to do it. You can also check the proposals quickly during a class period. For this lab we suggest you use Investigation Proposal B.
- Allow the students to become familiar with the friction block apparatus as part of the tool talk before they begin to design their investigation. This gives students a chance to see what they can and cannot do with the equipment.
- If you'd like to provide more test options for your students, consider covering a portion of your lab stations with sandpaper or wax paper to provide a different test surface for the friction blocks.
- Using a force sensor in place of a spring scale also works well for this lab. The data output for the force sensor will allow students to measure and determine the force of static friction and kinetic friction for the surface combinations they try.

Topic Connections

Table 12.2 provides an overview of the scientific practices, crosscutting concepts, disciplinary core ideas, and supporting ideas at the heart of this lab investigation. In addition, it lists NOS and NOSI concepts for the explicit and reflective discussion. Finally, it lists literacy and mathematics skills (*CCSS ELA* and *CCSS Mathematics*) that are addressed during the investigation.

TABLE 12.2

Lab 12 alignment with standards

Scientific practices	• Asking questions and defining problems • Planning and carrying out investigations • Analyzing and interpreting data • Using mathematics and computational thinking • Constructing explanations and designing solutions • Engaging in argument from evidence • Obtaining, evaluating, and communicating information
Crosscutting concepts	• Patterns • Stability and change
Core idea	• PS2.A: Forces and motion
Supporting ideas	• Balanced and unbalanced forces • Static friction and kinetic friction • Velocity and acceleration • Newton's second law of motion
NOS and NOSI concepts	• Difference between data and evidence • Methods used in scientific investigations
Literacy connections (*CCSS ELA*)	• *Reading*: Key ideas and details, craft and structure, integration of knowledge and ideas • *Writing*: Text types and purposes, production and distribution of writing, research to build and present knowledge, range of writing • *Speaking and listening*: Comprehension and collaboration, presentation of knowledge and ideas
Mathematics connections (*CCSS Mathematics*)	• Reason abstractly and quantitatively • Construct viable arguments and critique the reasoning of others

LAB 12

Lab Handout

Lab 12. Unbalanced Forces

How Does Surface Area Influence Friction and the Motion of an Object?

Introduction

The motion of an object is determined by the combination of all the forces acting on that object. Those forces may come in the form of a simple push or pull that result from contact forces, such as pushing a box with your hand, or non-contact forces, such as gravity pulling an object toward Earth. When forces are acting on an object from the same direction, the influence of each force is added together. Think of two people pushing a box across the floor. If they push from the same side, the forces acting on the box are unbalanced, meaning that there is more force on one side of the box than the other. When the forces acting on the box are unbalanced, the box moves in the direction of that force. However, if the two people stand on opposite sides of the box and push again, their forces are working against each other (see Figure L12.1). One person is trying to push the box to the right and the other person is trying to push the box to the left. In this case, their forces are balanced and cancel each other out. When the forces acting on an object are balanced, the motion of the object does not change.

FIGURE L12.1

Two people pushing on opposite sides of a box

When two people push on a box in opposite directions, it's easy to see that there are opposing forces acting on the box, but sometimes there are opposing forces acting on an object that are less obvious. If you give a book a push and let it slide across a table, it will eventually come to a stop. The book stops because there is an opposite force acting on it, just like the person pushing on the opposite side of the box. The force that causes the book to come to a stop is called friction. Friction is a force that occurs when two surfaces are in contact with each other. The force of friction always works on an object in a direction that is opposite of the motion of the object. Therefore, friction is always trying to slow an object down or keep an object from moving. Trying to slide a heavy object is often difficult due to friction. Placing a heavy object on a cart with wheels can make it easier to move. The amount of friction is greatly reduced by the wheels, so it is easier to move the heavy object.

The amount of friction between two objects depends on several factors. One important factor is the specific surfaces involved. The amount of friction between two objects as they slide past one another depends on the specific surfaces that are rubbing against each other. For example, a cardboard box sliding on a wooden floor has a different amount of friction than the same cardboard box sliding on a carpet floor. Different combinations of surface materials will result in different amounts of friction. Another factor that influences the amount of friction is the weight of the object that is moving. The heavier the object, the more friction there will be that opposes the motion of the object.

Reducing friction is often an important goal when there is work to be done. There are many different ways that someone can try to reduce the amount of friction when moving an object. One strategy is to change the surface that an object is sliding on. For example, there is very little friction between ice and other materials, therefore objects slide over ice very easily. Another strategy for reducing the amount of friction when trying to slide an object would be reducing the mass of the object; then there would be less force between the moving object and the surface it is sliding on. A third strategy would be to change the shape of the object and change the amount of contact between the two surfaces that are rubbing together. For example, instead of sliding two separate boxes side by side, perhaps stacking them one on top of the other would reduce the amount of friction and make them easier to move.

Your Task

Use what you know about forces and motion, patterns, and stability and change to design and carry out an investigation that will allow you to test how changing the amount of surface area between two objects influences the amount of friction between the objects and how changing the amount of surface area influences the motion of the object.

The guiding question of this investigation is, **How does surface area influence friction and the motion of an object?**

Materials

You may use any of the following materials during your investigation:

- Friction block set
- 20 N spring scale
- 10 N spring scale
- 5 N spring scale
- Mass set
- Meterstick
- Stopwatch
- Safety glasses or goggles

Safety Precautions

Follow all normal lab safety rules. In addition, take the following safety precautions:

1. Wear sanitized safety glasses or goggles during lab setup, hands-on activity, and takedown.
2. Wash hands with soap and water after completing the lab activity.

Investigation Proposal Required? ☐ Yes ☐ No

Getting Started

To answer the guiding question, you will need to design and conduct an investigation to measure the amount of friction between the friction block and the surface of your table and test how changes in the amount of surface area influence the motion of the block. To accomplish this task, you must determine what type of data you need to collect, how you will collect it, and how you will analyze it.

To determine *what type of data you need to collect,* think about the following questions:

- How will you determine the force due to friction?
- What information or measurements will you need to record?
- How will you determine the surface area of the block?
- Will you test different surface combinations?

To determine *how you will collect your data,* think about the following questions:

- What equipment will you need to collect the data you need?
- How will you make sure that your data are of high quality (i.e., how will you reduce error)?
- How will you keep track of the data you collect?
- How will you organize your data?

To determine *how you will analyze your data,* think about the following questions:

- How will you determine the influence of surface area on friction and motion of the block?
- What type of table or graph could you create to help make sense of your data?

Connections to Crosscutting Concepts, the Nature of Science, and the Nature of Scientific Inquiry

As you work through your investigation, be sure to think about

- why it is important to understand patterns and their causes,
- the importance of understanding stability and change within systems,
- the difference between observations and inferences, and
- the role of different methods in science.

Initial Argument

Once your group has finished collecting and analyzing your data, your group will need to develop an initial argument. Your initial argument needs to include a *claim*, *evidence* to support your claim, and a *justification* of the evidence. The claim is your group's answer to the guiding question. The evidence is an analysis and interpretation of your data. Finally, the justification of the evidence is why your group thinks the evidence matters. The justification of the evidence is important because scientists can use different kinds of evidence to support their claims. Your group will create your initial argument on a whiteboard. Your whiteboard should include all the information shown in Figure L12.2.

FIGURE L12.2

Argument presentation on a whiteboard

The Guiding Question:	
Our Claim:	
Our Evidence:	Our Justification of the Evidence:

Argumentation Session

The argumentation session allows all of the groups to share their arguments. One member of each group will stay at the lab station to share that group's argument, while the other members of the group go to the other lab stations to listen to and critique the arguments developed by their classmates. This is similar to how scientists present their arguments to other scientists at conferences. If you are responsible for critiquing your classmates' arguments, your goal is to look for mistakes so these mistakes can be fixed and they can make their argument better. The argumentation session is also a good time to think about ways you can make your initial argument better. Scientists must share and critique arguments like this to develop new ideas.

To critique an argument, you might need more information than what is included on the whiteboard. You will therefore need to ask the presenter lots of questions. Here are some good questions to ask:

- How did you collect your data? Why did you use that method? Why did you collect those data?
- What did you do to make sure the data you collected are reliable? What did you do to decrease measurement error?
- How did your group analyze the data? Why did you decide to do it that way? Did you check your calculations?
- Is that the only way to interpret the results of your analysis? How do you know that your interpretation of your analysis is appropriate?
- Why did your group decide to present your evidence in that way?
- What other claims did your group discuss before you decided on that one? Why did your group abandon those alternative ideas?

- How confident are you that your claim is valid? What could you do to increase your confidence?

Once the argumentation session is complete, you will have a chance to meet with your group and revise your initial argument. Your group might need to gather more data or design a way to test one or more alternative claims as part of this process. Remember, your goal at this stage of the investigation is to develop the most acceptable and valid answer to the research question!

Report

Once you have completed your research, you will need to prepare an *investigation report* that consists of three sections. Each section should provide an answer to the following questions:

1. What question were you trying to answer and why?
2. What did you do to answer your question and why?
3. What is your argument?

Your report should answer these questions in two pages or less. This report must be typed, and any diagrams, figures, or tables should be embedded into the document. Be sure to write in a persuasive style; you are trying to convince others that your claim is acceptable and valid!

Checkout Questions

Lab 12. Unbalanced Forces

How Does Surface Area Influence Friction and the Motion of an Object?

1. Jared knows that trucks and cars driving down the road normally have good traction between the rubber tires and the asphalt of the road. But he is unsure why trucks and cars begin to slide if they hit a patch of ice. Using what you know about balanced and unbalanced forces, explain to Jared why truck and car wheels slide on ice rather than roll like they do on normal roads.

2. A student in physical science class was conducting an investigation by sliding different blocks across her wooden lab table. Each block was launched by a rubber band with the same force. She measured how far the block traveled and obtained the following results:

	Distance traveled		
Surface of block	**Trial 1**	**Trial 2**	**Trial 3**
Wood	57 cm	66 cm	52 cm
Aluminum foil	64 cm	71 cm	69 cm
Carpet	32 cm	40 cm	45 cm

Use what you know about balanced and unbalanced forces to generate an argument (including a claim, evidence, and justification) that explains the results the student obtained.

3. When scientists make observations, they are more certain than when they make inferences.

   a. I agree with this statement.
   b. I disagree with this statement.

   Explain your answer, using an example from your investigation on unbalanced forces.

4. When several scientists are investigating the same thing, they all use the same methods so that they get the same answer.

   a. I agree with this statement.
   b. I disagree with this statement.

   Explain your answer, using an example from your investigation on unbalanced forces.

5. Scientists often look for patterns in nature or within the data that they collect during an investigation. Using an example from your investigation about unbalanced forces, explain why it is important to understand patterns and their causes within science.

6. In physics, there are times when scientists study many variables to learn how a system works. Using an example from your investigation on unbalanced forces, explain why it is important for scientists to understand what causes systems to be stable or change.

SECTION 4
Physical Science Core Idea 3

Energy

Introduction Labs

Teacher Notes

Lab 13. Kinetic Energy

How Do the Mass and Velocity of an Object Affect Its Kinetic Energy?

Purpose

The purpose of this lab is to *introduce* students to the relationships between an object's mass and velocity, and the impact these values have on the kinetic energy of that object. This lab gives students the opportunity to investigate cause-and-effect relationships and the ways in which scientists track how energy moves into, out of, and within systems. Students will also learn about the difference between laws and theories in science and how scientists use creativity and imagination in their work.

The Content

There are several types of energy that exist in the world, and energy can be readily transferred from one type to another. The *law of conservation of energy* states that within a given system the total amount of energy always stays the same. Essentially, this means that energy is neither created nor destroyed, but rather transferred from one type to another. The law of conservation of energy, like all scientific laws, *describes a specific relationship* that exists in the natural world—in this case, the relationship that exists among the many different forms of energy present in the world. In contrast, a scientific theory *provides an explanation* for why those things relate to each other. In a more practical sense, laws tell us how things relate, while theories tell us why they do.

Kinetic energy, which is also known as energy of motion, is the type that is measured in this lab. When stored *potential energy* (energy of position) is transformed into motion, it becomes kinetic energy. The car crash example discussed in the Lab Handout is a concrete illustration of *translational* kinetic energy, which is the energy an object has when it is moving from one place to another in a straight line. Other types of kinetic energy include rotational and vibrational kinetic energy.

The amount of translational kinetic energy an object has is dependent on two things: the object's mass and its velocity. Mass is the amount of matter in an object. A more massive object requires more energy to move and thus, when moving, will have more kinetic energy than an object of lesser mass moving at the same speed. Conversely, a less massive object moving quickly can potentially have more kinetic energy than a more massive object that is moving slowly. The formula that models this relationship mathematically is $KE = \frac{1}{2} mv^2$, where KE is kinetic energy, m is mass, and v is velocity.

Damage done in a car accident is caused by the translational kinetic energy of the vehicles involved. A large semitrailer traveling at 10 mph will cause much more significant damage than a small sedan traveling the same speed because the semitrailer has more kinetic energy available to transfer to the vehicle it is crashing into. That translational kinetic energy, when transferred, does the work to deform the vehicle or to push it a greater distance.

Timeline

The instructional time needed to complete this lab investigation is 200–230 minutes. Appendix 2 (p. 411) provides options for implementing this lab investigation over several class periods. Option C (230 minutes) should be used if students are unfamiliar with scientific writing, because this option provides extra instructional time for scaffolding the writing process. You can scaffold the writing process by modeling, providing examples, and providing hints as students write each section of the report. Option F (200 minutes) should be used if students are familiar with scientific writing and have developed the skills needed to write an investigation report on their own. In option F, students complete stage 6 (writing the investigation report) and stage 8 (revising the investigation report) as homework.

Materials and Preparation

The materials needed to implement this investigation are listed in Table 13.1 (p. 238). Much of the equipment can be purchased from general retail stores. The more specialized equipment can be purchased from a science supply company such as Carolina, Flinn Scientific, or Ward's Science.

Please ensure that the large plastic tub filled with flour is deep and wide enough that its size does not affect the movement of the flour when the racquetball makes contact. Cornstarch may also be used in lieu of flour. Be sure to cut the slits into the racquetballs beforehand. Also, ensure that both the slits and the funnels are large enough for the beans, rice, or other material you are using to increase the mass of the balls to pass through.

We recommend that you use a set routine for distributing and collecting the materials during the lab investigation. For example, the consumables and equipment for each group can be set up at each group's lab station before class begins, or one member from each group can collect them from a table or a cart when needed during class.

Safety Precautions and Laboratory Waste Disposal

Remind students to follow all normal lab safety rules. In addition, tell students to take the following safety precautions:

1. Wear sanitized safety glasses or goggles during lab setup, hands-on activity, and takedown.
2. Never put consumables in their mouth.
3. Sweep up flour off the floor to avoid a slip or fall hazard.
4. Wash hands with soap and water after completing the lab activity.

TABLE 13.1

Materials list for Lab 13

Item	Quantity
Safety glasses or goggles	1 per student
Large plastic tub filled with flour	1 per group
Racquetball with slit	1 per group
Funnel large enough for rice or beans	1 per group
Rice, beans, or other material to fill racquetball	1 cup per group
Meterstick	1 per group
Ruler	1 per group
String	20 cm per group
Stopwatch	1 per group
Electronic or triple beam balance	1 per group
Excel, graphing calculator, or other mathematical software (optional)	1 per group
Investigation Proposal A (optional)	1 per group
Whiteboard, 2' × 3'*	1 per group
Lab Handout	1 per student
Peer-review guide	1 per student
Checkout Questions	1 per student

*As an alternative, students can use computer and presentation software, such as Microsoft PowerPoint or Apple Keynote, to create their arguments.

There is no laboratory waste associated with this activity. The materials for this laboratory investigation can be stored and reused.

Topics for the Explicit and Reflective Discussion

Concepts That Can Be Used to Justify the Evidence

To provide an adequate justification of their evidence, students must explain why they included the evidence in their arguments and make the assumptions underlying their analysis and interpretation of the data explicit. In this investigation, students can use the following concepts to help justify their evidence:

- The law of conservation of energy

- The relationship between kinetic energy, mass, and velocity
- The transfer of energy when objects collide

We recommend that you review these concepts during the explicit and reflective discussion to help students make this connection.

How to Design Better Investigations

It is important for students to reflect on the strengths and weaknesses of the investigation they designed during the explicit and reflective discussion. Students should therefore be encouraged to discuss ways to eliminate potential flaws, measurement errors, or sources of bias in their investigations. To help students be more reflective about the design of their investigation, you can ask the following questions:

1. What were some of the strengths of your investigation? What made it scientific?
2. What were some of the weaknesses of your investigation? What made it less scientific?
3. If you were to do this investigation again, what would you do to address the weaknesses in your investigation? What could you do to make it more scientific?

Crosscutting Concepts

This investigation is well aligned with two crosscutting concepts found in *A Framework for K–12 Science Education,* and you should review these concepts during the explicit and reflective discussion.

- *Cause and effect: Mechanism and explanation:* One of the main objectives of science is to identify and establish relationships between a cause and an effect. In this lab students will examine the effects of energy transfer on objects.
- *Energy and matter: Flows, cycles, and conservation:* In science it is important to track how energy and matter move into, out of, and within systems. In this lab students will investigate how energy is transferred during a collision.

The Nature of Science and the Nature of Scientific Inquiry

This investigation is well aligned with two important concepts related to the *nature of science* (NOS) and the *nature of scientific inquiry* (NOSI), and you should review these concepts during the explicit and reflective discussion.

- *The difference between laws and theories in science:* A scientific law describes the behavior of a natural phenomenon or a generalized relationship under certain conditions; a scientific theory is a well-substantiated explanation of some aspect of the natural world. Theories do not become laws even with additional evidence;

they explain laws. However, not all scientific laws have an accompanying explanatory theory. It is also important for students to understand that scientists do not discover laws or theories; the scientific community develops them over time.

- *The importance of imagination and creativity in science:* Students should learn that developing explanations for or models of natural phenomena and then figuring out how they can be put to the test of reality is as creative as writing poetry, composing music, or designing video games. Scientists must also use their imagination and creativity to figure out new ways to test ideas and collect or analyze data.

Hints for Implementing the Lab

- Allowing students to design their own procedures for collecting data gives students an opportunity to try, to fail, and to learn from their mistakes. However, you can scaffold students as they develop their procedure by having them fill out an investigation proposal. These proposals provide a way for you to offer students hints and suggestions without telling them how to do it. You can also check the proposals quickly during a class period. For this lab we suggest you use Investigation Proposal A.
- Allow students to become familiar with the variable-mass racquetball as part of the tool talk before they begin to design their investigation. This gives students a chance to see what they can and cannot do with the equipment.
- Be sure that students record actual values (e.g., mass, velocity, size of depression left by the racquetball).

Topic Connections

Table 13.2 provides an overview of the scientific practices, crosscutting concepts, disciplinary core ideas, and supporting ideas at the heart of this lab investigation. In addition, it lists NOS and NOSI concepts for the explicit and reflective discussion. Finally, it lists literacy and mathematics skills (*CCSS ELA* and *CCSS Mathematics*) that are addressed during the investigation.

TABLE 13.2

Lab 13 alignment with standards

Scientific practices	• Asking questions and defining problems • Developing and using models • Planning and carrying out investigations • Analyzing and interpreting data • Using mathematics and computational thinking • Constructing explanations and designing solutions • Engaging in argument from evidence • Obtaining, evaluating, and communicating information
Crosscutting concepts	• Cause and effect: Mechanism and explanation • Energy and matter: Flows, cycles, and conservation
Core ideas	• PS2.B: Types of interactions • PS3.B: Conservation of energy and energy transfer • PS3.C: Relationship between energy and forces
Supporting ideas	• Law of conservation of energy • Relationship between kinetic energy, mass, and velocity • Transfer of energy
NOS and NOSI concepts	• Scientific laws and theories • Imagination and creativity in science
Literacy connections (*CCSS ELA*)	• *Reading:* Key ideas and details, craft and structure, integration of knowledge and ideas • *Writing:* Text types and purposes, production and distribution of writing, research to build and present knowledge, range of writing • *Speaking and listening:* Comprehension and collaboration, presentation of knowledge and ideas
Mathematics connections (*CCSS Mathematics*)	• Reason abstractly and quantitatively • Construct viable arguments and critique the reasoning of others • Model with mathematics • Use appropriate tools strategically

LAB 13

Lab Handout

Lab 13. Kinetic Energy

How Do the Mass and Velocity of an Object Affect Its Kinetic Energy?

Introduction

When law enforcement officials investigate car crashes (see Figure L13.1), it can sometimes be difficult to determine who is at fault and what laws were broken, especially when there is no footage of the crash. To re-create the crash scene, investigators use physics concepts to determine the specifics of a crash, including the speed and direction (together, the velocity) in which a car was traveling and when the driver attempted to stop. These figures can then be used to help determine who is at fault in a crash and the laws that person broke.

FIGURE L13.1

A police officer investigates a crash scene.

You already know that energy is conserved and transferred within and between systems, not created or destroyed. This is also true in car crashes. A traveling car has a certain amount of kinetic energy, and when that car hits another car or a different object, some of that energy is transformed into heat or sound, but most is used to do the work that deforms the car or object it crashes into. So, when cars collide, the transfer of their kinetic energy is responsible for the resulting damage. The damage done to the cars can be used, along with other pieces of evidence, to determine the velocity of the car.

In this investigation, you will be applying the same physics concepts to determine the mass or velocity of an object that has been dropped onto an inelastic surface (one that does not return to the same shape after impact). You will be creating a collision between a ball of variable mass and a large container of flour to investigate this relationship.

Your Task

Use what you know about force and motion, patterns, and causal relationships to design and carry out an investigation that will allow you to create a mathematical model explaining the relationship between mass, velocity, and force of impact.

The guiding question of this investigation is, **How do the mass and velocity of an object affect its kinetic energy?**

Materials

You may use any of the following materials during your investigation:

- Large plastic container filled with flour
- Racquetball with slit
- Funnel
- Rice or beans
- Meterstick
- Ruler
- String
- Stopwatch
- Electronic or triple beam balance
- Excel, graphing calculator, or other mathematical software (optional)
- Safety glasses or goggles

Safety Precautions

Follow all normal lab safety rules. In addition, take the following safety precautions:

1. Wear sanitized indirectly vented chemical-splash safety goggles during lab setup, hands-on activity, and takedown.
2. Never put consumables in your mouth.
3. Sweep up flour off the floor to avoid a slip or fall hazard.
4. Do not throw objects.
5. Wash hands with soap and water after completing the lab activity.

Investigation Proposal Required? ☐ Yes ☐ No

Getting Started

The first step in developing your mathematical model is to design and carry out an investigation to determine how mass and velocity affect the kinetic energy of the variable-mass ball. Because the kinetic energy of the ball will deform the flour surface during an impact, the extent of this deformation can be used to determine the kinetic energy of the ball at impact.

To determine *what type of data you need to collect,* think about the following questions:

- What information do you need to create your mathematical model?
- What measurements will you take during your investigation?
- How will you determine the velocity of the ball?
- In what way will you vary the mass of the ball, if at all?
- How will you know how much kinetic energy has been absorbed by the flour surface?

To determine *how you will collect your data,* think about the following questions:

- What equipment will you need to collect the data you need?

- How will you make sure that your data are of high quality (i.e., how will you reduce error)?
- How will you keep track of the data you collect?
- How will you organize your data?

To determine *how you will analyze your data,* think about the following questions:

- What type of calculations will you need to make?
- What type of table or graph could you create to help make sense of your data?

Once you have carried out your investigations, your group will need to use the data you collected to develop a mathematical model that can be used to help explain the relationship between the mass of an object, the velocity of the object, and the kinetic energy of the object. You should be able to use your mathematical model to predict any one of the three values when given the other two. For example, if given the kinetic energy measurement and the object's mass, you should be able to use your mathematical model to determine the velocity of the object at impact.

The last step in this investigation is to test your mathematical model. To accomplish this goal, you can predict the kinetic energy measurement of a ball with a given mass and velocity. If you are able to use your mathematical model to make accurate predictions about the object's mass, velocity, and kinetic energy, you will be able to generate the evidence you need to convince others that the mathematical model you developed is valid.

Connections to Crosscutting Concepts, the Nature of Science, and the Nature of Scientific Inquiry

As you work through your investigation, be sure to think about

- the importance of identifying cause and effect;
- how scientists often need to track how energy moves into, out of, and within a system;
- the difference between laws and theories in science; and
- how scientists must use imagination and creativity when developing models and explanations.

Initial Argument

Once your group has finished collecting and analyzing your data, your group will need to develop an initial argument. Your initial argument needs to include a *claim, evidence* to support your claim, and a *justification* of the evidence. The claim is your group's answer to the guiding question. The evidence is an analysis and interpretation of your data. Finally, the justification of the evidence is why your group thinks the evidence matters. The justification of the evidence is important because scientists can use different kinds of evidence to support their claims. Your group will create your initial argument on a whiteboard. Your whiteboard should include all the information shown in Figure L13.2.

FIGURE L13.2

Argument presentation on a whiteboard

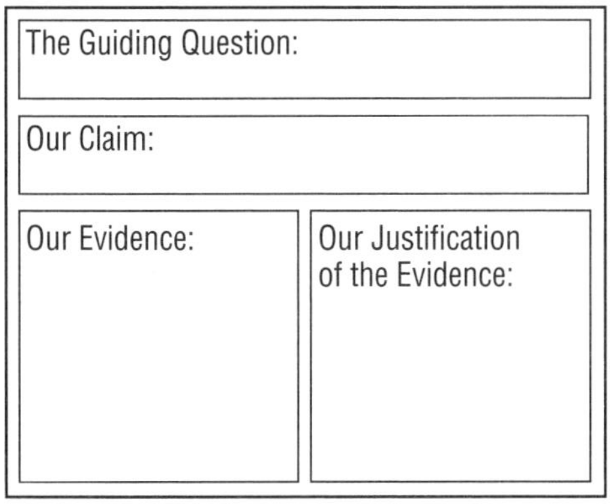

The Guiding Question:	
Our Claim:	
Our Evidence:	Our Justification of the Evidence:

Argumentation Session

The argumentation session allows all of the groups to share their arguments. One member of each group will stay at the lab station to share that group's argument, while the other members of the group go to the other lab stations to listen to and critique the arguments developed by their classmates. This is similar to how scientists present their arguments to other scientists at conferences. If you are responsible for critiquing your classmates' arguments, your goal is to look for mistakes so these mistakes can be fixed and they can make their argument better. The argumentation session is also a good time to think about ways you can make your initial argument better. Scientists must share and critique arguments like this to develop new ideas.

To critique an argument, you might need more information than what is included on the whiteboard. You will therefore need to ask the presenter lots of questions. Here are some good questions to ask:

- How did you collect your data? Why did you use that method? Why did you collect those data?
- What did you do to make sure the data you collected are reliable? What did you do to decrease measurement error?
- How did your group analyze the data? Why did you decide to do it that way? Did you check your calculations?
- Is that the only way to interpret the results of your analysis? How do you know that your interpretation of your analysis is appropriate?
- Why did your group decide to present your evidence in that way?
- What other claims did your group discuss before you decided on that one? Why did your group abandon those alternative ideas?
- How confident are you that your claim is valid? What could you do to increase your confidence?

Once the argumentation session is complete, you will have a chance to meet with your group and revise your initial argument. Your group might need to gather more data or design a way to test one or more alternative claims as part of this process. Remember, your goal at this stage of the investigation is to develop the most acceptable and valid answer to the research question!

Report

Once you have completed your research, you will need to prepare an *investigation report* that consists of three sections. Each section should provide an answer to the following questions:

1. What question were you trying to answer and why?
2. What did you do to answer your question and why?
3. What is your argument?

Your report should answer these questions in two pages or less. This report must be typed, and any diagrams, figures, or tables should be embedded into the document. Be sure to write in a persuasive style; you are trying to convince others that your claim is acceptable and valid!

Checkout Questions

Lab 13. Kinetic Energy: How Do the Mass and Velocity of an Object Affect Its Kinetic Energy?

1. Malik is throwing rocks into a lake. He throws a 1 kg rock that travels 5 m/s immediately after it left his hand. He also throws a 0.5 kg rock that travels 8 m/s immediately after it left his hand. Which rock did Malik throw the hardest?

   Explain your answer. How do you know which rock Malik threw the hardest?

2. Jerilyn dropped three spheres with masses of 0.25 kg, 0.75 kg, and 1 kg from equal height into a tub of flour. After dropping the spheres, her lab partner, Evan, put the spheres away before she recorded her data. Jerilyn and Evan are now unsure which sphere created which crater. A side view of the craters is shown below. Use what you know about mass, velocity, and kinetic energy to select which sphere created each crater.

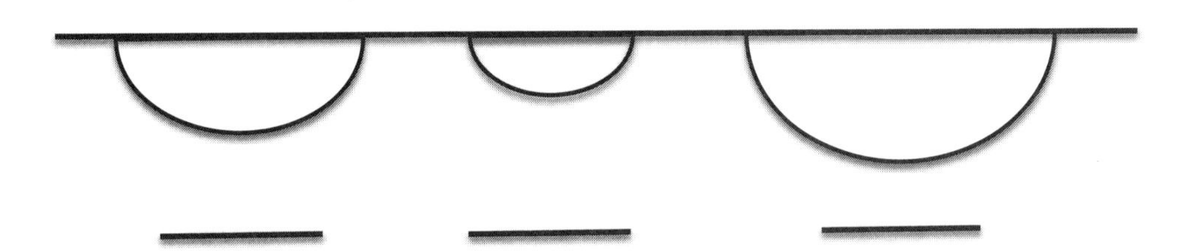

Explain your answer. How do you know which sphere made which crater? Use examples from your investigation about kinetic energy to support your answer.

3. Thinking creatively in science will lead to work that is less scientific and valid.

   a. I agree with this statement.
   b. I disagree with this statement.

   Explain your answer, using an example from your investigation about kinetic energy.

4. In science, theories and laws describe the same thing.

   a. I agree with this statement.
   b. I disagree with this statement.

   Explain your answer, using an example from your investigation about kinetic energy.

5. Understanding the ways in which the transfer of energy affects objects within a system is an important aspect of science and engineering. Using an example from your investigation about kinetic energy, explain why tracking the transfer of energy within a system is important.

6. Identifying cause-and-effect relationships in nature can help scientists make predictions about the behavior of objects. What cause-and-effect relationships did you observe in your investigation about kinetic energy, and how do these relationships allow you to make accurate predictions?

Teacher Notes

Lab 14. Potential Energy

How Can You Make an Action Figure Jump Higher?

Purpose

The purpose of this lab is to *introduce* students to the types of energy, specifically potential energy. Through this activity, students will have an opportunity to explore the crosscutting concepts of the importance of using and defining models to make sense of phenomena and how scientists focus on tracking the movement of energy through a system. Students will also learn about the difference between laws and theories and how scientists use multiple methods to investigate the natural world.

The Content

The *law of conservation of energy* states that within a given system the total amount of energy always stays the same. Essentially, this means that energy is neither created nor destroyed, but rather transferred from one type to another. Remember that scientific laws are used to describe specific relationships that exist in the natural world, whereas scientific theories provide broad-based explanations for different phenomena. In a more practical sense, laws tell us how things relate, while theories tell us why they do. In this case, the law of conservation of energy simply describes the relationship that exists among the many different types of energy present in the world.

There are several common forms of energy that exist in the world. Two of the most fundamental types of energy are potential and kinetic energy. When energy is stored in one form or another, it is called *potential energy*. Potential energy can be stored in the chemical bonds between atoms in a molecule and in the nuclei of atoms. Energy can also be stored based on the position of an object. Indeed, potential energy can be referred to as energy of position. The amount of potential energy an object has depends on the system being explored. In this use, a *system* refers to a specified collection of objects and their interactions. A ball on the floor has potential energy with respect to a desk in the same room, which can be called the ball-desk system. However, the potential energy of the ball is different if we are considering the ball-tree system, which includes a tree that exists outside of the room. Similarly, the amount of energy available and the different forms present will depend on the specific system that is being studied.

When potential energy is transformed into motion, it becomes *kinetic energy,* which can be detected when objects move. Kinetic energy is known as energy of motion. Kinetic energy is more obvious to identify, because it is the form of energy that does work on an object in a system. Other basic forms of energy include thermal energy (heat), chemical energy, electromagnetic energy, and nuclear energy. Some of these forms actually represent a mixture of potential and kinetic energies in more specific systems. More recognizable forms of energy, such as light and sound, also represent combinations of kinetic and potential energy.

As an example, think about climbing a hill. When you are at the bottom of a hill, you have low potential energy based on your position in the "hill-person" system. To increase your potential energy, you climb to the top of the hill. As you are climbing, you are moving, so you are using kinetic energy; you are transforming kinetic energy into increased potential energy; and you are changing position. Since you have climbed higher, you have greater potential energy. Now, you may wonder where the kinetic energy to climb the hill came from. That energy ultimately came from the energy stored in molecules that your body used to move your muscles.

Timeline

The instructional time needed to complete this lab investigation is 170–230 minutes. Appendix 2 (p. 411) provides options for implementing this lab investigation over several class periods. Option C (230 minutes) should be used if students are unfamiliar with scientific writing, because this option provides extra instructional time for scaffolding the writing process. You can scaffold the writing process by modeling, providing examples, and providing hints as students write each section of the report. Option D (170 minutes) should be used if students are familiar with scientific writing and have developed the skills needed to write an investigation report on their own. In option D, students complete stage 6 (writing the investigation report) and stage 8 (revising the investigation report) as homework.

Materials and Preparation

The materials needed to implement this investigation are listed in Table 14.1 (p. 252). The equipment can be purchased from a science supply company such as Carolina, Flinn Scientific, or Ward's Science. The clay and the action figures can be purchased at a toy store or general retail store.

We recommend that you use a set routine for distributing and collecting the materials during the lab investigation. For example, the equipment for each group can be set up at each group's lab station before class begins, or one member from each group can collect them from a table or a cart when needed during class.

TABLE 14.1

Materials list for Lab 14

Item	Quantity
Safety glasses or goggles	1 per student
Ruler	1 per group
Meterstick	1 per group
Electronic or triple beam balance	1 per group
Pencil	1 per group
Clay	100 g per group
Action figures	2–3 per group
Investigation Proposal C (optional)	1 per group
Whiteboard, 2' × 3'*	1 per group
Lab Handout	1 per student
Peer-review guide	1 per student
Checkout Questions	1 per student

*As an alternative, students can use computer and presentation software, such as Microsoft PowerPoint or Apple Keynote, to create their arguments.

Safety Precautions and Laboratory Waste Disposal

Follow all normal lab safety rules. In addition, tell students to take the following safety precautions:

1. Wear sanitized safety glasses or goggles during lab setup, hands-on activity, and take down.
2. Sweep clay up off the floor to avoid a slip or fall hazard.
3. Do not allow the action figures to jump too far from the work area.
4. Remove any fragile items from the work area.
5. Wash hands with soap and water after completing the lab activity.

There is no laboratory waste associated with this activity. The materials for this laboratory investigation can be stored and reused.

Topics for the Explicit and Reflective Discussion

Concepts That Can Be Used to Justify the Evidence

To provide an adequate justification of their evidence, students must explain why they included the evidence in their arguments and make the assumptions underlying their analysis and interpretation of the data explicit. In this investigation, students can use the following concepts to help justify their evidence:

- Law of conservation of energy
- Potential energy
- Kinetic energy
- Transformation of energy

We recommend that you review these concepts during the explicit and reflective discussion to help students make this connection.

How to Design Better Investigations

It is important for students to reflect on the strengths and weaknesses of the investigation they designed during the explicit and reflective discussion. Students should therefore be encouraged to discuss ways to eliminate potential flaws, measurement errors, or sources of bias in their investigations. To help students be more reflective about the design of their investigation, you can ask the following questions:

1. What were some of the strengths of your investigation? What made it scientific?
2. What were some of the weaknesses of your investigation? What made it less scientific?
3. If you were to do this investigation again, what would you do to address the weaknesses in your investigation? What could you do to make it more scientific?

Crosscutting Concepts

This investigation is well aligned with two crosscutting concepts found in *A Framework for K–12 Science Education*, and you should review these concepts during the explicit and reflective discussion.

- *System and system models:* Defining a system under study and making a model of it are tools for developing a better understanding of natural phenomena in science. In this lab students will investigate a system that can be used to convert potential energy to kinetic energy.

- *Energy and matter: Flows, cycles, and conservation:* In science it is important to track how energy and matter move into, out of, and within systems. In this lab students will investigate the conversion of energy from one type to another.

The Nature of Science and the Nature of Scientific Inquiry

This investigation is well aligned with two important concepts related to the *nature of science* (NOS) and the *nature of scientific inquiry* (NOSI), and you should review these concepts during the explicit and reflective discussion.

- *The difference between laws and theories in science:* A scientific law describes the behavior of a natural phenomenon or a generalized relationship under certain conditions; a scientific theory is a well-substantiated explanation of some aspect of the natural world. Theories do not become laws even with additional evidence; they explain laws. However, not all scientific laws have an accompanying explanatory theory. It is also important for students to understand that scientists do not discover laws or theories; the scientific community develops them over time.
- *Methods used in scientific investigations*: Examples of methods include experiments, systematic observations of a phenomenon, literature reviews, and analysis of existing data sets; the choice of method depends on the objectives of the research. There is no universal step-by step scientific method that all scientists follow; rather, different scientific disciplines (e.g., chemistry vs. physics) and fields within a discipline (e.g., organic vs. physical chemistry) use different types of methods, use different core theories, and rely on different standards to develop scientific knowledge.

Hints for Implementing the Lab

- Allowing students to design their own procedures for collecting data gives students an opportunity to try, to fail, and to learn from their mistakes. However, you can scaffold students as they develop their procedure by having them fill out an investigation proposal. These proposals provide a way for you to offer students hints and suggestions without telling them how to do it. You can also check the proposals quickly during a class period. For this lab we suggest you use Investigation Proposal C.
- Suggest that students use a small amount of clay to stick the pencil to the ruler when they construct their teeterboard.
- Have students focus on changing one characteristic of the system at a time. They should not change the mass of the dropped clay while also changing the height they drop it from.

- Encourage students to think of a way they could mathematically represent the relationships they find in this investigation.
- Action figures should not be too large, so that they can actually be launched using the ruler apparatus. We have had success using small, plastic army action figures that can be purchased in large quantities. Be sure to test your action figures with the equipment to determine if they are appropriate.

Topic Connections

Table 14.2 provides an overview of the scientific practices, crosscutting concepts, disciplinary core ideas, and supporting ideas at the heart of this lab investigation. In addition, it lists the NOS and NOSI concepts for the explicit and reflective discussion. Finally, it lists literacy and mathematics skills (*CCSS ELA* and *CCSS Mathematics*) that are addressed during the investigation.

TABLE 14.2

Lab 14 alignment with standards

Scientific practices	• Asking questions and defining problems • Planning and carrying out investigations • Analyzing and interpreting data • Using mathematics and computational thinking • Constructing explanations and designing solutions • Engaging in argument from evidence • Obtaining, evaluating, and communicating information
Crosscutting concepts	• Systems and system models • Energy and matter
Core ideas	• PS3.A: Definitions of energy • PS3.B: Conservation of energy and energy transfer
Supporting ideas	• Law of conservation of energy • Potential energy • Kinetic energy • Transformation of energy
NOS and NOSI concepts	• Scientific laws and theories • Methods used in scientific investigations
Literacy connections (*CCSS ELA*)	• *Reading*: Key ideas and details, craft and structure, integration of knowledge and ideas • *Writing*: Text types and purposes, production and distribution of writing, research to build and present knowledge, range of writing • *Speaking and listening*: Comprehension and collaboration, presentation of knowledge and ideas
Mathematics connections (*CCSS Mathematics*)	• Reason abstractly and quantitatively • Construct viable arguments and critique the reasoning of others • Use appropriate tools strategically • Attend to precision

Lab Handout

Lab 14. Potential Energy

How Can You Make an Action Figure Jump Higher?

Introduction

Teeterboards are typical pieces of equipment found on many playgrounds around the country. They are often used in shows that focus on gymnastic tricks. The picture in Figure L14.1 shows a circus act involving a performer launching another performer high into the air. It is easy to observe how the activity of a teeterboard involves objects' motion. However, that activity also involves energy shifting between forms.

FIGURE L14.1

Circus performers on a teeterboard

The law of conservation of energy states that within a given system the total amount of energy always stays the same—it is neither created nor destroyed; instead, energy is transformed from one form to another. When energy is stored in one form or another, it is called potential energy. Potential energy can be stored in the chemical bonds between atoms in a molecule and in the nuclei of atoms. Energy can also be stored based on the position of an object. Indeed, potential energy can be referred to as energy of position. When potential energy is transformed into motion, it becomes kinetic energy. Kinetic energy can be detected when objects move. Kinetic energy is known as energy of motion.

For an example, think about climbing a hill. When you are at the bottom of a hill, you have low potential energy based on your position. To increase your potential energy, you climb to the top of the hill. As you are climbing, you are moving, so you are using kinetic energy; you are transforming kinetic energy into increased potential energy; and you are changing position. Since you have climbed higher, you have greater potential energy. In this investigation you will explore the relationship between potential energy and kinetic energy as you try to make an action figure jump using a teeterboard.

The Task

Use what you know about the conservation of energy and models to design and carry out an investigation that will allow you to develop a rule that explains how an action figure can be made to jump lower or higher on a teeterboard.

The guiding question of this investigation is, **How can you make an action figure jump higher?**

Materials

You may use any of the following materials during your investigation:

- Ruler
- Meterstick
- Electronic or triple beam balance
- Pencil
- Clay (100 g)
- Action figures
- Safety glasses or goggles

Safety Precautions

Follow all normal lab safety rules. In addition, take the following safety precautions:

1. Wear sanitized safety glasses or goggles during lab setup, hands-on activity, and takedown.
2. Sweep clay up off the floor to avoid a slip or fall hazard.
3. Do not allow the action figure to jump too far from your work area.
4. Remove any fragile items from the work area.
5. Wash hands with soap and water after completing the lab activity.

Investigation Proposal Required? ☐ Yes ☐ No

Getting Started

To answer the guiding question, you will need to design and conduct an investigation that explores changing the potential energy of an action figure. To accomplish this task, you must determine what type of data you need to collect, how you will collect it, and how you will analyze it.

To determine *what type of data you need to collect,* think about the following questions:

- How will you test the ability to make the action figure jump higher?
- How will you measure the height of the jump?
- What type of measurements or observations will you need to record during your investigation?

To determine *how you will collect your data,* think about the following questions:

- How often will you collect data and when will you do it?

- How will you make sure that your data are of high quality (i.e., how will you reduce error)?
- How will you keep track of the data you collect and how will you organize it?

To determine *how you will analyze your data,* think about the following questions:

- What type of calculations will you need to make?
- What type of graph could you create to help make sense of your data?

Connections to Crosscutting Concepts, the Nature of Science, and the Nature of Scientific Inquiry

As you work through your investigation, be sure to think about

- how defining systems and models provides tools for understanding and testing of ideas;
- why it is important to track how energy and matter flows into, out of, and within a system;
- the difference between laws and theories in science; and
- the different forms of scientific investigation, including experiments, systematic observations, and analysis of data sets.

Initial Argument

Once your group has finished collecting and analyzing your data, your group will need to develop an initial argument. Your initial argument needs to include a *claim, evidence* to support your claim, and a *justification* of the evidence. The claim is your group's answer to the guiding question. The evidence is an analysis and interpretation of your data. Finally, the justification of the evidence is why your group thinks the evidence matters. The justification of the evidence is important because scientists can use different kinds of evidence to support their claims. Your group will create your initial argument on a whiteboard. Your whiteboard should include all the information shown in Figure L14.2.

FIGURE L14.2

Argument presentation on a whiteboard

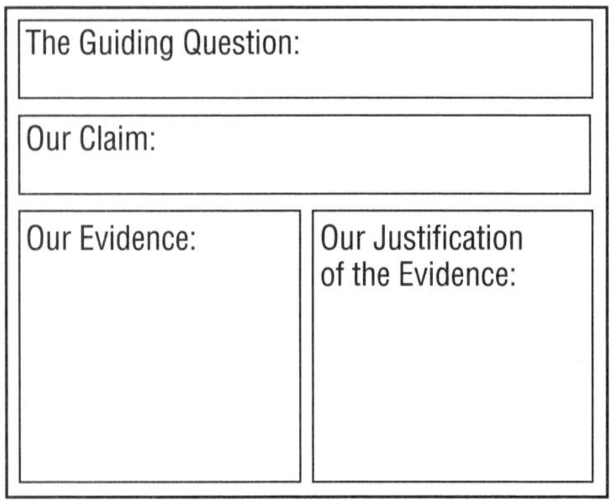

The Guiding Question:	
Our Claim:	
Our Evidence:	Our Justification of the Evidence:

Argumentation Session

The argumentation session allows all of the groups to share their arguments. One member of each group will stay at the lab station to share that group's argument, while the other members of the group go to the other lab stations to listen to and critique the arguments developed by their classmates. This is similar to how scientists present their arguments to other scientists at conferences. If you are responsible for critiquing your classmates' arguments, your goal is to look for mistakes so these mistakes can be fixed and they can make their argument better. The argumentation session is also a good time to think about ways you can make your initial argument better. Scientists must share and critique arguments like this to develop new ideas.

To critique an argument, you might need more information than what is included on the whiteboard. You will therefore need to ask the presenter lots of questions. Here are some good questions to ask:

- How did you collect your data? Why did you use that method? Why did you collect those data?
- What did you do to make sure the data you collected are reliable? What did you do to decrease measurement error?
- How did your group analyze the data? Why did you decide to do it that way? Did you check your calculations?
- Is that the only way to interpret the results of your analysis? How do you know that your interpretation of your analysis is appropriate?
- Why did your group decide to present your evidence in that way?
- What other claims did your group discuss before you decided on that one? Why did your group abandon those alternative ideas?
- How confident are you that your claim is valid? What could you do to increase your confidence?

Once the argumentation session is complete, you will have a chance to meet with your group and revise your initial argument. Your group might need to gather more data or design a way to test one or more alternative claims as part of this process. Remember, your goal at this stage of the investigation is to develop the most acceptable and valid answer to the research question!

Report

Once you have completed your research, you will need to prepare an *investigation report* that consists of three sections. Each section should provide an answer to the following questions:

1. What question were you trying to answer and why?
2. What did you do to answer your question and why?
3. What is your argument?

Your report should answer these questions in two pages or less. This report must be typed, and any diagrams, figures, or tables should be embedded into the document. Be sure to write in a persuasive style; you are trying to convince others that your claim is acceptable and valid!

Checkout Questions

Lab 14. Potential Energy

How Can You Make an Action Figure Jump Higher?

1. What is potential energy?

2. What is kinetic energy?

3. A student is trying to get a cart to reach the wall at the end of the system pictured below. He uses a ramp to get the cart some energy to cover that distance. However, as shown below, using the ramp as constructed, he was not able to reach the wall.

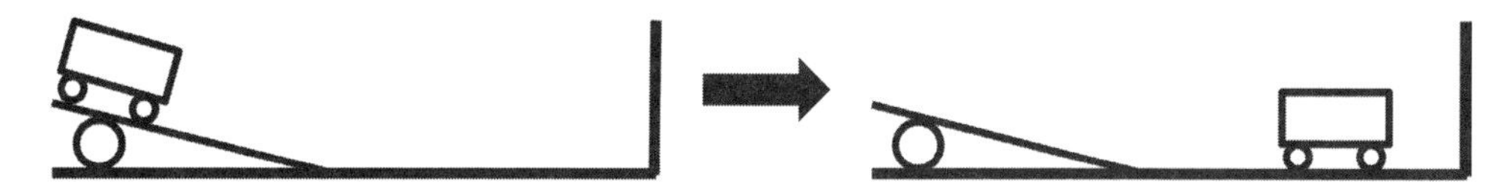

   a. What can the student change to get the cart to reach the wall?

   b. How do you know?

4. The law of conservation of energy describes how energy exists in physical systems but not why it acts in certain ways.

   a. I agree with this statement.

   b. I disagree with this statement.

   Explain your answer, using an example from your investigation about potential energy.

5. Science only relies on experiments to understand the physical world.

   a. I agree with this statement.

   b. I disagree with this statement.

   Explain your answer, using an example from your investigation about potential energy.

6. Scientists often have to define the boundaries of physical systems and use them to create models to test ideas. Explain why defining systems and models is important in science, using an example from your investigation about potential energy.

7. It is important to track how energy flows into, out of, and within a system during an investigation. Explain why it is important to keep track of energy when studying a system, using an example from your investigation about potential energy.

Teacher Notes

Lab 15. Thermal Energy and Specific Heat

Which Material Has the Greatest Specific Heat?

Purpose

The purpose of this lab is to *introduce* students to the concepts of specific heat, the conservation of energy, and heat transfer. In addition, students will have an opportunity to explore the crosscutting concepts of systems and the flow and conservation of energy. Students will also learn about the differences between observations and inferences and how scientists use different methods to answer different types of questions.

The Content

All matter is made up of atoms or molecules that are constantly in motion. The atoms or molecules vibrate (wiggle about a fixed position), move from one location to another, or rotate about an imaginary axis. These atoms or molecules, as a result, have kinetic energy. *Temperature* is a measure of the average amount of kinetic energy of all the atoms or molecules in a sample of matter. The temperature of an object goes up when the average kinetic energy of the atoms or molecules that make up that object increases and goes down when the average kinetic energy of the atoms or molecules decreases. *Heat,* in contrast, is the total amount of kinetic energy in a sample of matter. Heat transfer is the flow of energy from a higher-temperature object to a lower-temperature object. Energy will move between two objects as a result of a difference in temperature and will continue moving until the two objects are at the same temperature.

A physical property of a substance is its ability to retain heat energy. Some substances, such as water, can gain a large amount of heat energy without a significant change in temperature. Other substances, such as lead, will change temperature quickly as they gain energy. This property is based mainly on the structure of the material, the size of the atoms and molecules, and the interactions between them. This property is known as the *specific heat* of the substance and is defined as the amount of heat energy that is required to raise the temperature of 1 gram of a substance by 1 degree Celsius. The SI unit for specific heat is J/g•°C. The specific heat value of an object indicates how much heat a substance will absorb before its temperature rises. It also indicates the ability of a substance to transfer heat to a cooler object.

The amount of heat delivered by a material (q) is equal to the mass of the material delivering the heat (m) multiplied by the specific heat of the material (s) multiplied by the resulting temperature change. The mathematical relationship between these three factors and the amount of heat transferred to an object is

$$q = m \times s \times \Delta T$$

In this investigation, the students will need to determine the specific heat of several different materials using a calorimeter. A calorimeter is a device used to measure heat flow, where the heat given off by a material is absorbed by the calorimeter and its contents. The contents of the calorimeter is usually water, which has a specific heat of 4.18 J/g•°C. You can make an inexpensive calorimeter using polystyrene cups (see Figure 15.1).

FIGURE 15.1

A basic calorimeter

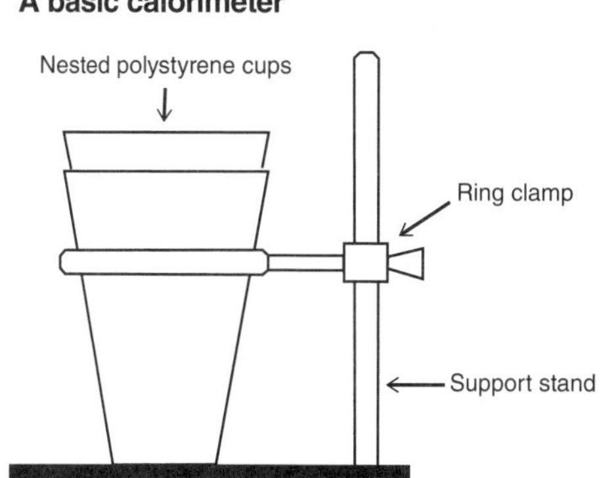

The law of conservation of energy states that energy is not created nor destroyed; it is only converted from one form to another. Therefore, the heat gained by the water in the calorimeter must be equal in magnitude (and opposite in sign) to the heat lost by the material that was added to the calorimeter. The specific heat of a material, as a result, can be calculated using the following equation:

$$m_{\text{water}} \times s_{\text{water}} \times \Delta T_{\text{water}} = -[m_{\text{material}} \times s_{\text{material}} \times \Delta T_{\text{material}}]$$

The basic procedure for determining the specific heat of a material using a calorimeter is as follows:

1. Weigh a sample of a material on a balance to the nearest tenth of a gram. For this investigation, we suggest that you use a 40–60 g sample. Record this value.
2. Place the sample in a boiling-water bath for 10–15 minutes to be sure that the temperature of the sample is 100°C.
3. Fill a calorimeter with 25–50 ml of room-temperature water. Be sure to determine the exact mass and temperature of the water. Record the mass and initial temperature of the water.
4. Using tongs, lift up the heated sample from the boiling-water bath and carefully place it into the calorimeter.
5. Stir the water in the calorimeter with a stirring rod for several minutes. Use a thermometer to monitor the temperature change of the water.
6. Once the temperature of the water stops increasing, record the final temperature of the water.
7. Calculate the heat gained by the water using the mass of the water, the temperature change of the water in degrees Celsius, and the specific heat of the water. The formula for this calculation is

$$q_{\text{water}} = m_{\text{water}} \times 4.18 \times \Delta T_{\text{water}}$$

8. Calculate the specific heat of the sample using the mass of the sample, the temperature change of the metal, and the heat gained by the water. The initial temperature of the metal is 100°C. The formula for this calculation is

$$s_{material} = q_{water} / m_{material} \times \Delta T_{material}$$

Timeline

The instructional time needed to complete this lab investigation is 230–280 minutes. Appendix 2 (p. 411) provides options for implementing this lab investigation over several class periods. Option A (280 minutes) should be used if students are unfamiliar with scientific writing, because this option provides extra instructional time for scaffolding the writing process. You can scaffold the writing process by modeling, providing examples, and providing hints as students write each section of the report. Option B (230 minutes) should be used if students are familiar with scientific writing and have developed the skills needed to write an investigation report on their own. In option B, students complete stage 6 (writing the investigation report) and stage 8 (revising the investigation report) as homework.

Materials and Preparation

The materials needed to implement this investigation are listed in Table 15.1. The consumables and equipment can be purchased from a science supply company such as Carolina, Flinn Scientific, or Ward's Science.

We recommend that the set samples include aluminum, copper, tin, zinc, glass, plastic, and wood. There are a number of options for creating a set of samples. You can purchase one or more different specific heat sets, density kits, or specific gravity sets from a science supply company and then use the materials from the kits to create your own sets. Good kits to use for this approach include the following (item numbers are accurate at the time of writing this book but may change in the future):

- Specific Heat Specimen Set from Carolina (item 753490)
- Deluxe Density Cube Set from Carolina (item 752476)
- Study of Density Kit from Carolina (item 840942)
- Density Rod Set from Carolina (item 752480)
- Specific Gravity Set, Square from Carolina (item 752491)
- Specific Heat Set from Flinn (item AP9220)
- Density Cube Set from Flinn Scientific (item AP6058)
- Specific Gravity Metal Specimens Set from Flinn Scientific (item AP9234)
- Investigating Measurement and Density Kit from Ward's Science (item 180222)
- Density Blocks Lab Activity from Ward's Science (item 366859)
- Specific Gravity Specimens Set from Ward's Science (Item 364001)

You can also purchase metal shot, such as aluminum, brass, copper, tin, and zinc, and some different plastic beads or blocks, and then use these materials to create a set of samples. Another option is to gather different materials from around the classroom to create your own sets.

TABLE 15.1

Materials list for Lab 15

Item	Quantity
Samples	
Set of substances (should include aluminum, copper, tin, zinc, glass, plastic, and wood)	1 per group
Consumables	
Water (in squirt bottle)	1 per group
Equipment and other materials	
Safety glasses or goggles	1 per student
Chemical-resistant apron	1 per student
Nonlatex gloves	1 pair per student
Graduated cylinder, 100 ml	1 per group
Beaker, 250 ml	2 per group
Polystyrene cup	2 per group
Ring clamp and support stand	1 per group
Thermometer (or temperature probe with interface)	1 per group
Hot plate	1 per group
Electronic or triple beam balance	1 per group
Tongs	1 per group
Stirring rod	1 per group
Investigation Proposal C (optional)	1 per group
Whiteboard, 2' × 3'*	1 per group
Lab Handout	1 per student
Peer-review guide	1 per student
Checkout Questions	1 per student

*As an alternative, students can use computer and presentation software, such as Microsoft PowerPoint or Apple Keynote, to create their arguments.

We recommend that you use a set routine for distributing and collecting the materials during the lab investigation. For example, the consumables and equipment for each group can be set up at each group's lab station before class begins, or one member from each group can collect them from a table or a cart when needed during class.

Safety Precautions and Laboratory Waste Disposal

Remind students to follow all normal lab safety rules. In addition, tell the students to take the following safety precautions:

1. Wear sanitized indirectly vented chemical-splash goggles and chemical-resistant nonlatex gloves and aprons during lab setup, hands-on activity, and takedown.
2. Use caution when working with hot plates, because they can burn skin and cause fires.
3. Hot plates also need to be kept away from water and other liquids.
4. Use only GFCI-protected electrical receptacles for hot plates.
5. Clean up any spilled liquid immediately to avoid a slip or fall hazard.
6. Handle all glassware with care.
7. Handle glass thermometers with care. They are fragile and can break, causing a sharp hazard that can cut or puncture skin.
8. Wash hands with soap and water after completing the lab activity.
9. The water can be disposed of down a drain if it is connected to a sanitation sewer system. The sets of samples can be cleaned, dried, and stored for later use.

Topics for the Explicit and Reflective Discussion

Concepts That Can Be Used to Justify the Evidence

To provide an adequate justification of their evidence, students must explain why they included the evidence in their arguments and make the assumptions underlying their analysis and interpretation of the data explicit. In this investigation, students can use the following concepts to help justify their evidence:

- Properties of matter
- Conservation of energy
- Heat and temperature
- Heat transfer and thermal equilibrium
- Heat capacity and specific heat

We recommend that you review these concepts during the explicit and reflective discussion to help students make this connection.

How to Design Better Investigations

It is important for students to reflect on the strengths and weaknesses of the investigation they designed during the explicit and reflective discussion. Students should therefore be encouraged to discuss ways to eliminate potential flaws, measurement errors, or sources of bias in their investigations. To help students be more reflective about the design of their investigation, you can ask the following questions:

1. What were some of the strengths of your investigation? What made it scientific?
2. What were some of the weaknesses of your investigation? What made it less scientific?
3. If you were to do this investigation again, what would you do to address the weaknesses in your investigation? What could you do to make it more scientific?

Crosscutting Concepts

This investigation is well aligned with two crosscutting concepts found in *A Framework for K–12 Science Education*, and you should review these concepts during the explicit and reflective discussion.

- *Systems and system models:* Defining a system under study and making a model of it are tools for developing a better understanding of natural phenomena in science. In this lab it is important to model and keep track of the system and surroundings during energy transfer.
- *Energy and matter: Flows, cycles, and conservation.* In science it is important to track how energy and matter move into, out of, and within systems. In this lab, for example, it is important to know that energy is transferred between different objects.

The Nature of Science and the Nature of Scientific Inquiry

This investigation is well aligned with two important concepts related to the *nature of science* (NOS) and the *nature of scientific inquiry* (NOSI), and you should review these concepts during the explicit and reflective discussion.

- *The difference between observations and inferences in science*: An observation is a descriptive statement about a natural phenomenon, whereas an inference is an interpretation of an observation. Students should also understand that current scientific knowledge and the perspectives of individual scientists guide both observations and inferences. Thus, different scientists can have different but equally valid interpretations of the same observations due to differences in their perspectives and background knowledge.

- *Methods used in scientific investigations*: Examples of methods include experiments, systematic observations of a phenomenon, literature reviews, and analysis of existing data sets; the choice of method depends on the objectives of the research. There is no universal step-by step scientific method that all scientists follow; rather, different scientific disciplines (e.g., chemistry vs. physics) and fields within a discipline (e.g., organic vs. physical chemistry) use different types of methods, use different core theories, and rely on different standards to develop scientific knowledge.

Hints for Implementing the Lab

- Allowing students to design their own procedures for collecting data gives students an opportunity to try, to fail, and to learn from their mistakes. However, you can scaffold students as they develop their procedure by having them fill out an investigation proposal. These proposals provide a way for you to offer students hints and suggestions without telling them how to do it. You can also check the proposals quickly during a class period. For this lab we suggest you use Investigation Proposal C.
- We recommend including 5–10 materials in the set samples to foster higher-quality argumentation during the lab. The more specific heat values that student groups have to determine, the more opportunities there are for variation among groups that can lead to critical questioning and discussion during the argumentation session. However, if necessary for time or scheduling issues, the number of objects can be decreased.
- If students are only seeing small temperature changes in the water, encourage them to use less water and more of the material they are testing.
- This is a good lab for students to make mistakes during the data collection stage. Students will quickly figure out what they did wrong during the argumentation session. It will also create an opportunity for students to reflect on and identify ways to improve the way they design investigations during the explicit and reflective discussion.
- Be sure to allow students to go back and re-collect data at the end of the argumentation session if needed. Students often realize that they made numerous mistakes when they were collecting data as a result of their discussions during the argumentation session. The students, as a result, will want a chance to re-collect data, and the re-collection of data should be encouraged when time allows. This also offers an opportunity to discuss what scientists do when they realize a mistake is made inside the lab.

Topic Connections

Table 15.2 provides an overview of the scientific practices, crosscutting concepts, disciplinary core ideas, and supporting ideas at the heart of this lab investigation. In addition, it lists the NOS and NOSI concepts for the explicit and reflective discussion. Finally, it lists literacy and mathematics skills (*CCSS ELA* and *CCSS Mathematics*) that are addressed during the investigation.

TABLE 15.2

Lab 15 alignment with standards

Scientific practices	• Asking questions and defining problems • Planning and carrying out investigations • Analyzing and interpreting data • Using mathematics and computational thinking • Constructing explanations and designing solutions • Engaging in argument from evidence • Obtaining, evaluating, and communicating information
Crosscutting concepts	• Systems and system models • Energy and matter: Flows, cycles, and conservation
Core ideas	• PS1.A: Structure and properties of matter • PS3.A: Definitions of energy • PS3.B: Conservation of energy and energy transfer
Supporting ideas	• Properties of matter • Conservation of energy • Heat and temperature • Heat transfer and thermal equilibrium • Heat capacity and specific heat
NOS and NOSI concepts	• Difference between observations and inferences • Methods used in scientific investigations
Literacy connections (*CCSS ELA*)	• *Reading*: Key ideas and details, craft and structure, integration of knowledge and ideas • *Writing*: Text types and purposes, production and distribution of writing, research to build and present knowledge, range of writing • *Speaking and listening*: Comprehension and collaboration, presentation of knowledge and ideas
Mathematics connections (*CCSS Mathematics*)	• Reason abstractly and quantitatively • Construct viable arguments and critique the reasoning of others • Use appropriate tools strategically • Attend to precision

Lab Handout

Lab 15. Thermal Energy and Specific Heat

Which Material Has the Greatest Specific Heat?

Introduction

Scientists are able to identify unknown substances based on their chemical and physical properties. A substance is a type of matter with a specific composition and specific properties. One physical property of a substance is the amount of energy it will absorb per unit of mass. This property is called specific heat (s). Specific heat is the amount of energy, measured in joules, that is needed to raise the temperature of 1 gram of the substance 1 degree Celsius. Scientists often need to know the specific heat of different substances when they attempt to track how energy moves into, out of, and within a system.

Chemists use a technique called calorimetry to determine the specific heat of a substance. Calorimetry, or the measurement of heat transfer, is based on the law of conservation of energy. This law states that energy is not created nor destroyed; it is only converted from one form to another. This fundamental law serves as the foundation for all the research that is done in the field of thermodynamics, which is the study of heat, temperature, and heat transfer. Heat is defined as the total kinetic energy of all the atoms or molecules that make up a substance. Temperature, in contrast, is defined as a measure of the average kinetic energy of the atoms or molecules that make up a substance.

Heat, or thermal energy, can be transferred through a substance and between two different objects. Scientists call this process conduction (see Figure L15.1). The transfer of heat energy through the process of conduction can be explained by thinking of the heat from a source causing the atoms of a substance to vibrate faster, which means they have greater kinetic energy. These atoms then cause the atoms next to them to vibrate faster by bumping into them, which means that the kinetic energy of the neighboring atoms increases as well. Over time, kinetic energy is transferred from one atom to the next. As more atoms in the substance gain kinetic energy over time, the temperature of the substance increases. This process is also how heat energy is able to transfer between two different objects that are in contact with each other.

The amount of heat (q) transferred to an object depends on three factors. The first is the mass (m) of the object. The second factor is the specific heat (s) value of object. This is important because an object will consist of a specific type of substance, and each type of substance has a unique specific heat value. The third factor is the resulting temperature change (ΔT). The mathematical relationship between these three factors and the amount of heat transferred to an object is

$$q = m \times s \times \Delta T$$

FIGURE L15.1

Thermal energy can transfer through a substance or from one substance to another by conduction.

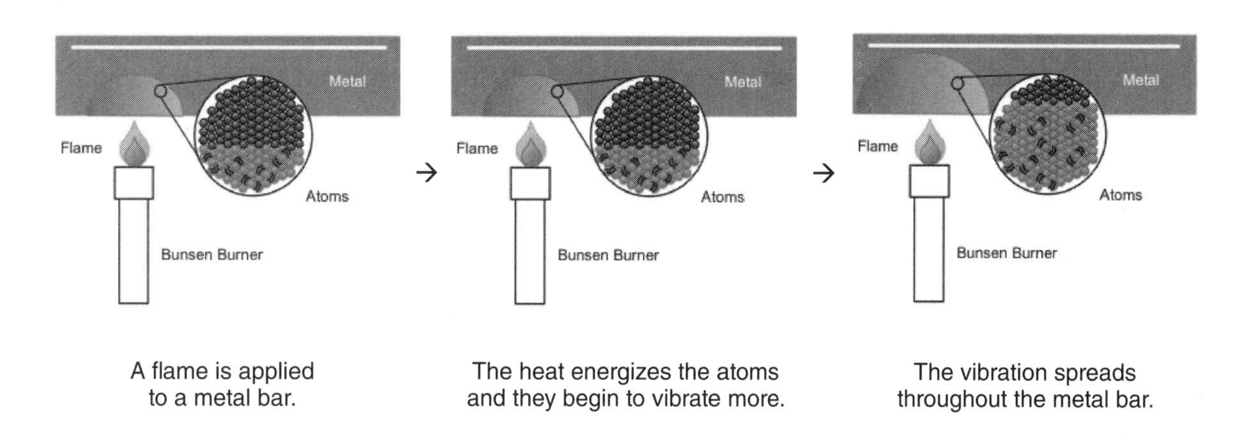

The materials that people use to build a new structure or to manufacture commercial goods have a wide range of specific heat values. Take concrete and wood as an example. Both of these materials can be used to build benches in parks or at bus stops for people to use. Wood, however, has a much higher specific heat than concrete. It therefore takes more heat energy to increase the temperature of a 10 kg piece of wood than it does to increase the temperature of a 10 kg piece of concrete. The piece of concrete, as a result, will get hotter faster than the piece of wood when it is exposed to the same amount of heat energy. This issue could be a potential problem in cities that tend to be hot and sunny most of the year. Engineers and manufacturers therefore need to know how to look up or determine the specific heat value of a potential building or manufacturing material before they decide to use it. In this investigation, you will have an opportunity to learn how to determine the specific heat value of a material using the process of calorimetry.

Your Task

Use what you know about heat, temperature, the conservation of energy, and defining systems to design and carry out an investigation to determine the specific heat values of several different materials.

The guiding question of this investigation is, **Which material has the greatest specific heat?**

Materials

You may use any of the following materials during your investigation:

Samples
- Aluminum (Al)
- Copper (Cu)
- Tin (Sn)
- Zinc (Zn)
- Glass
- Plastic
- Wood

Consumables
- Water (in squirt bottle)

Equipment
- Graduated cylinder (100 ml)
- 2 Beakers (each 250 ml)
- 2 Polystyrene cups
- Ring clamp and support stand
- Thermometer or temperature probe
- Hot plate
- Electronic or triple beam balance
- Tongs
- Stirring rod
- Safety glasses or goggles
- Chemical-resistant apron
- Nonlatex gloves

Safety Precautions

Follow all normal lab safety rules. In addition, take the following safety precautions:

1. Wear sanitized indirectly vented chemical-splash goggles and chemical-resistant nonlatex gloves and aprons during lab setup, hands-on activity, and takedown.
2. Use caution when working with hot plates, because they can burn skin and cause fires.
3. Hot plates also need to be kept away from water and other liquids.
4. Use only GFCI-protected electrical receptacles for hot plates.
5. Clean up any spilled liquid immediately to avoid a slip or fall hazard.
6. Handle all glassware with care.
7. Handle glass thermometers with care. They are fragile and can break, causing a sharp hazard that can cut or puncture skin.
8. Wash hands with soap and water after completing the lab activity.

Investigation Proposal Required? ☐ Yes ☐ No

Getting Started

To calculate the specific heat of a material, you will need to determine how much energy the material is able to transfer to a sample of water using a calorimeter. A calorimeter is used to prevent heat loss to the surroundings (see Figure L15.2). The heat gained by the water in a calorimeter is therefore equal in magnitude (but opposite in sign) to the heat lost by the material:

$$q_{water} = -q_{material}$$

The amount of heat gained by the water is calculated using the mass of water used, the specific heat of water (4.18 J/g•°C), and the difference between the final and initial temperature of the water in the calorimeter. The amount of water used for calorimetry varies, but most people use between 10 and 50 ml because water has such a high specific heat. The equation for calculating the amount of heat gained by the water is

$$q_{\text{water}} = m_{\text{water}} \times s_{\text{water}} \times \Delta T_{\text{water}}$$

FIGURE L15.2

A basic calorimeter

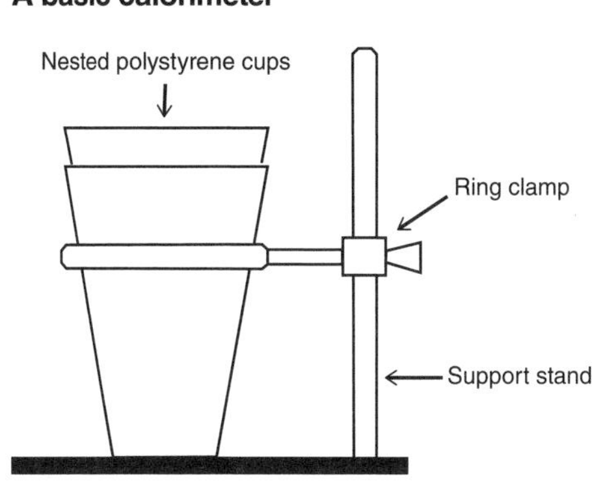

The amount of heat lost by a material once it is added to the water is calculated using the mass of the material, the specific heat of that material, and the difference between the material's final temperature and its initial temperature. The final temperature of the material is assumed to be the same as the final temperature of the water in the cup. The initial temperature of the material will be 100°C. To ensure that the initial temperature of the material will be 100°C before you add it to the water in the calorimeter, you can place the material in a boiling-water bath for 10–15 minutes. The equation for calculating the amount of heat lost by a metal is

$$-q_{metal} = m_{\text{material}} \times s_{\text{material}} \times \Delta T_{material}$$

Now that you understand the basics of calorimetry, you must determine what data you need to collect, how you will collect it, and how you will analyze it in order to answer the guiding question.

To determine *what data you will need to collect,* think about the following questions:

- How will you know how much thermal energy has been transferred from a material to the water in a calorimeter?
- What information do you need to calculate the specific heat of material once you know how much thermal energy has been transferred from a material to the water in a calorimeter?

To determine *how you will collect your data,* think about the following questions:

- What equipment will you need to collect the data you need?
- How will you make sure that your data are of high quality (i.e., how will you reduce error)?
- How will you keep track of the data you collect?
- How will you organize your data?

To determine *how you will analyze your data,* think about the following questions:

- What type of calculations will you need to make?
- What type of graph could you create to help make sense of your data?

Connections to Crosscutting Concepts, the Nature of Science, and the Nature of Scientific Inquiry

As you work through your investigation, be sure to think about

- the importance of defining a system under study;
- how scientists often need track how energy moves into, out of, and within a system;
- the difference between observations and inferences in science; and
- how scientists use different methods to answer different types of questions.

Initial Argument

Once your group has finished collecting and analyzing your data, your group will need to develop an initial argument. Your initial argument needs to include a *claim, evidence* to support your claim, and a *justification* of the evidence. The claim is your group's answer to the guiding question. The evidence is an analysis and interpretation of your data. Finally, the justification of the evidence is why your group thinks the evidence matters. The justification of the evidence is important because scientists can use different kinds of evidence to support their claims. Your group will create your initial argument on a whiteboard. Your whiteboard should include all the information shown in Figure L15.3.

FIGURE L15.3

Argument presentation on a whiteboard

The Guiding Question:	
Our Claim:	
Our Evidence:	Our Justification of the Evidence:

Argumentation Session

The argumentation session allows all of the groups to share their arguments. One member of each group will stay at the lab station to share that group's argument, while the other members of the group go to the other lab stations to listen to and critique the arguments developed by their classmates. This is similar to how scientists present their arguments to other scientists at conferences. If you are responsible for critiquing your classmates' arguments, your goal is to look for mistakes so these mistakes can be fixed and they can make their argument better. The argumentation session is also a good time to think about ways you can make your initial argument better. Scientists must share and critique arguments like this to develop new ideas.

To critique an argument, you might need more information than what is included on the whiteboard. You will therefore need to ask the presenter lots of questions. Here are some good questions to ask:

- How did you collect your data? Why did you use that method? Why did you collect those data?
- What did you do to make sure the data you collected are reliable? What did you do to decrease measurement error?
- How did your group analyze the data? Why did you decide to do it that way? Did you check your calculations?
- Is that the only way to interpret the results of your analysis? How do you know that your interpretation of your analysis is appropriate?
- Why did your group decide to present your evidence in that way?
- What other claims did your group discuss before you decided on that one? Why did your group abandon those alternative ideas?
- How confident are you that your claim is valid? What could you do to increase your confidence?

Once the argumentation session is complete, you will have a chance to meet with your group and revise your initial argument. Your group might need to gather more data or design a way to test one or more alternative claims as part of this process. Remember, your goal at this stage of the investigation is to develop the most acceptable and valid answer to the research question!

Report

Once you have completed your research, you will need to prepare an *investigation report* that consists of three sections. Each section should provide an answer to the following questions:

1. What question were you trying to answer and why?
2. What did you do to answer your question and why?
3. What is your argument?

Your report should answer these questions in two pages or less. This report must be typed, and any diagrams, figures, or tables should be embedded into the document. Be sure to write in a persuasive style; you are trying to convince others that your claim is acceptable and valid!

Checkout Questions

Lab 15. Thermal Energy and Specific Heat

Which Material Has the Greatest Specific Heat?

1. The diagrams below show a 50 g piece of iron and a 50 g piece of tin being added to 50 ml of water in two different calorimeters. The initial temperature of each piece of metal is 100°C. The initial temperature of the water in each calorimeter is 25°C.

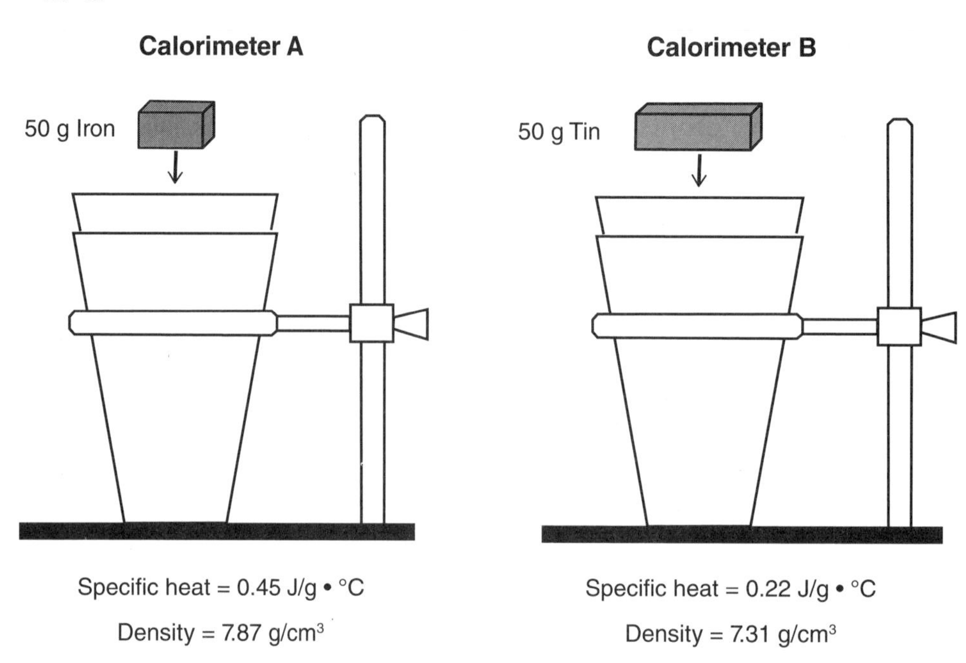

Specific heat = 0.45 J/g • °C
Density = 7.87 g/cm^3

Specific heat = 0.22 J/g • °C
Density = 7.31 g/cm^3

What do you think will happen to the temperature of the water in each calorimeter?

a. The temperature of the water in calorimeter A will increase more than it will in calorimeter B.

b. The temperature of the water in calorimeter B will increase more than it will in calorimeter A.

c. The temperature of the water in calorimeters A and B will go up by the same amount.

d. Unsure

How do you know?

2. "Heat from the metal transferred into the water" is an example of an observation.

    a. I agree with this statement.

    b. I disagree with this statement.

    Explain your answer, using an example from your investigation about specific heat.

3. Investigations are only scientific if someone designs and then carries out an experiment.

   a. I agree with this statement.
   b. I disagree with this statement.

   Explain your answer, using an example from your investigation about specific heat.

4. Scientists often need to define a system before they attempt to study it. Explain what it means to define a system and then explain why it is important in science, using an example from your investigation about specific heat.

5. Scientists often need to track how energy or matter moves into, out of, or within systems to explain a natural phenomenon. Explain why tracking energy or matter is so useful in science, using an example from your investigation about specific heat.

LAB 16

Teacher Notes

Lab 16. Electrical Energy and Lightbulbs

How Does the Arrangement of Lightbulbs That Are Connected to a Battery Affect the Brightness of a Single Bulb in That Circuit?

Purpose

The purpose of this lab is to *introduce* students to electrical circuits arranged in series and in parallel. In addition, this lab gives students an opportunity to track how energy and matter move through an electrical system and to examine the relationship between the structure and function of electrical circuits. Students will also learn about the difference between data and evidence, and the nature and role of experiments.

The Content

A basic circuit has three components: a voltage source (in this investigation, that is the battery), a resistor (the lightbulb), and a wire or wires connecting the resistor to the voltage source. When the voltage source (i.e., battery) and resistor (i.e., lightbulb) are connected to complete the circuit, an electrical current flows through the circuit and the resistor can then use the electrical energy (the voltage) to do work. In the case of a lightbulb, this means that the voltage, or electrical energy, provided by the battery can be used by the lightbulb to produce light. The mathematical relationship between these three components of the circuit is $V = IR$, where V is the voltage supplied by the battery (units of volts, V), I is the current in the circuit (units of amperes, A) and R is the resistance of the bulb (units of ohms, Ω).

The guiding question asks about the brightness of a lightbulb. The term *brightness* is often ambiguous, because it can refer to the amount of light given off by a source (e.g., one lightbulb is brighter than another) but can also be used colloquially to mean a different shade of color (e.g., a bright pink T-shirt). In this lab, *brightness* refers to the total amount of energy given off by a lightbulb. The brightness of a lightbulb (or the total energy given off) is a function of two factors: the current flowing through the circuit and the voltage drop across the lightbulb ("voltage drop" is just a more technical way of saying the voltage used by the bulb). If all bulbs in a circuit have the same resistance, then the brightness of the bulb and, by extension, the energy used by the bulb, is just a function of the voltage drop across the bulb. For a simple circuit with only one bulb, the brightness is directly related to the voltage of the battery, because the single bulb will use all of the voltage. Thus, to increase the brightness of the single bulb, increase the voltage of the battery.

For circuits that contain two or more bulbs, the situation is a bit more complex. When a number of lightbulbs are connected in series (see Figure 16.1), the total voltage used by the circuit is still equal to the voltage supplied by the battery. However, the total resistance of the circuit increases and the current flowing through the circuit decreases. If *all bulbs are of the same resistance*, then the voltage drop across each bulb will be shared equally. For example, if the battery in Figure 16.1 is 9 V, then each bulb will have a voltage drop of 3 V. If a fourth bulb of the same resistance were added to the circuit, then each bulb would have a voltage drop of 2.25 V. In this case, adding a fourth bulb would cause the brightness of each bulb to drop. The brightness would decrease because adding a fourth bulb in series would decrease the current flowing through each bulb. Thus, the decrease in brightness by adding another bulb in series is not a linear relationship.

FIGURE 16.1

Lightbulbs connected in series

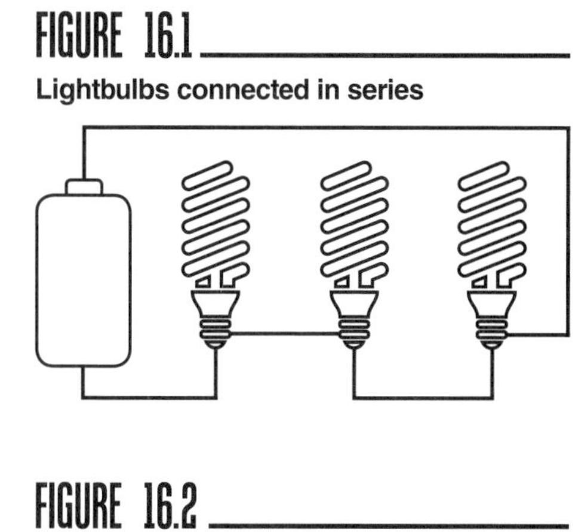

When lightbulbs are arranged in parallel (see Figure 16.2), the total resistance of the circuit is found by the following formula: $\frac{1}{R_{total}} = \frac{1}{R_1} + \frac{1}{R_2} + \frac{1}{R_3} + \cdots$ If *all of the bulbs are of the same resistance,* then the total resistance in the circuit is just R/x, where R is the resistance of each bulb and x is the number of bulbs. In comparison with a series circuit with the same number of bulbs, in a parallel circuit the resistance decreases and the current increases. Furthermore, each path in a parallel circuit must use all of the voltage supplied by the battery. This means that the voltage drop across each battery is equal to the total voltage supplied by the battery. Thus, the brightness of each lightbulb in the parallel circuit will be brighter than in the series circuit. This is because each bulb will have a higher voltage drop and a higher current flowing through it.

FIGURE 16.2

Lightbulbs connected in parallel

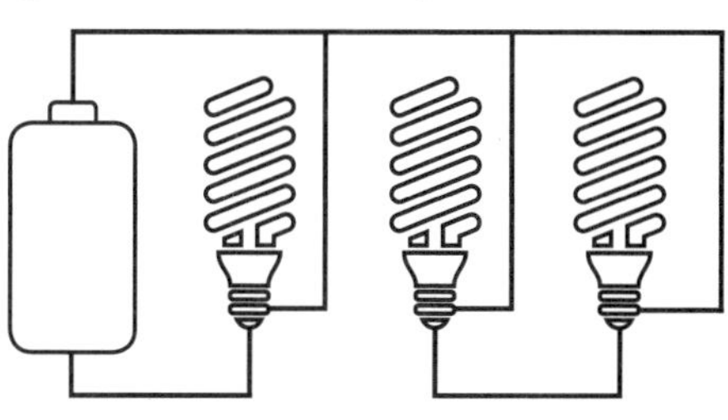

Timeline

The instructional time needed to complete this lab investigation is 170–230 minutes. Appendix 2 (p. 411) provides options for implementing this lab investigation over several class periods. Option C (230 minutes) should be used if students are unfamiliar with scientific writing, because this option provides extra instructional time for scaffolding the writing process. You can scaffold the writing process by modeling, providing examples, and providing hints as students write each section of the report. Option D (170 minutes) should be used if students are familiar with scientific writing and have developed the skills needed to write an investigation report on their own. In option D, students complete stage 6 (writing the investigation report) and stage 8 (revising the investigation report) as homework.

Materials and Preparation

The materials needed to implement this investigation are listed in Table 16.1 (p. 284). The equipment can be purchased from a science supply company such as Carolina, Flinn Scientific, or Ward's Science.

TABLE 16.1

Materials list for Lab 16

Item	Quantity
Safety glasses or goggles	1 per student
Size D batteries	2 per group
Battery holders	2 per group
Small lightbulbs	8 per group
Lightbulb holders	8 per group
Electrical wire	12 wires per group
Light sensor with interface	1 per group
Investigation Proposal C (optional)	1 per group
Whiteboard, 2' × 3'*	1 per group
Lab Handout	1 per student
Peer-review guide	1 per student

* As an alternative, students can use computer and presentation software, such as Microsoft PowerPoint or Apple Keynote, to create their arguments.

We recommend that you use a set routine for distributing and collecting the materials during the lab investigation. For example, the equipment for each group can be set up at each group's lab station before class begins, or one member from each group can collect them from a table or a cart when needed during class.

Safety Precautions and Laboratory Waste Disposal

Remind students to follow all normal lab safety rules. In addition, tell students to take the following safety precautions:

1. Wear sanitized safety glasses or goggles during lab setup, hands-on activity, and takedown.
2. Use caution when handling bulbs, wires, and batteries. They can get hot and burn skin.
3. Never put batteries in their mouth or on their tongue.
4. Use caution in handling wire ends. They are sharp and can cut or puncture skin.
5. Lightbulbs are made of glass. Be careful handling them. If they break, clean them up immediately and place in a broken glass box.

6. Wash hands with soap and water after completing the lab activity.
7. Batteries, lightbulbs, and wire may be stored for future use. When batteries need replacing, dispose of old batteries according to manufacturer's recommendations. The box of broken glass should be discarded in accordance with local policies.

Topics for the Explicit and Reflective Discussion

Concepts That Can Be Used to Justify the Evidence

To provide an adequate justification of their evidence, students must explain why they included the evidence in their arguments and make the assumptions underlying their analysis and interpretation of the data explicit. In this investigation, students can use the following concepts to help justify their evidence:

- Electrical circuits
- Series and parallel circuits
- Light energy
- Conservation of energy

We recommend that you review these concepts during the explicit and reflective discussion to help students make this connection.

How to Design Better Investigations

It is important for students to reflect on the strengths and weaknesses of the investigation they designed during the explicit and reflective discussion. Students should therefore be encouraged to discuss ways to eliminate potential flaws, measurement errors, or sources of bias in their investigations. To help students be more reflective about the design of their investigation, you can ask the following questions:

1. What were some of the strengths of your investigation? What made it scientific?
2. What were some of the weaknesses of your investigation? What made it less scientific?
3. If you were to do this investigation again, what would you do to address the weaknesses in your investigation? What could you do to make it more scientific?

Crosscutting Concepts

This investigation is well aligned with two crosscutting concepts found in *A Framework for K–12 Science Education,* and you should review these concepts during the explicit and reflective discussion.

- *Energy and matter: Flows, cycles, and conservation:* In science it is important to track how energy and matter move into, out of, and within systems. In this lab students will track how electrical energy flows through a circuit and how that energy is transformed into light energy.
- *Structure and function:* The way an object is shaped or structured determines many of its properties and functions. In this lab students will explore how the structure of a circuit influences the brightness of the bulbs within the circuit.

The Nature of Science and the Nature of Scientific Inquiry

This investigation is well aligned with two important concepts related to the *nature of science* (NOS) and the *nature of scientific inquiry* (NOSI), and you should review these concepts during the explicit and reflective discussion.

- *The difference between data and evidence in science:* Data are measurements, observations, and findings from other studies that are collected as part of an investigation. Evidence, in contrast, is analyzed data and an interpretation of the analysis.
- *The nature and role of experiments:* Scientists use experiments to test the validity of a hypothesis (i.e., a tentative explanation) for an observed phenomenon. Experiments include a test and the formulation of predictions (expected results) if the test is conducted and the hypothesis is valid. The experiment is then carried out and the predictions are compared with the observed results of the experiment. If the observed results match the predictions, then the hypothesis is supported. If the observed results do not match the predictions, then the hypothesis is not supported. A signature feature of an experiment is the control of variables to help eliminate alternative explanations for observed results.

Hints for Implementing the Lab

- Allowing students to design their own procedures for collecting data gives students an opportunity to try, to fail, and to learn from their mistakes. However, you can scaffold students as they develop their procedure by having them fill out an investigation proposal. These proposals provide a way for you to offer students hints and suggestions without telling them how to do it. You can also check the proposals quickly during a class period. For this lab we suggest using Investigation Proposal C.
- We strongly recommend using lightbulbs that have the same resistance. This way, the resistance of the bulb does not affect the students' results.
- Allow the students to become familiar with the light sensor as part of the tool talk before they begin to design their investigation. This gives students a chance to see what they can and cannot do with the equipment.

- We recommend that students use a test bulb to determine a baseline measurement. It is also useful to test how the arrangement of lightbulbs near the light sensor influences the measurements.

Topic Connections

Table 16.2 provides an overview of the scientific practices, crosscutting concepts, disciplinary core ideas, and supporting ideas at the heart of this lab investigation. In addition, it lists NOS and NOSI concepts for the explicit and reflective discussion. Finally, it lists literacy and mathematics skills (*CCSS ELA* and *CCSS Mathematics*) that are addressed during the investigation.

TABLE 16.2

Lab 16 alignment with standards

Scientific practices	• Asking questions and defining problems • Planning and carrying out investigations • Analyzing and interpreting data • Using mathematics and computational thinking • Constructing explanations and designing solutions • Engaging in argument from evidence • Obtaining, evaluating, and communicating information
Crosscutting concepts	• Energy and matter: Flows, cycles, and conservation • Structure and function
Core idea	• This lab does not address a disciplinary core idea from the *NGSS.* However, many non-*NGSS* states do address circuits in their standards.
Supporting ideas	• Electrical circuits • Series and parallel circuits • Light energy • Conservation of energy
NOS and NOSI concepts	• Difference between data and evidence • Nature and role of experiments
Literacy connections (*CCSS ELA*)	• *Reading*: Key ideas and details, craft and structure, integration of knowledge and ideas • *Writing*: Text types and purposes, production and distribution of writing, research to build and present knowledge, range of writing • *Speaking and listening*: Comprehension and collaboration, presentation of knowledge and ideas
Mathematics connections (*CCSS Mathematics*)	• Reason abstractly and quantitatively • Construct viable arguments and critique the reasoning of others • Use appropriate tools strategically

Lab Handout

Lab 16. Electrical Energy and Lightbulbs

How Does the Arrangement of Lightbulbs That Are Connected to a Battery Affect the Brightness of a Single Bulb in That Circuit?

Introduction

Scientists and historians generally agree that the approximately 250-year period from the 1550s to about 1800 was one of the most influential periods in history. During this time period—often referred to as the Scientific Revolution—science became increasingly important, and many of the people and developments of this period still influence our society today. For example, scientists such as Copernicus, Galileo, Kepler, and Newton published their most influential works during this time period.

While the ideas of Copernicus, Galileo, Kepler, and Newton (among others) are no doubt important and still remain influential, the most important development from this period may have come from a series of debates between Robert Boyle and Thomas Hobbes. Boyle was an important chemist and inventor. Hobbes was an influential philosopher. Boyle was a member of The Royal Society (along with Isaac Newton, Nicholas Mercator, and Edmond Halley, among many others), a scientific group that began in London during the 1600s (and the oldest scientific group still in existence). Hobbes was not a member of The Royal Society. Boyle and Hobbes had different views on how science should be conducted. Hobbes felt that science should be based on logic and reason, by which he meant that scientists should think about their questions and use philosophical approaches to answer those questions. This, said Hobbes, was how science had been done dating back to Aristotle. Boyle, on the other hand, suggested that science should be based on empirical results (a fancy term for evidence) and scientists should use rigorous investigative methods to answer their questions. Boyle also put forth the idea that scientists need to control for all the potential factors that might affect the outcome of an investigation (a more scientific way of saying that they should account for a factor by keeping it the same across conditions that are being tested). This, said Boyle, was how science should be done in the future, despite how it had been done in the past. After a series of debates and demonstrations, most of the members of The Royal Society sided with Boyle. The reliance on empirical support is what many scientists and historians say is the most important development of the Scientific Revolution.

Another outcome of the Scientific Revolution was the development of new questions that scientists could investigate and attempt to answer. These questions gave rise to entire fields of science, such as microbiology and geology. A third field to develop during this time was the study of electricity—a field that is very much with us today. The study of electricity has led to the development of many important technologies that we still use today. In the year 1800, Alessandro Volta invented the battery. Another important invention due to the study of electricity is the lightbulb, invented by Thomas Edison in 1879.

One of the most difficult aspects of inventing a reliable lightbulb was identifying the best material to use for the filament in the lightbulb. The filament is a small wire inside the lightbulb that the electricity must pass through. When the electricity passes through the filament there is a lot of resistance, meaning it is difficult for the electric current to pass through the small wire. As the electric current moves through the filament, it generates heat energy due to the resistance of the wire (similarly to how your hands generate heat energy when you rub them together quickly) and that heat energy causes the wire of the filament to glow. During this process electrical energy is converted to radiant energy (or light energy).

Since the lightbulb and battery were invented, people have been investigating their behavior when they are connected as part of an electric circuit in many different ways. An electric circuit is a continuous path that allows electricity to leave a source (such as a battery), travel through wires and other objects (such as lightbulbs), and then return to the source. Research on batteries, lightbulbs, and circuits has shown that there are two general categories of ways to arrange lightbulbs in an electric circuit and connect them to a battery in a way that will still allow the bulbs to light. The two categories are called series circuits and parallel circuits. When lightbulbs are arranged in series, such as is shown in Figure L16.1(a), each bulb is connected to the next bulb and so forth. When lightbulbs are connected in parallel, such as is shown in Figure L16.1(b), each bulb is connected directly to the battery. The amount of light given off by a lightbulb is influenced in part by the strength of the battery (or other source of electricity) and the ways the lightbulbs are connected together. Scientists have investigated what happens to the brightness of the light emitted by the lightbulb when they are connected in series and parallel. In this investigation, you will have an opportunity to examine how the arrangement of bulbs connected to a battery in an electric circuit affects the brightness of a specific bulb in the circuit.

FIGURE L16.1

Bulbs in series (a) and in parallel (b)

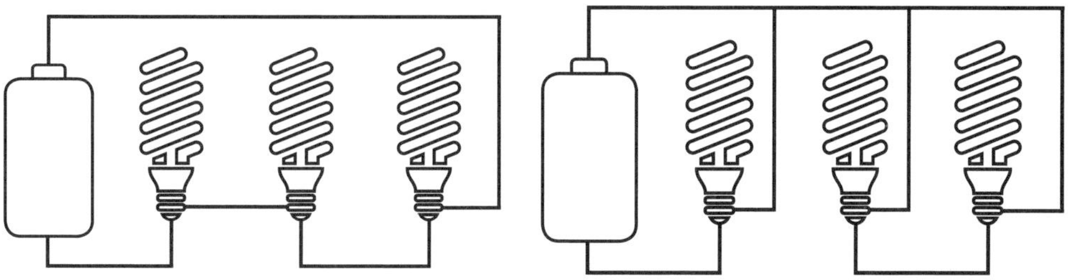

(a)

(b)

LAB 16

Your Task

Use what you know about circuits, the relationship between structure and function, and how to design and carry out an investigation to develop a rule that will allow you to predict the brightness of a bulb based on how it is arranged in an electric circuit. During this investigation, you will want to keep in mind the ideas of Robert Boyle—that scientific rules need empirical support and it is important to control for all the factors that might influence your results during an investigation. Once you develop your rule, you will need to test it to determine if it allows you to predict the brightness of a bulb in a wide range of different circuits.

The guiding question of this investigation is, **How does the arrangement of lightbulbs that are connected to a battery affect the brightness of a single bulb in that circuit?**

Materials

You may use any of the following materials during your investigation:

- Size D batteries
- Battery holders
- Small lightbulbs
- Lightbulb holders
- Electrical wire
- Light sensor with interface
- Safety glasses or goggles

Safety Precautions

Follow all normal lab safety rules. In addition, take the following safety precautions:

1. Wear sanitized safety glasses or goggles during lab setup, hands-on activity, and takedown.
2. Use caution when handling bulbs, wires, and batteries. They can get hot and burn skin.
3. Never put batteries in your mouth or on your tongue.
4. Use caution in handling wire ends. They are sharp and can cut or puncture skin.
5. Lightbulbs are made of glass. Be careful handling them. If they break, clean them up immediately and place in a broken glass box.
6. Wash hands with soap and water after completing the lab activity.

Investigation Proposal Required? ☐ Yes ☐ No

Getting Started

The first step in this investigation is to learn more about how the number and arrangement of bulbs in a circuit affect the brightness of a specific bulb in that circuit. To accomplish this task, you must determine what type of data you need to collect, how you will collect it, and how you will analyze it before you begin.

To determine *what type of data you need to collect,* think about the following questions:

- How will you determine brightness?
- What other factors, besides the type of circuit, could affect the brightness of a lightbulb?
- How will you control for those factors?

To determine *how you will collect your data,* think about the following questions:

- What equipment will you need to collect the data you need?
- How will you make sure that your data are of high quality (i.e., how will you reduce error)?
- How will you keep track of the data you collect?
- How will you organize your data?

To determine *how you will analyze your data,* think about the following questions:

- What factors will you compare to generate your rule?
- What type of table or graph could you create to help make sense of your data?

The second step in this investigation is to develop a rule that you can use to predict the brightness of a bulb in a circuit. Once you have your rule, you will need to test it to determine if it allows you to accurately predict the brightness of a bulb in several new circuits (ones that you did not use to develop your rule). It is important for you to test your rule, because the results of your test will not only allow you to demonstrate that your rule is valid but also will allow you to show that it is a useful way to predict the behavior of a lightbulb when it is connected to a battery and one or more other bulbs. Be sure to modify your rule as needed if it does not allow you to accurately predict the brightness of a bulb in a particular circuit.

Connections to Crosscutting Concepts, the Nature of Science, and the Nature of Scientific Inquiry

As you work through your investigation, be sure to think about

- the importance of tracking how energy and matter move within electrical systems,
- the relationship between structure and function in nature,
- the difference between data and evidence in science, and
- the nature and role of experiments in science.

Initial Argument

Once your group has finished collecting and analyzing your data, your group will need to develop an initial argument. Your initial argument needs to include a *claim*, *evidence* to support your claim, and a *justification* of the evidence. The claim is your group's answer to the guiding question. The evidence is an analysis and interpretation of your data. Finally, the justification of the evidence is why your group thinks the evidence matters. The justification of the evidence is important because scientists can use different kinds of evidence to support their claims. Your group will create your initial argument on a whiteboard. Your whiteboard should include all the information shown in Figure L16.2.

FIGURE L16.2

Argument presentation on a whiteboard

The Guiding Question:	
Our Claim:	
Our Evidence:	Our Justification of the Evidence:

Argumentation Session

The argumentation session allows all of the groups to share their arguments. One member of each group will stay at the lab station to share that group's argument, while the other members of the group go to the other lab stations to listen to and critique the arguments developed by their classmates. This is similar to how scientists present their arguments to other scientists at conferences. If you are responsible for critiquing your classmates' arguments, your goal is to look for mistakes so these mistakes can be fixed and they can make their argument better. The argumentation session is also a good time to think about ways you can make your initial argument better. Scientists must share and critique arguments like this to develop new ideas.

To critique an argument, you might need more information than what is included on the whiteboard. You will therefore need to ask the presenter lots of questions. Here are some good questions to ask:

- How did you collect your data? Why did you use that method? Why did you collect those data?
- What did you do to make sure the data you collected are reliable? What did you do to decrease measurement error?
- How did your group analyze the data? Why did you decide to do it that way? Did you check your calculations?
- Is that the only way to interpret the results of your analysis? How do you know that your interpretation of your analysis is appropriate?
- Why did your group decide to present your evidence in that way?
- What other claims did your group discuss before you decided on that one? Why did your group abandon those alternative ideas?

- How confident are you that your claim is valid? What could you do to increase your confidence?

Once the argumentation session is complete, you will have a chance to meet with your group and revise your initial argument. Your group might need to gather more data or design a way to test one or more alternative claims as part of this process. Remember, your goal at this stage of the investigation is to develop the most acceptable and valid answer to the research question!

Report

Once you have completed your research, you will need to prepare an *investigation report* that consists of three sections. Each section should provide an answer to the following questions:

- What question were you trying to answer and why?
- What did you do to answer your question and why?
- What is your argument?

Your report should answer these questions in two pages or less. This report must be typed, and any diagrams, figures, or tables should be embedded into the document. Be sure to write in a persuasive style; you are trying to convince others that your claim is acceptable and valid!

Checkout Questions

Lab 16. Electrical Energy and Lightbulbs

How Does the Arrangement of Lightbulbs That Are Connected to a Battery Affect the Brightness of a Single Bulb in That Circuit?

1. In which circuit will lightbulb A be brightest: circuit 1, circuit 2, or circuit 3?

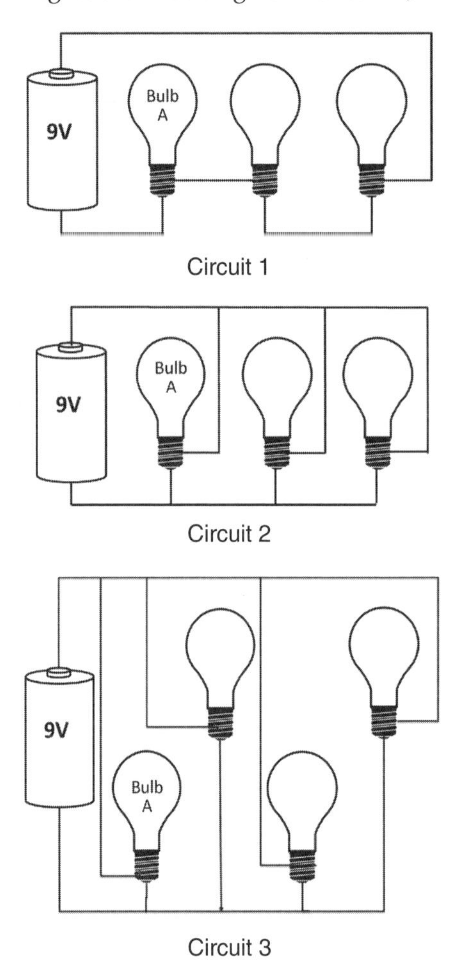

Explain your answer. Why do you think that bulb will be brightest?

2. If you want to make a lightbulb less bright, what are some things you could do to the circuit to make the bulb decrease in brightness? How do you know those things will work?

3. In science, there is no difference between data and evidence.

   a. I agree with this statement.
   b. I disagree with this statement.

   Explain your answer, using an example from your investigation about series and parallel circuits.

4. No matter what is being investigated, conducting an experiment is the best way to develop scientific knowledge.

   a. I agree with this statement.
   b. I disagree with this statement.

Explain your answer, using an example from your investigation about series and parallel circuits.

5. Often, the structure of a system influences the function of that system. In other words, how something is structured influences how it works. Using an example from your investigation about series and parallel circuits, explain how the structure of a system can influence how it functions.

6. Scientists often need to keep track of the movement of energy into, out of, and within systems. Using an example from your investigation about series and parallel circuits, explain why it is important to track how input of energy into a system affects how it behaves.

Application Labs

Teacher Notes

Lab 17. Rate of Energy Transfer

How Does the Surface Area of a Substance Affect the Rate at Which Thermal Energy Is Transferred From One Substance to Another?

Purpose

The purpose of this lab is for students to *apply* their knowledge of thermal energy transfer and kinetic molecular theory to the relationship between surface area and thermal energy transfer. This lab gives students an opportunity to further their understanding of why it is important to track how energy moves within systems and how the structure and shape of an object can influence how it interacts with other matter. Students will also learn about the difference between observations and inferences and the different methods used in scientific investigations.

The Content

According to kinetic-molecular theory, *thermal energy* is the energy an object has due to the kinetic energy of the particles that make up the object or substance. *Temperature* is related to thermal energy, in that temperature is a measure of the average kinetic energy of the particles in an object or substance. *Heat* is also related to thermal energy, in that heat is a measure of the transfer of thermal energy from one object to another object or substance.

In everyday use, the terms *heat* and *thermal energy* are often used interchangeably to mean the same thing (we often use *heat* when we mean *thermal energy*). In science, we define *heat* as only the transfer of thermal energy. This means that an object or substance *does not have heat* as a property itself. Instead, an object has *thermal energy* and, through heat transfer, it can lose thermal energy to other objects or substances that have less thermal energy.

The transfer of thermal energy is dependent on several properties of substances. The first property is temperature. Under normal circumstances, thermal energy is always transferred from higher temperature to lower temperatures. If you put a cold piece of ice in warm water, thermal energy will transfer from the water (higher temperature) to the ice (lower temperature). This also means that when you put water in the freezer to make ice, the water loses thermal energy to the surroundings (this is important because in normal conversation we say that the freezer transfers cold to the ice, which is not correct).

The second property of a substance that affects heat transfer is the *specific heat* of the two substances. Specific heat is a measure of how much thermal energy must be added to raise the temperature of 1 gram of a substance by 1 degree Celsius. All substances have a different specific heat. This means that different substances will see a different temperature change when the same amount of thermal energy is added. For example, the specific heat of water is 4.18 J/g • °C. This means that to raise the temperature of 1 g of water by 1°C, 4.18 J of energy must be added. The specific heat of aluminum is 0.9 J/g • °C. Thus, it takes much less energy to raise the temperature of aluminum by 1°C than to raise the temperature of water by 1°C.

Thermal energy is transferred between two substances because particles of the substance with the higher temperature collide with the particles of the substance with the lower temperature. In other words, thermal energy is transferred when the two substances are in contact. When these microscopic collisions occur, the particle with the higher kinetic energy will lose kinetic energy to the particle with the lower kinetic energy. This occurs when two solids at different temperatures are in contact, when a solid is in contact with a liquid, when a solid is in contact with a gas, when two liquids are mixed together, when a liquid is in contact in a gas, or when two gases are mixed together.

As an example, when liquid water is put into a freezer, the water molecules have more kinetic energy than the air molecules in the freezer. The water molecules on the surface of the water will collide with the gaseous air molecules, and the water molecules will lose kinetic energy to the air molecules in the freezer. This will lower the temperature of the water, and the water will eventually freeze.

Thus, a third factor influencing the transfer of thermal energy is the ability of molecules of one substance to come in contact with the molecules in the other substance. The more often these particles can collide, the more quickly thermal energy will be transferred. This means that the greater the surface area, the quicker thermal energy will be transferred.

Using a block of ice as an example, a 100 g block of ice will take much longer to melt than will 100 pieces of ice (1 g each), even though it is the same mass of ice and they both start at the same temperature. This is because the surface area to mass ratio for the small pieces is greater than with the one large block of ice.

The same principle is at the heart of this lab; you will be using small pieces of metal and placing them in a beaker of water to heat the water. Holding mass and temperature of both the metal and the water constant, the smaller the pieces of metal are, the more quickly thermal energy will be transferred from the metal to the water.

Heat transfer stops when the two substances reach an equal temperature. This temperature, which is called the *equilibrium temperature,* will be in between the initial temperatures of the substances. However, the final temperature does not necessarily have to be at the midpoint between the two substances. The exact equilibrium temperature is dependent on the mass of the two samples and the specific heat of the two samples.

Timeline

The instructional time needed to complete this lab investigation is 230–280 minutes. Appendix 2 (p. 411) provides options for implementing this lab investigation over several class periods. Option A (280 minutes) should be used if students are unfamiliar with scientific writing, because this option provides extra instructional time for scaffolding the writing process. You can scaffold the writing process by modeling, providing examples, and providing hints as students write each section of the report. Option B (230 minutes) should be used if students are familiar with scientific writing and have developed the skills needed to write an investigation report on their own. In option B, students complete stage 6 (writing the investigation report) and stage 8 (revising the investigation report) as homework.

Materials and Preparation

The materials needed to implement this investigation are listed in Table 17.1. The mesh bags can be purchased from craft stores such as Michaels. The rest of the equipment can be purchased from a science supply company such as Carolina, Flinn Scientific, or Ward's Science. For the metal samples, we recommend using fishing weights or metal shot (made of lead or stainless steel) that can be purchased at a sporting goods or outdoor supply store, such as Wal-Mart or Outdoor World Sporting Goods. Some density kits from science suppliers also have various-size metal samples that are suitable.

We recommend that you use a set routine for distributing and collecting the materials during the lab investigation. For example, the consumables and equipment for each group can be set up at each group's lab station before class begins, or one member from each group can collect them from a table or a cart when needed during class.

Safety Precautions and Laboratory Waste Disposal

Remind students to follow all normal lab safety rules. In addition, tell students to take the following safety precautions:

1. Wear sanitized indirectly vented chemical-splash goggles and chemical-resistant nonlatex gloves and aprons during lab setup, hands-on activity, and takedown.
2. Use caution when working with hot plates, because they can burn skin and cause fires.
3. Hot plates also need to be kept away from water and other liquids.
4. Use only GFCI-protected electrical receptacles for hot plates.
5. Use caution when working with hot water, because it can burn skin.
6. Clean up any spilled liquid immediately to avoid a slip or fall hazard.
7. Never put consumables in their mouth.
8. Students should always use tongs to move the heated metal.
9. Handle all glassware with care.
10. Handle glass thermometers with care. They are fragile and can break, causing a sharp hazard that can cut or puncture skin.
11. Never return the consumables to stock bottles.

12. Wash hands with soap and water after completing the lab activity
13. Water can be poured down any normal drain. We recommend letting the water reach room temperature first.

TABLE 17.1

Materials list for Lab 17

Item	Quantity
Consumable	
Water	As needed per group
Equipment and other materials	
Safety glasses or goggles	1 per student
Chemical-resistant apron	1 per student
Nonlatex gloves	1 pair per student
Beaker, 1,000 ml	1 per group
Graduated cylinder, 100 ml	1 per group
Tongs	1 per group
Mesh bags with drawstrings	3–5 per group
Styrofoam cups	5 per group
Electronic or triple beam balance	1 per group
Thermometer (or temperature probe with interface)	1 per group
Ruler	1 per group
Stopwatch	1 per group
Investigation Proposal B (optional)	1 per group
Whiteboard, 2' × 3'*	1 per group
Lab Handout	1 per student
Peer-review guide	1 per student
Checkout Questions	1 per student

* As an alternative, students can use computer and presentation software, such as Microsoft PowerPoint or Apple Keynote, to create their arguments.

Topics for the Explicit and Reflective Discussion

Concepts That Can Be Used to Justify the Evidence

To provide an adequate justification of their evidence, students must explain why they included the evidence in their arguments and make the assumptions underlying their analysis and interpretation of the data explicit. In this investigation, students can use the following concepts to help justify their evidence:

- Thermal energy versus heat energy
- Transfer of energy
- Kinetic-molecular theory of matter
- The law of conservation of energy

We recommend that you review these concepts during the explicit and reflective discussion to help students make this connection.

How to Design Better Investigations

It is important for students to reflect on the strengths and weaknesses of the investigation they designed during the explicit and reflective discussion. Students should therefore be encouraged to discuss ways to eliminate potential flaws, measurement errors, or sources of bias in their investigations. To help students be more reflective about the design of their investigation, you can ask the following questions:

1. What were some of the strengths of your investigation? What made it scientific?
2. What were some of the weaknesses of your investigation? What made it less scientific?
3. If you were to do this investigation again, what would you do to address the weaknesses in your investigation? What could you do to make it more scientific?

Crosscutting Concepts

This investigation is well aligned with two crosscutting concepts found in *A Framework for K–12 Science Education,* and you should review these concepts during the explicit and reflective discussion:

- *Energy and matter: Flows, cycles, and conservation:* In science it is important to track how energy and matter move into, out of, and within systems. In this lab students will investigate how energy is transferred between different substances.
- *Structure and function:* The way an object is shaped or structured determines many of its properties and functions. In this lab students will investigate how

the structure of a substance, specifically surface area, influences the rate of heat transfer.

The Nature of Science and the Nature of Scientific Inquiry

This investigation is well aligned with two important concepts related to the *nature of science* (NOS) and the *nature of scientific inquiry* (NOSI), and you should review these concepts during the explicit and reflective discussion.

- *The difference between observations and inferences in science:* An observation is a descriptive statement about a natural phenomenon, whereas an inference is an interpretation of an observation. Students should also understand that current scientific knowledge and the perspectives of individual scientists guide both observations and inferences. Thus, different scientists can have different but equally valid interpretations of the same observations due to differences in their perspectives and background knowledge.
- *Methods used in scientific investigations:* Examples of methods include experiments, systematic observations of a phenomenon, literature reviews, and analysis of existing data sets; the choice of method depends on the objectives of the research. There is no universal step-by-step scientific method that all scientists follow; rather, different scientific disciplines (e.g., chemistry vs. biology) and fields within a discipline (e.g., organic vs. physical chemistry) use different types of methods and core theories and rely on different standards to develop scientific knowledge.

Hints for Implementing the Lab

- We recommend that you attempt this investigation on your own before the first class, so that you can gauge the time needed to sufficiently heat the metal pieces.
- It is important for you to know how to use the equipment so you can help students when technical issues arise.
- Allowing students to design their own procedures for collecting data gives students an opportunity to try, to fail, and to learn from their mistakes. However, you can scaffold students as they develop their procedure by having them fill out an investigation proposal. These proposals provide a way for you to offer students hints and suggestions without telling them how to do it. You can also check the proposals quickly during a class period. For this lab we suggest using Investigation Proposal B.
- Allow the students to become familiar with the thermometer or temperature probe as part of the tool talk before they begin to design their investigation. This gives students a chance to see what they can and cannot do with the equipment.
- If time is limited, students can be given multiple sets of equipment so that each group can collect data on different surface areas at the same time.

- When students are heating several small pieces of metal, it is useful to place them in a mesh bag with a drawstring that allows all the objects to be placed into and removed from the water bath at the same time. Small bags used for holding loose candy work well and can be purchased from most craft stores.
- Be sure that students record actual values (e.g., temperature or time).

Topic Connections

Table 17.2 provides an overview of the scientific practices, crosscutting concepts, disciplinary core ideas, and supporting ideas at the heart of this lab investigation. In addition, it lists NOS and NOSI concepts for the explicit and reflective discussion. Finally, it lists literacy and mathematics skills (*CCSS ELA* and *CCSS Mathematics*) that are addressed during the investigation.

TABLE 17.2

Lab 17 alignment with standards

Scientific practices	• Asking questions and defining problems • Planning and carrying out investigations • Analyzing and interpreting data • Using mathematics and computational thinking • Constructing explanations and designing solutions • Engaging in argument from evidence • Obtaining, evaluating, and communicating information
Crosscutting concepts	• Energy and matter: Flows, cycles, and conservation • Structure and function
Core ideas	• PS1.A: Structure and properties of matter • PS3.A: Definitions of energy
Supporting ideas	• Thermal energy versus heat energy • Transfer of energy • Kinetic-molecular theory of matter • Law of conservation of energy
NOS and NOSI concepts	• Observations and inferences • Methods used in scientific investigations
Literacy connections (*CCSS ELA*)	• *Reading*: Key ideas and details, craft and structure, integration of knowledge and ideas • *Writing*: Text types and purposes, production and distribution of writing, research to build and present knowledge, range of writing • *Speaking and listening*: Comprehension and collaboration, presentation of knowledge and ideas
Mathematics connections (*CCSS Mathematics*)	• Reason abstractly and quantitatively • Construct viable argument and critique the reasoning of others • Model with mathematics

Lab Handout

Lab 17. Rate of Energy Transfer

How Does the Surface Area of a Substance Affect the Rate at Which Thermal Energy Is Transferred From One Substance to Another?

Introduction

Understanding how energy is transferred from one object or substance to another is an important concept within science. Many common events involve a transfer of energy, such as heating a pot of water on the stove or when a car engine burns gasoline. In the first example, heat energy is transferred from the stove to the pot of water, and the water absorbs the heat energy and will eventually begin to boil. In a car, the chemical energy stored in the gasoline is released when it is burned in the engine; that chemical energy is ultimately converted to kinetic energy and results in the motion of the car.

The law of conservation of energy indicates that energy is not created or destroyed, only converted from one form to another. There are many different types of energy that can be transferred between objects. When two objects are at different temperatures, it is possible for heat or thermal energy to transfer from one object to the other. If you place a cold pot of water on a hot stove burner, for example, thermal energy will transfer from the stove to the water and the water will get warmer. Heat energy always moves from objects with a high temperature toward objects with a lower temperature.

When we measure the temperature of an object, we often use the Celsius scale. On the Celsius scale water freezes at 0°C and boils at 100°C. The temperature of a substance is a measure of the average kinetic energy of the particles of that substance. Water molecules in a cold sample of water at 10°C have less kinetic energy than water molecules in a sample of hot water at 50°C. In this example, the water molecules at 50°C will be moving faster. If these two samples of water with different temperatures are mixed together, the fast- and slow-moving particles will transfer energy until eventually the molecules all have similar amounts of kinetic energy. When molecules with higher kinetic energy bump into molecules with lower kinetic energy, the faster-moving particles transfer kinetic energy to the slower particles. The transfer of energy will result in the water mixture having a temperature that is in the middle of the starting temperatures, which is called an equilibrium temperature—in this example, perhaps about 30°C.

Whenever substances at different temperatures come into contact with one another, thermal energy will be transferred from the hotter object to the cooler object until an equilibrium temperature is reached. The rate at which that thermal energy is transferred from one substance to another, however, is based on several factors. Some of those factors include the properties of the specific substances, the amount of the substances involved, the starting temperatures, and the size and shape of the objects (see Figure L17.1 on page 306 for three samples of metal with equal mass but different surface areas). In this activity,

LAB 17

you will investigate how surface area affects the rate of heat transfer.

FIGURE L17.1

Three samples of metal with equal mass but different surface areas

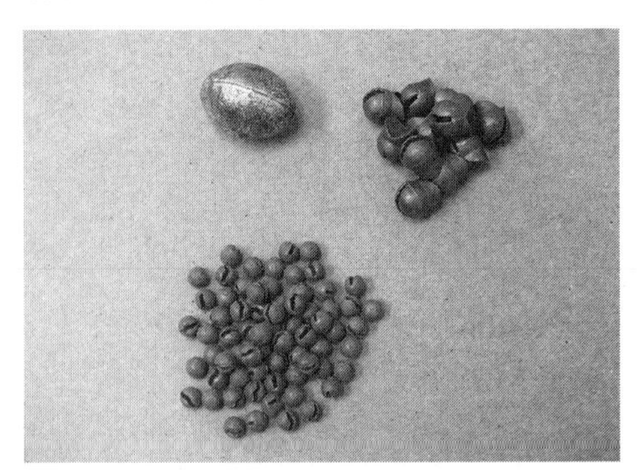

Your Task

Use what you know about thermal energy, tracking energy, and the relationship between structure and function to design and carry out an investigation that will allow you to test how the surface area of a hot object affects the rate at which thermal energy is transferred from that object to water. To complete this task, you will need to heat up several objects with different surface areas and then place them into room-temperature water. It is up to your group to determine how much and at what rate thermal energy is transferred to the room-temperature water.

The guiding question of this investigation is, **How does the surface area of a substance affect the rate at which thermal energy is transferred from one substance to another?**

Materials

You may use any of the following materials during your investigation:

Consumable
- Water

Equipment
- Metal samples
- Hot plate
- Beaker (1,000 ml)
- Graduated cylinder (100 ml)
- Tongs
- Mesh bags
- Styrofoam cups
- Electronic or triple beam balance
- Thermometer or temperature probe
- Ruler
- Stopwatch
- Safety glasses or goggles
- Chemical-resistant apron
- Nonlatex gloves

Safety Precautions

Follow all normal lab safety rules. In addition, take the following safety precautions:

1. Wear sanitized indirectly vented chemical-splash goggles and chemical-resistant nonlatex gloves and aprons during lab setup, hands-on activity, and takedown.
2. Use caution when working with hot plates, because they can burn skin and cause fires.

3. Hot plates also need to be kept away from water and other liquids.
4. Use caution when working with hot water, because it can burn skin.
5. Only use GFCI-protected electrical receptacles for hot plates.
6. Clean up any spilled liquid immediately to avoid a slip or fall hazard.
7. Never put consumables in your mouth.
8. Always use tongs to move the heated metal.
9. Handle all glassware with care.
10. Handle glass thermometers with care. They are fragile and can break, causing a sharp hazard that can cut or puncture skin.
11. Never return the consumables to stock bottles.
12. Wash hands with soap and water after completing the lab activity.

Investigation Proposal Required? ☐ Yes ☐ No

Getting Started

To answer the guiding question, you will need to design and conduct an investigation to measure the rate at which thermal energy is transferred to the water. To accomplish this task, you must determine what type of data you need to collect, how you will collect it, and how you will analyze it.

To determine *what type of data you need to collect,* think about the following questions:

- How will you determine the amount of energy transferred?
- What information or measurements will you need to record?
- How will you know when the equilibrium temperature is achieved?
- How will you measure the surface area of the different samples?
- What variables will you control from one sample to the next?

To determine *how you will collect your data,* think about the following questions:

- What equipment will you need to collect the data you need?
- How will you make sure that your data are of high quality (i.e., how will you reduce error)?
- Are there different ways you can measure the amount of energy transferred?
- How will you keep track of the data you collect?
- How will you organize your data?

To determine *how you will analyze your data,* think about the following questions:

- How will you determine the rate of heat transfer?
- What type of table or graph could you create to help make sense of your data?

Connections to Crosscutting Concepts, the Nature of Science, and the Nature of Scientific Inquiry

As you work through your investigation, be sure to think about

- why it is important to track how energy and matter move into, out of, and within systems;
- how the structure or shape of something can influence how it functions and places limits on what it can and cannot do;
- the difference between observations and inferences in science; and
- the different methods used in science.

Initial Argument

Once your group has finished collecting and analyzing your data, your group will need to develop an initial argument. Your initial argument needs to include a *claim, evidence* to support your claim, and a *justification* of the evidence. The claim is your group's answer to the guiding question. The evidence is an analysis and interpretation of your data. Finally, the justification of the evidence is why your group thinks the evidence matters. The justification of the evidence is important because scientists can use different kinds of evidence to support their claims. Your group will create your initial argument on a whiteboard. Your whiteboard should include all the information shown in Figure L17.2.

FIGURE L17.2

Argument presentation on a whiteboard

The Guiding Question:	
Our Claim:	
Our Evidence:	Our Justification of the Evidence:

Argumentation Session

The argumentation session allows all of the groups to share their arguments. One member of each group will stay at the lab station to share that group's argument, while the other members of the group go to the other lab stations to listen to and critique the arguments developed by their classmates. This is similar to how scientists present their arguments to other scientists at conferences. If you are responsible for critiquing your classmates' arguments, your goal is to look for mistakes so these mistakes can be fixed and they can make their argument better. The argumentation session is also a good time to think about ways you can make your initial argument better. Scientists must share and critique arguments like this to develop new ideas.

To critique an argument, you might need more information than what is included on the whiteboard. You will therefore need to ask the presenter lots of questions. Here are some good questions to ask:

- How did you collect your data? Why did you use that method? Why did you collect those data?
- What did you do to make sure the data you collected are reliable? What did you do to decrease measurement error?
- How did your group analyze the data? Why did you decide to do it that way? Did you check your calculations?
- Is that the only way to interpret the results of your analysis? How do you know that your interpretation of your analysis is appropriate?
- Why did your group decide to present your evidence in that way?
- What other claims did your group discuss before you decided on that one? Why did your group abandon those alternative ideas?
- How confident are you that your claim is valid? What could you do to increase your confidence?

Once the argumentation session is complete, you will have a chance to meet with your group and revise your initial argument. Your group might need to gather more data or design a way to test one or more alternative claims as part of this process. Remember, your goal at this stage of the investigation is to develop the most acceptable and valid answer to the research question!

Report

Once you have completed your research, you will need to prepare an *investigation report* that consists of three sections. Each section should provide an answer to the following questions:

1. What question were you trying to answer and why?
2. What did you do to answer your question and why?
3. What is your argument?

Your report should answer these questions in two pages or less. This report must be typed, and any diagrams, figures, or tables should be embedded into the document. Be sure to write in a persuasive style; you are trying to convince others that your claim is acceptable and valid!

Checkout Questions

Lab 17. Rate of Energy Transfer

How Does the Surface Area of a Substance Affect the Rate at Which Thermal Energy Is Transferred From One Substance to Another?

1. Denise was conducting an investigation on how long it would take ice to melt. She decided to test ice cubes versus crushed ice. She put 500 ml of 25°C water into two cups, and then she put 200 g of ice into each cup. One cup had ice cubes and the other cup had small crushed ice. She recorded the time it took for the ice in each cup to melt; the setup and results are shown below.

200 g cubed ice before melting

Time to melt: 14 minutes

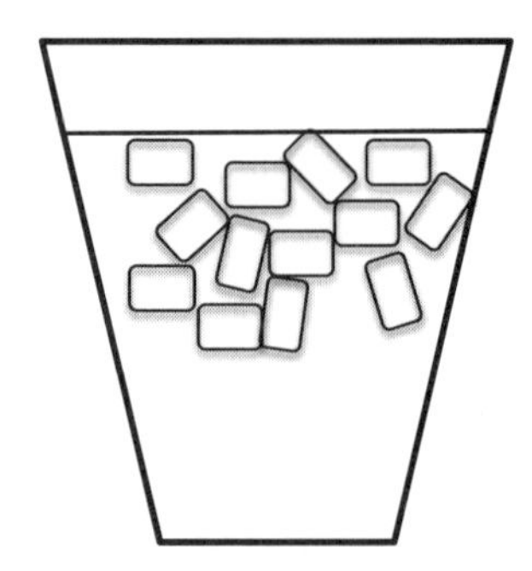

200 g crushed ice before melting

Time to melt: 9 minutes

Use what you know about energy transfer to explain the results that Denise obtained.

2. An engineer needs to put a hot piece of metal and a cold piece of metal together in a way that makes them reach their equilibrium temperature the fastest. Below are four options that she has come up with; in each option, the gray bar is hot and the black bar is cold. Which option would you recommend?

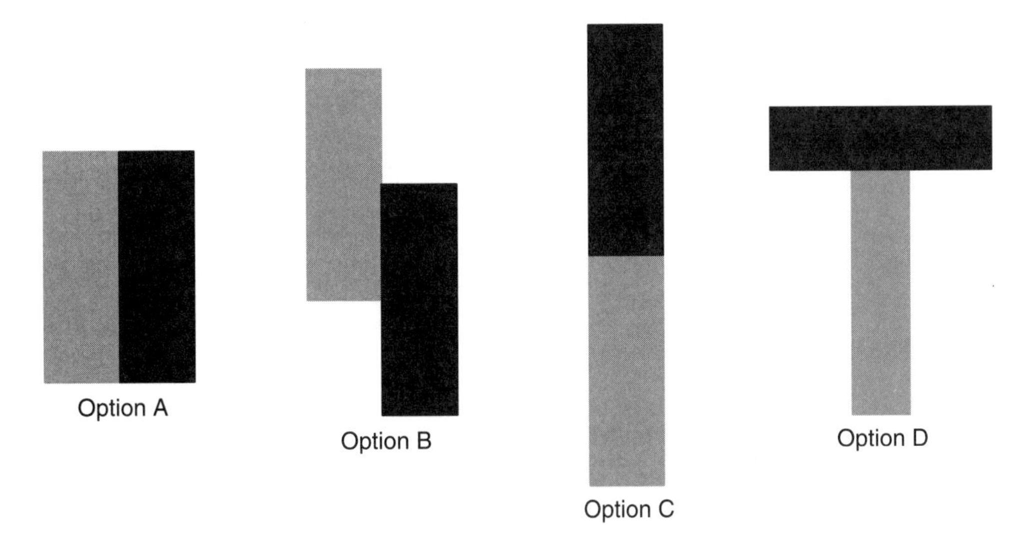

Explain why you chose that option to recommend to the engineer.

3. It is more important for scientists to make observations than inferences.

   a. I agree with this statement.

   b. I disagree with this statement.

   Explain your answer, using an example from your investigation on energy transfer.

4. Different scientists may use different procedures to investigate the same question.

   a. I agree with this statement.

   b. I disagree with this statement.

   Explain your answer, using an example from your investigation on energy transfer.

5. Understanding how systems work is an important aspect of science and engineering. Use an example from your investigation about energy transfer to help explain why it is important to track how energy and matter move into, out of, and within systems.

6. Scientists often study the structure of objects because the structure can provide useful clues about the function of that object. Explain why it is important for scientists to understand the connection between structure and function, using an example from your investigation on energy transfer.

Teacher Notes

Lab 18. Radiation and Energy Transfer

What Color Should We Paint a Building to Reduce Cooling Costs?

Purpose

The purpose of this lab is for students to *apply* what they know about energy transfer, the electromagnetic spectrum, and radiation to the problem of energy cost reduction in building design. In addition, this lab gives students an opportunity to describe the relationship between cause and effect by providing an explanation for the effect of variation of temperature between canisters of different colors. Students will also learn about the difference between laws and theories in science as well as the nature and role of experiments in scientific investigations.

The Content

Radiant energy, or *electromagnetic radiation*, can be conceptualized as being transported by *electromagnetic waves* or by their quanta, which are called photons. Each conceptualization is equally valid. For the purposes of this lab and its explanation, we will use the electromagnetic wave conceptualization. It is important to remember that waves, including electromagnetic waves, are not energy themselves. Rather, waves transport energy. Radiant energy is energy emitted when any charged particle is accelerated. When this occurs, an electromagnetic wave is generated. Electromagnetic waves are composed of alternating oscillating electric and magnetic fields. Because they require no medium through which to travel, electromagnetic waves can propagate through space and other vacuums.

Radiant energy encompasses all energy within the electromagnetic spectrum. Thus, visible light is radiant energy, but so is heat emitted from an infrared lamp, microwaves emitted from an antenna, and x-rays emitted from a radiograph machine.

Electromagnetic waves are characterized by the amount of energy they are carrying. The electromagnetic spectrum ranges from low-energy radio waves to high-energy gamma radiation (see Figure 18.1).

Visible light, a small portion of the electromagnetic spectrum, is also characterized by the amount of energy waves carry. Waves of various wavelengths within the visible light portion of the electromagnetic spectrum are perceived as different colors (see Figure 18.2; a full-color version of this figure is available at *www.nsta.org/adi-physicalscience*). Wavelengths perceived as red are longest and carry the least amount of energy; violet wavelengths are the shortest and carry the greatest amount of energy. White light, which is not really white, is a combination of all of the wavelengths of light.

The electrons of atoms within objects have a natural vibrational frequency. When light that matches that natural frequency contacts the electrons in the object, they absorb the light's energy. However, when light of a frequency that does not match the natural

FIGURE 18.1

The electromagnetic spectrum

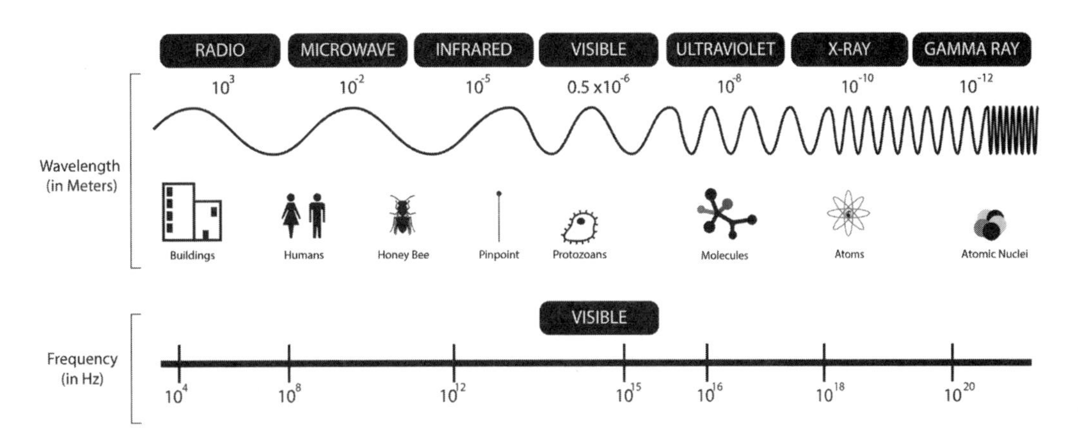

FIGURE 18.2

Wavelengths of visible light

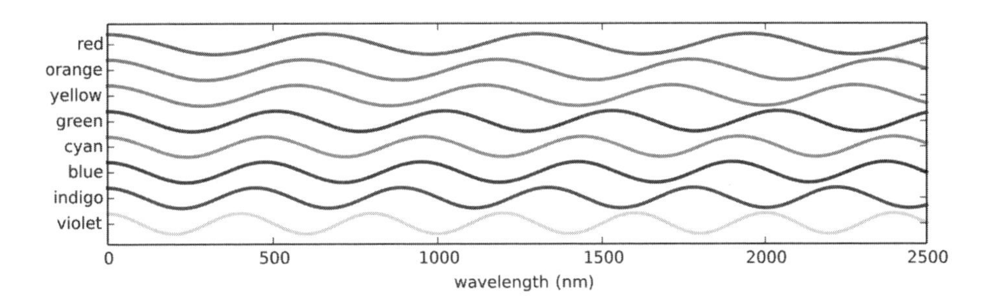

vibrational frequency of the electrons contacts the object, that specific wavelength of light is reflected off the object. For example, an object that appears indigo is reflecting indigo light while absorbing other colors in the spectrum. A white object reflects all colors, and a black object absorbs them. Thus, a red object will absorb more radiant energy than a violet object because the red object will reflect lower-energy red light and absorb higher-energy violet light, whereas the opposite is true for the violet object.

The *law of conservation of energy* states that energy within a system cannot be created or destroyed, only transformed into different types of energy. Because radiant energy is carried by electromagnetic waves, it does not require a medium to transfer energy to an object, unlike the processes of conduction and convection for thermal energy transfer. When an object absorbs radiant energy, the *kinetic energy* of its atoms (or molecules, depending on the object) increases, which increases the average temperature of the object. The greater the amount of radiant energy transferred, the greater the increase in kinetic energy, and the hotter the object will become. If left to absorb radiant energy for the same amount of time, a red object will be hotter than a violet object, because it will have absorbed more radiant energy. A white object left to absorb radiant energy will be cooler than an object of any other color, because it will not absorb wavelengths on the visible spectrum. However, it will not reflect all wavelengths of light on the electromagnetic spectrum, and will thus still increase in temperature.

Timeline

The instructional time needed to complete this lab investigation is 230–280 minutes. Appendix 2 (p. 411) provides options for implementing this lab investigation over several class periods. Option E (280 minutes) should be used if students are unfamiliar with scientific writing, because this option provides extra instructional time for scaffolding the writing process. You can scaffold the writing process by modeling, providing examples, and providing hints as students write each section of the report. Option B (230 minutes) should be used if students are familiar with scientific writing and have developed the skills needed to write an investigation report on their own. In option B, students complete stage 6 (writing the investigation report) and stage 8 (revising the investigation report) as homework.

Materials and Preparation

The materials needed to implement this investigation are listed in Table 18.1. The materials required for constructing the color canisters used in this lab can be readily obtained at most grocery or general stores. Instructions for making the canisters are provided below. Other required equipment can be purchased from a science supply company such as Carolina, Flinn Scientific, or Ward's Science.

You will need to make the colored canisters before the first class using this lab. These canisters can be easily made with chip containers and spray paint and are durable enough to be shared between classrooms and used from year to year. Small, cylindrical chip canisters work best because they are shorter than most lab thermometers, allowing the thermometer to be inserted and read without removing the top. Obtain enough chip canisters so that you can make at least three of each color (white, black, orange, green, and blue work well). Colors can be substituted as long as white and black canisters are available and the other colors vary enough in shade that the canisters' temperature difference will be detectable by the lab equipment. If you have the resources and time, a full set of colors can be made for each student group. You will also need masking tape, a utility knife, and a clean, well-ventilated area in which to spray-paint the canisters.

Remove the tops from the canisters and tape over the openings to avoid overspray into the canisters' interior. Following the paint manufacturer's instructions, use spray paint to cover the exterior of the canisters. While the canisters dry, use the utility knife to cut a

TABLE 18.1

Materials list for Lab 18

Item	Quantity
Safety glasses or goggles	1 per student
Colored canisters (see instructions in text)	Class set
Heat lamp	1 per group
Stopwatch	1 per group
Alcohol thermometer (or temperature probe with interface)	At least 1 per group
Investigation Proposal B (optional)	1 per group
Whiteboard, 2' × 3'*	1 per group
Lab Handout	1 per student
Peer-review guide	1 per student
Checkout Questions	1 per student

* As an alternative, students can use computer and presentation software, such as Microsoft PowerPoint or Apple Keynote, to create their arguments.

small "+" shape, like those on fast-food soft drink lids, into the center of each plastic top. The opening should be just large enough for your lab thermometers. The canisters are ready to use once the paint has dried.

We recommend that you use a set routine for distributing and collecting the materials during the lab investigation. For example, the consumables and equipment for each group can be set up at each group's lab station before class begins, or one member from each group can collect them from a table or a cart when needed during class.

Safety Precautions and Laboratory Waste Disposal

Remind students to follow all normal lab safety rules. In addition, tell students to take the following safety precautions:

1. Wear sanitized safety goggles or glasses during lab setup, hands-on activity, and takedown.
2. Use caution when working with heat lamps and metal containers, because they get hot and can burn skin.

3. Clamp lamps to a secure structure out of the way of foot traffic, and avoid touching parts of the lamp other than the power switch until it has cooled. Only use GFCI-protected electrical receptacles for the heat lamp.
4. Handle glass thermometers with care. They are fragile and can break, causing a sharp hazard that can cut or puncture skin.
5. Wash hands with soap and water after completing the lab activity.

No waste disposal is needed in this lab investigation.

Topics for the Explicit and Reflective Discussion

Concepts That Can Be Used to Justify the Evidence

To provide an adequate justification of their evidence, students must explain why they included the evidence in their arguments and make the assumptions underlying their analysis and interpretation of the data explicit. In this investigation, students can use the following concepts to help justify their evidence:

- Electromagnetic spectrum
- Relationship between wavelength and energy for electromagnetic waves
- Reflection and absorption of light waves based on an object's color
- Transformation of radiant energy to thermal energy
- Law of conservation of energy

We recommend that you review these concepts during the explicit and reflective discussion to help students make this connection.

How to Design Better Investigations

It is important for students to reflect on the strengths and weaknesses of the investigation they designed during the explicit and reflective discussion. Students should therefore be encouraged to discuss ways to eliminate potential flaws, measurement errors, or sources of bias in their investigations. To help students be more reflective about the design of their investigation, you can ask the following questions:

1. What were some of the strengths of your investigation? What made it scientific?
2. What were some of the weaknesses of your investigation? What made it less scientific?
3. If you were to do this investigation again, what would you do to address the weaknesses in your investigation? What could you do to make it more scientific?

Crosscutting Concepts

This investigation is well aligned with two crosscutting concepts found in *A Framework for K–12 Science Education,* and you should review these concepts during the explicit and reflective discussion.

- *Cause and effect: Mechanism and explanation:* A major goal of science is to determine or describe the underlying cause of natural phenomena. Some causes are simple and some are multifaceted, so it is important for scientists to develop and test potential explanations for what is observed. In this lab students will investigate the effect of color on energy transfer.
- *Energy and matter: Flows, cycles, and conservation:* In science it is important to track how energy and matter move into, out of, and within systems. In this lab students will investigate how energy transfer is influenced by properties of an object.

The Nature of Science and the Nature of Scientific Inquiry

This investigation is well aligned with two important concepts related to the *nature of science* (NOS) and the *nature of scientific inquiry* (NOSI), and you should review these concepts during the explicit and reflective discussion.

- *The difference between laws and theories in science:* A scientific law describes the behavior of a natural phenomenon or a generalized relationship under certain conditions; a scientific theory is a well-substantiated explanation of some aspect of the natural world. For example, the first law of thermodynamics (law of conservation of energy) does not explain why energy is not created or destroyed; it simply describes what happens. Theories do not become laws even with additional evidence; they explain laws. However, not all scientific laws have an accompanying explanatory theory. It is also important for students to understand that scientists do not discover laws or theories; the scientific community develops them over time.
- *The nature and role of experiments:* Scientists use experiments to test the validity of a hypothesis (i.e., a tentative explanation) for an observed phenomenon. Experiments include a test and the formulation of predictions (expected results) if the test is conducted and the hypothesis is valid. The experiment is then carried out and the predictions are compared with the observed results of the experiment. If the observed results match the predictions, then the hypothesis is supported. If the observed results do not match the predictions, then the hypothesis is not supported. A signature feature of an experiment is the control of variables to help eliminate alternative explanations for observed results.

Hints for Implementing the Lab

- If using a temperature probe and accompanying software, learn how to use it before the lab begins. It is important for you to know how to use the equipment so you can help students when technical issues arise.
- Allowing students to design their own procedures for collecting data gives students an opportunity to try, to fail, and to learn from their mistakes. However, you can scaffold students as they develop their procedure by having them fill out an investigation proposal. These proposals provide a way for you to offer students hints and suggestions without telling them how to do it. You can also check the proposals quickly during a class period. For this lab we suggest using Investigation Proposal B.
- Allow the students to become familiar with the temperature probe and software as part of the tool talk before they begin to design their investigation. This gives students a chance to see what they can and cannot do with the equipment.
- Be sure that the heat lamps are clamped securely, if applicable, and are located in areas where students will not be walking frequently. Try to place heat lamps such that cords are off of the floor and out of the way of students.

Topic Connections

Table 18.2 provides an overview of the scientific practices, crosscutting concepts, disciplinary core ideas, and supporting ideas at the heart of this lab investigation. In addition, it lists NOS and NOSI concepts for the explicit and reflective discussion. Finally, it lists literacy and mathematics skills (*CCSS ELA* and *CCSS Mathematics*) that are addressed during the investigation.

TABLE 18.2

Lab 18 alignment with standards

Scientific practices	• Asking questions and defining problems • Planning and carrying out investigations • Analyzing and interpreting data • Using mathematics and computational thinking • Constructing explanations and designing solutions • Engaging in argument from evidence • Obtaining, evaluating, and communicating information
Crosscutting concepts	• Cause and effect: Mechanism and explanation • Energy and matter: Flows, cycles, and conservation
Core ideas	• PS3.B: Conservation of energy and energy transfer • PS4.A: Wave properties • PS4.B: Electromagnetic radiation
Supporting ideas	• Wave properties • The electromagnetic spectrum • Conservation of energy • Properties of matter • Heat transfer
NOS and NOSI concepts	• Scientific laws and theories • Nature and role of experiments
Literacy connections (*CCSS ELA*)	• *Reading*: Key ideas and details, craft and structure, integration of knowledge and ideas • *Writing*: Text types and purposes, production and distribution of writing, research to build and present knowledge, range of writing • *Speaking and listening*: Comprehension and collaboration, presentation of knowledge and ideas
Mathematics connections (*CCSS Mathematics*)	• Reason abstractly and quantitatively • Construct viable arguments and critique the reasoning of others • Look for and make use of structure • Look for and express regularity in repeated reasoning

Lab Handout

Lab 18. Radiation and Energy Transfer

What Color Should We Paint a Building to Reduce Cooling Costs?

Introduction

Radiant energy is the energy transported by electromagnetic waves. Electromagnetic waves transport many different types of energy (see Figure L18.1). The microwaves that warm up your food when you place it into a microwave oven are electromagnetic waves, as are the x-rays that a doctor or dentist uses to take pictures of your bones. In fact, everything you see is also due to electromagnetic waves. Visible light, the light that humans can see, travels in waves. Each color has its own wavelength, which corresponds to a different amount of energy. When those waves reach our eyes, they can then be processed and perceived as color. Certain properties of an object cause it to reflect one wavelength of light and absorb others. For example, the reason an object appears blue is because it absorbs all other wavelengths and reflects a blue wavelength of light, which your eyes receive and, along with your brain, process as the color blue.

FIGURE L18.1

The electromagnetic spectrum

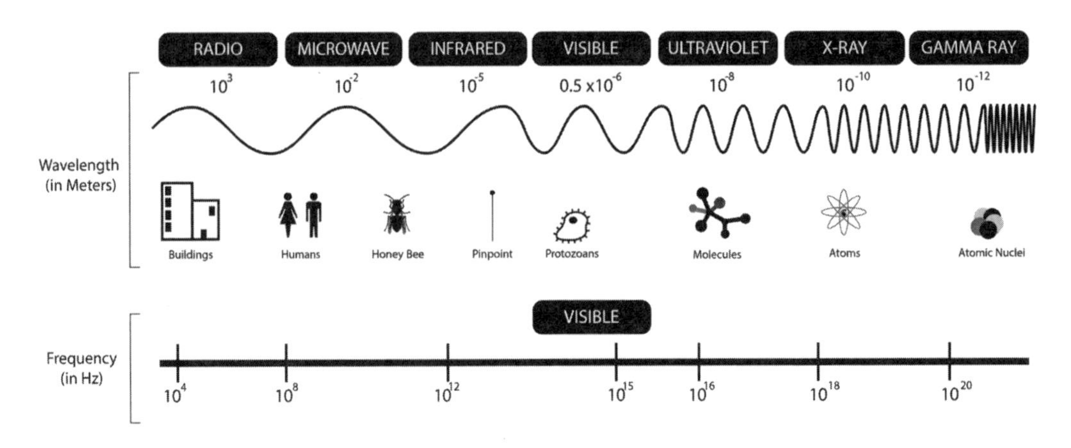

As is true for all energy, radiant energy can be transferred into other forms but cannot be created or destroyed. However, radiant energy is different in that it does not need a medium (matter), such as air or metal, to travel. Radiant energy can travel through a vacuum, such as space. The Sun emits radiant energy that travels through space. Some of that energy reaches Earth. When radiant energy reaches an object, it increases the rate of vibration of the atoms and/or molecules in that object, raising its overall temperature.

Radiant energy from the Sun raises the temperature of nearly everything on Earth, but some things are more affected than others.

When a new building is designed, its architects take into account the future energy costs of the building. Energy-efficient buildings are cheaper and more efficient to own and operate, and are also better for the environment. Some energy-saving or energy-storing measures, such as advanced heating and cooling systems, are expensive, but other measures, such as insulation or paint color, are simpler and less expensive. However, each step taken to increase the energy efficiency of a building is beneficial, not only for those who will own and use the building but for everyone, because we all benefit from the reduced use of energy resources.

Your Task

Use what you know about electromagnetic waves, visible light, and energy transfer to design and conduct an experiment to determine which paint color keeps a building the coolest, reducing its cooling costs.

The guiding question of this investigation is, **What color should we paint a building to reduce cooling costs?**

Materials

You may use any of the following materials during your investigation:

- Canisters with various paint colors
- Heat lamp
- Stopwatch
- Thermometer or temperature probe
- Safety glasses or goggles

Safety Precautions

Follow all normal lab safety rules. In addition, take the following safety precautions:

- Wear sanitized safety glasses or goggles during lab setup, hands-on activity, and takedown.
- Use caution when working with heat lamps and metal containers, because they get hot and can burn skin.
- Clamp lamps to a secure structure out of the way of foot traffic, and avoid touching parts of the lamp other than the power switch until it has cooled.
- Only use GFCI-protected electrical receptacles for the heat lamp.
- Handle glass thermometers with care. They are fragile and can break, causing a sharp hazard that can cut or puncture skin.
- Wash hands with soap and water after completing the lab activity.

Investigation Proposal Required? ☐ Yes ☐ No

FIGURE L18.2

Heating canisters with a heat lamp

Getting Started

To answer the guiding question, you will need to design an experiment that will allow you to determine which exterior paint color is associated with the lowest average canister temperature. To accomplish this task, you can heat canisters that are painted different colors using a heat lamp (see Figure L18.2). Before you can begin heating different canisters, you must first determine what type of data you need to collect, how you will collect it, and how you will analyze it.

To determine *what type of data you need to collect,* think about the following questions:

- What will serve as your independent variable in the investigation?
- What will serve as your dependent variable in the investigation?
- What types of measurements will you need to make?

To determine *how you will collect your data,* think about the following questions:

- What equipment will you use to take measurements?
- When will you make the measurements that you need?
- What other factors will you need to control during your experiment?
- How will you make sure that your data are of high quality (i.e., how will you reduce error)?
- How will you keep track of the data you collect?
- How will you organize your data?

To determine *how you will analyze your data,* think about the following questions:

- What type of calculations will you need to make?
- What type of table or graph could you create to help make sense of your data?

Connections to Crosscutting Concepts, the Nature of Science, and the Nature of Scientific Inquiry

As you work through your investigation, be sure to think about

- how scientists work to explain the relationships between causes and effects;
- the importance of tracking how energy moves into, out of, and within systems;

- the difference between scientific laws and scientific theories; and
- the nature and role of experiments in science.

Initial Argument

Once your group has finished collecting and analyzing your data, your group will need to develop an initial argument. Your initial argument needs to include a *claim, evidence* to support your claim, and a *justification* of the evidence. The claim is your group's answer to the guiding question. The evidence is an analysis and interpretation of your data. Finally, the justification of the evidence is why your group thinks the evidence matters. The justification of the evidence is important because scientists can use different kinds of evidence to support their claims. Your group will create your initial argument on a whiteboard. Your whiteboard should include all the information shown in Figure L18.3.

FIGURE L18.3

Argument presentation on a whiteboard

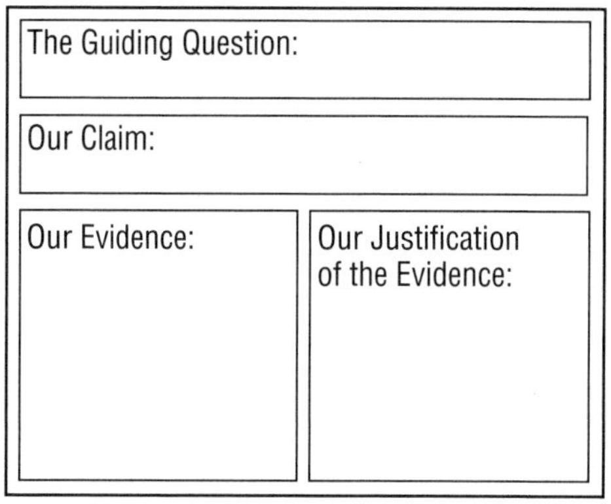

The Guiding Question:	
Our Claim:	
Our Evidence:	Our Justification of the Evidence:

Argumentation Session

The argumentation session allows all of the groups to share their arguments. One member of each group will stay at the lab station to share that group's argument, while the other members of the group go to the other lab stations to listen to and critique the arguments developed by their classmates. This is similar to how scientists present their arguments to other scientists at conferences. If you are responsible for critiquing your classmates' arguments, your goal is to look for mistakes so these mistakes can be fixed and they can make their argument better. The argumentation session is also a good time to think about ways you can make your initial argument better. Scientists must share and critique arguments like this to develop new ideas.

To critique an argument, you might need more information than what is included on the whiteboard. You will therefore need to ask the presenter lots of questions. Here are some good questions to ask:

- How did you collect your data? Why did you use that method? Why did you collect those data?
- What did you do to make sure the data you collected are reliable? What did you do to decrease measurement error?
- How did your group analyze the data? Why did you decide to do it that way? Did you check your calculations?
- Is that the only way to interpret the results of your analysis? How do you know that your interpretation of your analysis is appropriate?

- Why did your group decide to present your evidence in that way?
- What other claims did your group discuss before you decided on that one? Why did your group abandon those alternative ideas?
- How confident are you that your claim is valid? What could you do to increase your confidence?

Once the argumentation session is complete, you will have a chance to meet with your group and revise your initial argument. Your group might need to gather more data or design a way to test one or more alternative claims as part of this process. Remember, your goal at this stage of the investigation is to develop the most acceptable and valid answer to the research question!

Report

Once you have completed your research, you will need to prepare an *investigation report* that consists of three sections. Each section should provide an answer to the following questions:

1. What question were you trying to answer and why?
2. What did you do to answer your question and why?
3. What is your argument?

Your report should answer these questions in two pages or less. This report must be typed, and any diagrams, figures, or tables should be embedded into the document. Be sure to write in a persuasive style; you are trying to convince others that your claim is acceptable and valid!

Checkout Questions

Lab 18. Radiation and Energy Transfer

What Color Should We Paint a Building to Reduce Cooling Costs?

1. Each of the objects below was left in sunlight for one hour. Each object reflects a different wavelength of light, as shown below, and absorbs the other wavelengths. At the end of the hour, which object would you expect to have the highest average temperature?

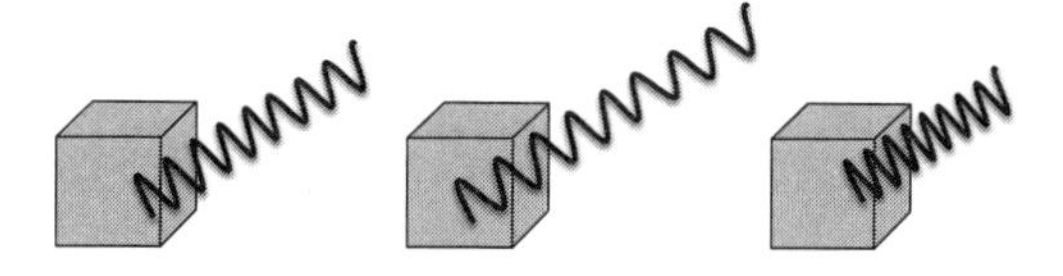

Explain your answer. Why do you think that object will have the highest average temperature at the end of the hour?

2. Shelby is choosing a solar panel to gather energy from the Sun for heating her home. There is one model available, but it comes in three colors. The first color reflects light with wavelengths between 350 and 400 nm, the second between 600 and 700 nm, and the third reflects no visible light. Which option is the best choice for Shelby if she wants her solar panel to absorb the most energy possible?

Explain your answer. Why did you recommend that option to Shelby?

3. In science, laws are more important than theories.

   a. I agree with this statement.

   b. I disagree with this statement.

   Explain your answer, using an example from your investigation about radiant energy transfer.

4. Regardless of the question you want to answer, an experiment is always the best way to conduct a scientific investigation.

   a. I agree with this statement.

   b. I disagree with this statement.

   Explain your answer, using an example from your investigation about radiant energy transfer.

5. Scientists often need to look for and understand the underlying cause of events in nature. Explain why it is important to be able to identify and understand cause-and-effect relationships in science, using an example from your investigation about radiant energy transfer.

6. Scientists often need to keep track of the flow of energy within systems. Using an example from your investigation about radiant energy transfer, explain why it is important to keep track of how energy moves into, out of, and within systems.

SECTION 5

Physical Science Core Idea 4

Waves and Their Applications in Technologies for Information Transfer

Introduction Labs

LAB 19

Teacher Notes

Lab 19. Wave Properties

How Do Frequency, Amplitude, and Wavelength of a Transverse Wave Affect Its Energy?

Purpose

The purpose of this lab is to *introduce* students to waves and their properties. This lab specifically investigates the properties of transverse waves. In addition, this lab gives students an opportunity to investigate the cause-and-effect relationships between wave components and how those various relationships influence the energy transferred by a wave. Students will also learn about the difference between data and evidence in science and how and why methods used in science may differ despite common goals.

The Content

A *wave* can be generally defined as some sort of disturbance that travels through space or matter. *Mechanical waves,* such as sound waves, seismic waves, water waves, or the waves created by moving the loose end of a rope up and down, require a physical medium, such as a liquid, a solid, or a gas, through which to travel. Despite traveling through matter, mechanical waves transfer energy, not particles. *Electromagnetic waves,* including radio waves, microwaves, infrared waves, visible light, ultraviolet waves, X-rays, and gamma rays (all of the waves along the electromagnetic spectrum), do not require a physical medium for travel and can thus move throughout the vacuum of space.

There are many different *waveforms* (shapes) that exist in this world. *Transverse* and *longitudinal* waves are the two most common waveforms. In transverse waves, the particles of the wave move perpendicular to the direction in which the wave propagates. Longitudinal wave particles move parallel to the propagation of a wave. The movement of particles in transverse and longitudinal waves is shown in Figure 19.1.

FIGURE 19.1

Transverse (a) and longitudinal (b) waveforms

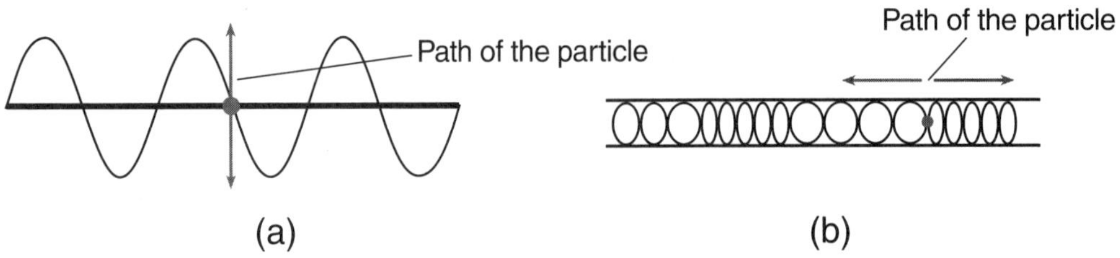

Path of the particle

Path of the particle

(a)

(b)

For the purposes of this lab, we will be investigating the motion of transverse waves. Longitudinal waves can also be represented as transverse waves. For example, an oscilloscope can be used to display the transverse waveform of a normally longitudinal sound wave. It does this by measuring the velocity and frequency of the displacement of a particle in the longitudinal wave and displaying its transverse equivalent. The wave created on a rope when a person moves one end of it up and then down is a transverse wave. Electromagnetic waves are also transverse waves.

The energy of a transverse wave can be described by the size of its components. The *wavelength* of a transverse wave is the distance between any two like points on two consecutive propagations of a wave. As wavelength decreases, the energy of the wave increases. Another component of a transverse wave is *amplitude,* which is the measure of displacement of a wave particle from average. Amplitude can be measured from the average line to the crest (top) or trough (bottom) of a wave (see Figure 19.2). The energy of a transverse wave is proportional to its amplitude squared. The number of wave cycles that occur within 1 second is the wave's frequency. This value is measured in hertz (Hz), which is equivalent to 1 cycle/second. For example, a wave that cycles three full times in 1 second has a frequency of 3 Hz. As frequency increases, the energy of the wave also increases. Increasing the frequency of a wave will also decrease its wavelength.

FIGURE 19.2

Components of a transverse wave

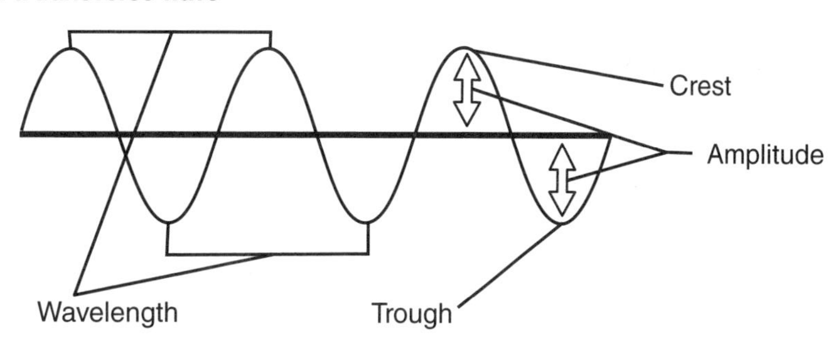

So, for a transverse wave created in a rope or string, if a person expends a lot of energy creating the wave, that wave is likely to have a higher amplitude, shorter wavelength, and higher frequency than a wave that was created with less energy. If less energy is used to create the wave, in general, the wave will have a smaller amplitude, longer wavelength, and lower frequency than a high-energy wave. It is possible that a high-energy wave and a low-energy wave have some similar components (i.e., the same amplitude, the same wavelength, or the same frequency), but because of the law of conservation of energy and the relationship between the components of waves, they will not have identical components if the total energy of the two waves is different.

LAB 19

Timeline

The instructional time needed to complete this lab investigation is 170–230 minutes. Appendix 2 (p. 411) provides options for implementing this lab investigation over several class periods. Option C (230 minutes) should be used if students are unfamiliar with scientific writing, because this option provides extra instructional time for scaffolding the writing process. You can scaffold the writing process by modeling, providing examples, and providing hints as students write each section of the report. Option D (170 minutes) should be used if students are familiar with scientific writing and have developed the skills needed to write an investigation report on their own. In option D, students complete stage 6 (writing the investigation report) and stage 8 (revising the investigation report) as homework.

Materials and Preparation

The materials needed to implement this investigation are listed in Table 19.1. This lab requires internet access to manipulate an online simulation. The *Wave on a String* simulation was developed by PhET Interactive Simulations, University of Colorado (*http://phet.colorado.edu*), and is available at *https://phet.colorado.edu/sims/html/wave-on-a-string/latest/wave-on-a-string_en.html*. It is free to use and can be run online using an internet browser on a school computer or tablet. You should access the website and learn how the simulation works before beginning the lab investigation. In addition, it is important to check if students can access and use the simulation from a school computer, because some schools have set up firewalls and other restrictions on web browsing.

TABLE 19.1

Materials list for Lab 19

Item	Quantity
Computer (or tablet) with internet access	1 per group
Investigation Proposal C (optional)	1 per group
Whiteboard, 2' × 3'*	1 per group
Lab Handout	1 per student
Peer-review guide	1 per student
Checkout Questions	1 per student

* As an alternative, students can use computer and presentation software, such as Microsoft PowerPoint or Apple Keynote, to create their arguments.

This simulation does not allow students to manipulate a button labeled "Energy." However, the frequency of the oscillator can be manipulated. Students can be left to determine that as frequency increases, so does energy, or they can be told this explicitly during the tool talk.

Safety Precautions and Laboratory Waste Disposal

Remind students to follow all normal lab safety rules. There is no laboratory waste associated with this activity.

Topics for the Explicit and Reflective Discussion

Concepts That Can Be Used to Justify the Evidence

To provide an adequate justification of their evidence, students must explain why they included the evidence in their arguments and make the assumptions underlying their analysis and interpretation of the data explicit. In this investigation, students can use the following concepts to help justify their evidence:

- Wave properties
- Conservation of energy
- Electromagnetic spectrum

We recommend that you review these concepts during the explicit and reflective discussion to help students make this connection.

How to Design Better Investigations

It is important for students to reflect on the strengths and weaknesses of the investigation they designed during the explicit and reflective discussion. Students should therefore be encouraged to discuss ways to eliminate potential flaws, measurement errors, or sources of bias in their investigations. To help students be more reflective about the design of their investigation, you can ask the following questions:

- What were some of the strengths of your investigation? What made it scientific?
- What were some of the weaknesses of your investigation? What made it less scientific?
- If you were to do this investigation again, what would you do to address the weaknesses in your investigation? What could you do to make it more scientific?

Crosscutting Concepts

This investigation is well aligned with two crosscutting concepts found in *A Framework for K–12 Science Education,* and you should review these concepts during the explicit and reflective discussion.

- *Cause and effect: Mechanism and explanation:* One of the main objectives of science is to identify and establish relationships between a cause and an effect. For example, the effect of increasing the frequency of waves is that wavelength decreases.
- *Energy and matter: Flows, cycles, and conservation:* In science it is important to track how energy and matter move into, out of, and within systems. In this lab students will investigate how changes in the properties of a wave influence the amount of energy that is transferred by the wave.

The Nature of Science and the Nature of Scientific Inquiry

This investigation is well aligned with two important concepts related to the *nature of science* (NOS) and the *nature of scientific inquiry* (NOSI), and you should review these concepts during the explicit and reflective discussion.

- *The difference between data and evidence in science:* Data are measurements, observations, and findings from other studies that are collected as part of an investigation. Evidence, in contrast, is analyzed data and an interpretation of the analysis.
- *Methods used in scientific investigations:* Examples of methods include experiments, systematic observations of a phenomenon, literature reviews, and analysis of existing data sets; the choice of method depends on the objectives of the research. There is no universal step-by step scientific method that all scientists follow; rather, different scientific disciplines (e.g., chemistry vs. biology) and fields within a discipline (e.g., organic vs. physical chemistry) use different types of methods, use different core theories, and rely on different standards to develop scientific knowledge.

Hints for Implementing the Lab

- Learn how to use the online simulation before the lab begins. It is important for you to know how to use the simulation so you can help students when they get stuck or confused.
- Allowing students to design their own procedures for collecting data gives students an opportunity to try, to fail, and to learn from their mistakes. However, you can scaffold students as they develop their procedure by having them fill out an investigation proposal. These proposals provide a way for you to offer students hints and suggestions without telling them how to do it. You can also check the proposals quickly during a class period. For this lab we suggest using Investigation Proposal C.
- A group of three students per computer or tablet tends to work well.
- Allow the students to play with the simulation as part of the tool talk before they begin to design their investigation. This gives students a chance to see what they

can and cannot do with the simulation. You can also show them the full-color version of the simulation screenshot that appears in black and white in the Lab Handout; the full-color version is available at *www.nsta.org/adi-physicalscience.*

- Be sure that the students record actual values (e.g., wavelength, amplitude, frequency) and are not just attempting to hand draw what they see on the computer screen.
- This is a good lab for students to make mistakes during the data collection stage. Students will quickly figure out what they did wrong during the argumentation session, and it will only take them a short period of time to re-collect data. It will also create an opportunity for students to reflect on and identify ways to improve the way they design investigations (especially how they attempt to control variables as part of an experiment) during the explicit and reflective discussion.

Topic Connections

Table 19.2 (p. 340) provides an overview of the scientific practices, crosscutting concepts, disciplinary core ideas, and supporting ideas at the heart of this lab investigation. In addition, it lists NOS and NOSI concepts for the explicit and reflective discussion. Finally, it lists literacy and mathematics skills (*CCSS ELA* and *CCSS Mathematics*) that are addressed during the investigation.

TABLE 19.2

Lab 19 alignment with standards

Scientific practices	• Asking questions and defining problems • Developing and using models • Planning and carrying out investigations • Analyzing and interpreting data • Using mathematics and computational thinking • Constructing explanations and designing solutions • Engaging in argument from evidence • Obtaining, evaluating, and communicating information
Crosscutting concepts	• Cause and effect: Mechanism and explanation • Energy and matter: Flows, cycles, and conservation
Core idea	• PS4.A Wave properties
Supporting ideas	• Wave properties • Law of conservation of energy • Electromagnetic spectrum
NOS and NOSI concepts	• Difference between data and evidence • Methods used in scientific investigations
Literacy connections (*CCSS ELA*)	• *Reading*: Key ideas and details, craft and structure, integration of knowledge and ideas • *Writing*: Text types and purposes, production and distribution of writing, research to build and present knowledge, range of writing • *Speaking and listening*: Comprehension and collaboration, presentation of knowledge and ideas
Mathematics connections (*CCSS Mathematics*)	• Reason abstractly and quantitatively • Construct viable arguments and critique the reasoning of others • Attend to precision

Lab Handout

Lab 19. Wave Properties

How Do Frequency, Amplitude, and Wavelength of a Transverse Wave Affect Its Energy?

Introduction

Energy can be transported by waves. There are many forms of waves that exist in the world. Mechanical waves, such as sound waves or water waves, must travel through a medium, or matter. For example, when you speak, you create a pressure disturbance in the air that travels as a wave through the air molecules. You can also create a wave in a rope or string by moving one end from side to side. In each case, the wave travels through the medium, the air or the rope (or string). Electromagnetic waves, such as radio, ultraviolet, and visible light waves, don't require a medium to travel. Instead, the vibrations of perpendicular electric and magnetic fields form these waves.

Although waves may travel differently, some mechanical and electromagnetic waves can be represented by the same basic shape, or waveform. Electromagnetic waves and the waves you might make in a rope or string, for example, are called transverse waves. A drawing of a transverse wave is shown in Figure L19.1. The highest point of the wave is called the crest, and the lowest point is called the trough. The wavelength of the wave, a measure of how long the wave is, can be found by measuring the distance between the same point on a wave and the wave in front of or behind it. Usually, this is done by measuring crest to crest or trough to trough. The amplitude of a wave is the distance from the resting position (the horizontal line) to the crest or trough. The frequency of a wave is hard to show in a picture. A wave's frequency is a measure of how many times a wave passes a certain point in a certain amount of time. To measure frequency, scientists measure the number of wave cycles (trough to trough or crest to crest) that occur in 1 second, and they measure this value in hertz (Hz). One cycle per second is 1 Hz, two per second is 2 Hz, and so on.

FIGURE L19.1

Transverse wave

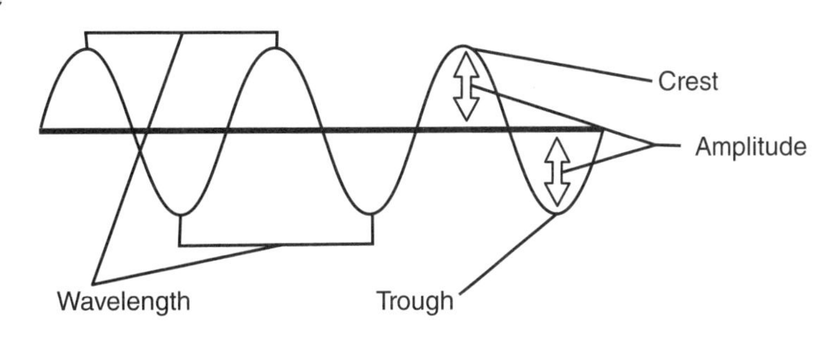

The properties of a wave contain information about the energy that wave is carrying and also determine its use. For example, electromagnetic radio waves are used to transmit the radio signals your car stereo picks up. Your favorite station numbers are actually measurements of the frequency at which that station broadcasts.

Your Task

Use what you know about waves and energy to design and carry out an investigation that will allow you to describe the relationship between a wave's energy and its amplitude, wavelength, and frequency.

The guiding question of this investigation is, **How do frequency, amplitude, and wavelength of a transverse wave affect its energy?**

Materials

You will use an online simulation called *Wave on a String* to conduct your investigation. You can access the simulation by going to the following website: *https://phet.colorado.edu/sims/html/wave-on-a-string/latest/wave-on-a-string_en.html.*

Safety Precautions

Follow all normal lab safety rules.

Investigation Proposal Required? ☐ Yes ☐ No

Getting Started

To answer the guiding question, you will need to design and carry out an experiment. To accomplish this task, you must determine what type of data you need to collect, how you will collect it, and how you will analyze it. The *Wave on a String* simulation allows you to propagate (start) and manipulate a wave on a virtual rope. The rope is shown as a series of red circles, with every ninth circle colored green (see Figure L19.2). This will make it easier for you to track and measure the properties of the various waves you create.

The upper-left-hand corner of the screen has a box with options for manual, oscillate, and pulse. These options allow you to choose how you will make the waves you will use for data. The "Manual" option requires that you move a wrench up and down to create a wave. The "Oscillate" option creates the wave for you, and you can adjust the frequency and amplitude of the waves using a slider that will appear at the bottom of the screen. Do not choose the "Pulse" option. Because this option only moves upward, it does not create a transverse wave and will not produce a wave that will be helpful for your investigation. You also have the option to use a rope with a fixed end, a loose end, or no end. The simulation provides rulers, a timer, and a reference line for you to use. To activate these tools, simply check the box next to each option. You can move the rulers by clicking

FIGURE L19.2

A screenshot of the *Wave on a String* simulation

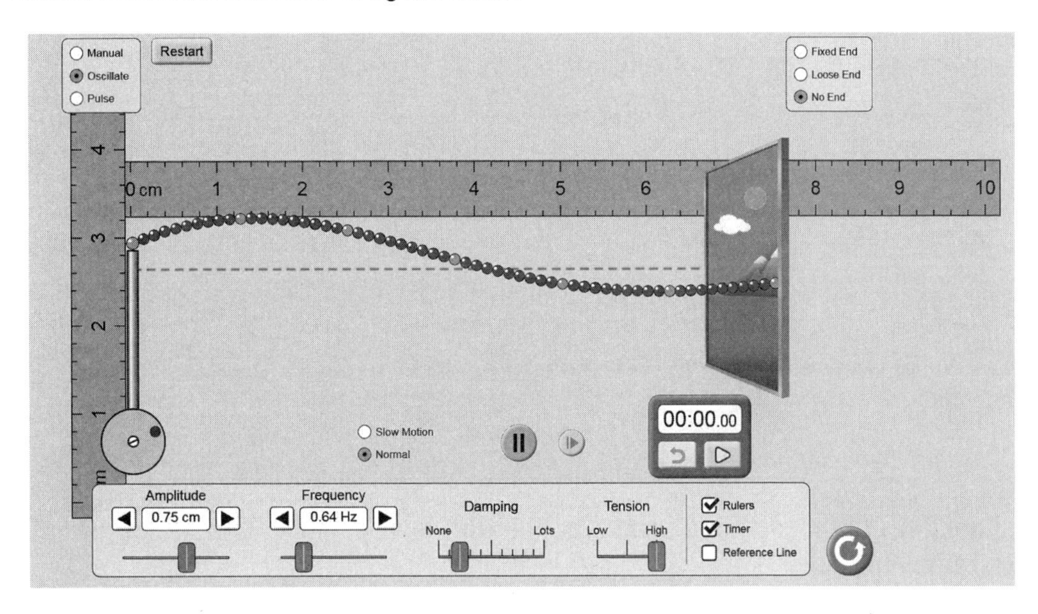

and dragging them to different locations. You may start and pause the simulation at any time by selecting the play/pause button at the bottom of the screen. You can also view the simulation in normal time or in slow motion.

You will need to design and carry out at least three different experiments using the *Wave on a String* simulation in order to determine the relationship between frequency, amplitude, wavelength, and energy. You will need to conduct at least three different experiments, because you will need to be able to answer three specific questions before you will be able to develop an answer to the guiding question:

- How does changing the frequency affect the energy of the wave?
- How does changing the amplitude affect the energy of the wave?
- How does changing the wavelength affect the energy of the wave?

It will be important for you to determine what type of data you need to collect, how to collect the data you need, and how you will need to analyze your data for each experiment, because each experiment is slightly different.

To determine *what type of data you need to collect,* think about the following questions:

- What will serve as your independent variable in the investigation?
- What will serve as your dependent variable(s) in the investigation?
- How will you define and determine the amount of energy being put into the waves?

- How will you measure the various properties of the waves?

To determine *how you will collect your data,* think about the following questions:

- What simulation settings will you use to collect the data you need?
- How will you make sure that your data are of high quality (i.e., how will you reduce error)?
- How will you keep track of the data you collect?
- How will you organize your data?

To determine *how you will analyze your data,* think about the following questions:

- What type of calculations will you need to make?
- What type of table or graph could you create to help make sense of your data?
- How will you determine if there is a relationship between different variables?

Connections to Crosscutting Concepts, the Nature of Science, and the Nature of Scientific Inquiry

As you work through your investigation, be sure to think about

- how scientists work to explain the relationships between causes and effects;
- the importance of tracking how energy moves into, out of, and within systems;
- the difference between data and evidence in science; and
- methods used in scientific investigations.

Initial Argument

Once your group has finished collecting and analyzing your data, your group will need to develop an initial argument. Your initial argument needs to include a *claim, evidence* to support your claim, and a *justification* of the evidence. The claim is your group's answer to the guiding question. The evidence is an analysis and interpretation of your data. Finally, the justification of the evidence is why your group thinks the evidence matters. The justification of the evidence is important because scientists can use different kinds of evidence to support their claims. Your group will create your initial argument on a whiteboard. Your whiteboard should include all the information shown in Figure L19.3.

Argumentation Session

The argumentation session allows all of the groups to share their arguments. One member of each group will stay at the lab station to share that group's argument, while the other members of the group go to the other lab stations to listen to and critique the arguments developed by their classmates. This is similar to how scientists present their arguments to other scientists at conferences. If you are responsible for critiquing your classmates' arguments, your goal is to look for mistakes so these mistakes can be fixed and they can make their argument better. The argumentation session is also a good time to think about ways you can make your initial argument better. Scientists must share and critique arguments like this to develop new ideas.

FIGURE L19.3

Argument presentation on a whiteboard

The Guiding Question:	
Our Claim:	
Our Evidence:	Our Justification of the Evidence:

To critique an argument, you might need more information than what is included on the whiteboard. You will therefore need to ask the presenter lots of questions. Here are some good questions to ask:

- How did you collect your data? Why did you use that method? Why did you collect those data?
- What did you do to make sure the data you collected are reliable? What did you do to decrease measurement error?
- How did your group analyze the data? Why did you decide to do it that way? Did you check your calculations?
- Is that the only way to interpret the results of your analysis? How do you know that your interpretation of your analysis is appropriate?
- Why did your group decide to present your evidence in that way?
- What other claims did your group discuss before you decided on that one? Why did your group abandon those alternative ideas?
- How confident are you that your claim is valid? What could you do to increase your confidence?

Once the argumentation session is complete, you will have a chance to meet with your group and revise your initial argument. Your group might need to gather more data or design a way to test one or more alternative claims as part of this process. Remember, your goal at this stage of the investigation is to develop the most acceptable and valid answer to the research question!

Report

Once you have completed your research, you will need to prepare an *investigation report* that consists of three sections. Each section should provide an answer to the following questions:

1. What question were you trying to answer and why?
2. What did you do to answer your question and why?

3. What is your argument?

Your report should answer these questions in two pages or less. This report must be typed, and any diagrams, figures, or tables should be embedded into the document. Be sure to write in a persuasive style; you are trying to convince others that your claim is acceptable and valid!

Checkout Questions

Lab 19. Wave Properties

How Do Frequency, Amplitude, and Wavelength of a Transverse Wave Affect Its Energy?

1. Order the transverse waves below from greatest to least energy carried:

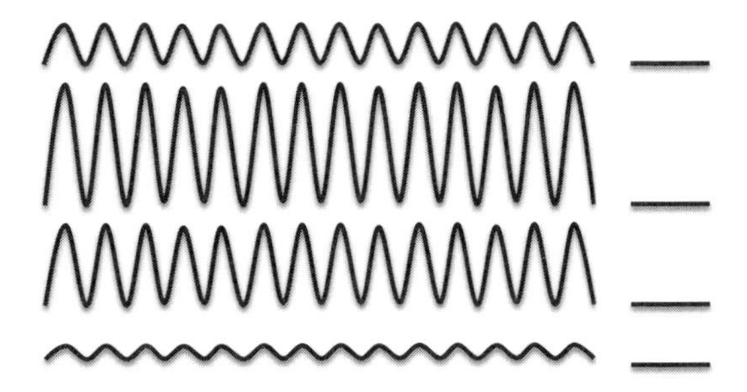

Explain your answer. Why do you think the order that you chose is correct?

2. Jalen moves the end of a rope to produce the wave shown below:

Without changing how far up or down he moves his arm, Jalen moves the end of the rope faster than before, working harder to move the rope. Which wave shown below looks like the wave Jalen is now making?

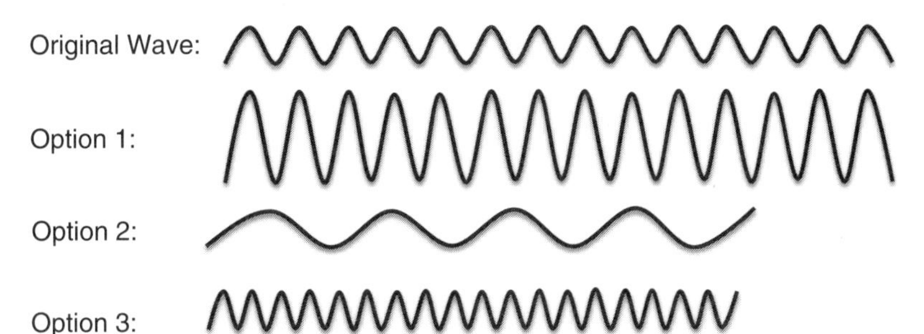

Explain your answer. Why do you think the wave Jalen is making looks like the option you chose?

3. In science, there is no difference between data and evidence.

   a. I agree with this statement.

   b. I disagree with this statement.

   Explain your answer, using an example from your investigation about the properties of transverse waves.

4. No matter what is being investigated, conducting an experiment is the best way to develop scientific knowledge.

   a. I agree with this statement.

   b. I disagree with this statement.

   Explain your answer, using an example from your investigation about the properties of transverse waves.

5. Often, changing one part of a system will cause another part of that system to change as well. Determining the cause of observed effects is an important pursuit in science. Using an example from your investigation about properties of transverse waves, explain why it is helpful for scientists to investigate cause-and-effect relationships in the natural world.

6. Scientists often need to keep track of the movement of energy into, out of, and within systems. Using an example from your investigation about properties of transverse waves, explain why it is important to track how input of energy into a system affects how it behaves.

LAB 20

Teacher Notes

Lab 20. Reflection and Refraction

How Can You Predict Where a Ray of Light Will Go When It Comes in Contact With Different Types of Transparent Materials?

Purpose

The purpose of this lab is to *introduce* students to the nature of light and the concepts of reflection and refraction. In addition, students will have an opportunity to explore the crosscutting concepts of looking for and explaining patterns and the role of models in science during this activity. Students will also learn about how scientific knowledge changes over time and the culture of science.

The Content

Light is energy carried in an electromagnetic wave. In a vacuum, light travels at a speed of 299,792,458 meters per second. Light is able to pass though some materials but not others. Materials that allow light to pass through it are called *transparent.* The speed of light varies in different types of transparent materials. In air, for example, the speed of light is slightly slower than it is in a vacuum, but in a solid transparent material, such as glass, the speed of light is about half of what it is in a vacuum. An important property of a transparent material is how it affects the speed of light as it travels through the material. The *index of refraction* for a transparent material is defined as the ratio of the speed of light in a vacuum to the speed of light in the material. The index of refraction for all transparent materials is greater than 1.0. The index of refraction of air is about 1.0003; the index of refraction of acrylic is 1.49.

When a ray of light passes between any two transparent materials, part of the ray is reflected and stays in the first material, while the rest of the ray is refracted and passes into the second material. Figure 20.1 shows a ray of light crossing the boundary between two transparent materials. The index of refraction for the first material is n_1 and for the second material is n_2. A line that runs perpendicular to the boundary is called the *surface normal,* and this line is used to measure the angles of the light rays. The angle the incident ray makes with the normal is called the *angle of incidence* (θ_i), the angle the reflected ray makes with the normal is called the *angle of reflection* (θ_r), and the angle the refracted ray makes with the normal is called the *angle of refraction* (θ_R).

FIGURE 20.1

A ray of light crossing the boundary between two transparent media when $n_1 < n_2$

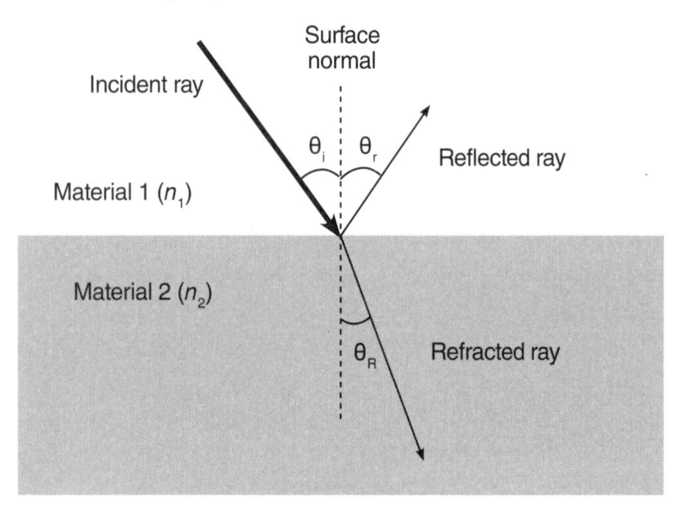

Scientists use the law of reflection and the law of refraction to describe and predict the path a ray of light will take when it reaches the boundary between two transparent materials. The *law of reflection* states that the angle of incidence (θ_i) equals the angle of reflection (θ_r), and the reflected ray will be on the opposite side of the normal from the incident ray. The *law of refraction* states that the sine of the angle of incidence (θ_i) and the sine of the angle of refraction (θ_R) are in a constant ratio to each other, and the refracted ray lies on the opposite side of the normal from the incident ray. The law of refraction is also known as Snell's law. The constant in Snell's law is the ratio of the indices of refraction for two materials, n_1 and n_2. The law of refraction is usually written as

$$\frac{\sin \theta_i}{\sin \theta_R} = \frac{n_2}{n_1}$$

When $n_1 < n_2$ (see Figure 20.1), a light ray passes from a medium or material with a low index of refraction, such as air, to a medium or material with a higher index of refraction, such as water, plastic, or glass. The reflected portion of the light remains in the first medium. This is called *external reflection.* Based on the law of reflection, the incident ray and the reflected ray both are at the same angle with respect to the normal ($\theta_i = \theta_r$). The refracted part of the light ray enters the material with the higher index of refraction at an angle that can be calculated using Snell's law. The angle of refraction is smaller than the incident angle ($\theta_R < \theta_i$) in this situation.

When $n_1 = n_2$, the angle of incidence will always equal the angle of refraction ($\theta_i = \theta_R$). This happens when light passes through a boundary between two materials with the same index of refraction or when it moves through a single material without a boundary. The path of the incident ray and the path of the refracted ray are therefore identical.

When $n_1 > n_2$ (see Figure 20.2, p. 352), a light ray moves from a medium or material with a higher index of refraction, such as water or plastic, to a medium or material that has a lower index of refraction, such as air. The reflected part of the light will stay inside the medium or material with the higher index of refraction. This is called *internal reflection.* Given the law of reflection, the angle of incidence (θ_i) will equal the angle of reflection (θ_r). The refracted part of the incident ray will leave the medium or material with the higher index of refraction and enter the medium or material with the lower index of refraction at an angle (θ_R) that can be predicted using Snell's law. The angle of refraction is larger than the incident angle ($\theta_R > \theta_i$) in this situation except when the incident angle is greater than the critical angle (θ_c). When the angle of incidence is greater than the critical angle ($\theta_i > \theta_c$), there is no refraction. Instead, the entire ray will reflect back inside the medium or material. This is called *total internal reflection.* The critical angle can be determined using the following formula:

$$\theta_c = \sin^{-1} \frac{n_1}{n_2}$$

FIGURE 20.2

A ray of light crossing the boundary between two transparent media when $n_1 > n_2$

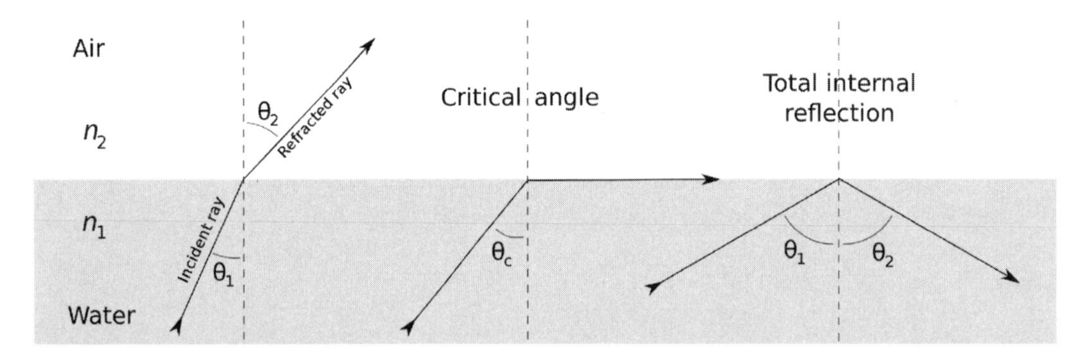

Timeline

The instructional time needed to complete this lab investigation is 200–230 minutes. Appendix 2 (p. 411) provides options for implementing this lab investigation over several class periods. Option C (230 minutes) should be used if students are unfamiliar with scientific writing, because this option provides extra instructional time for scaffolding the writing process. You can scaffold the writing process by modeling, providing examples, and providing hints as students write each section of the report. Option F (200 minutes) should be used if students are familiar with scientific writing and have developed the skills needed to write an investigation report on their own. In option F, students complete stage 6 (writing the investigation report) and stage 8 (revising the investigation report) as homework.

Materials and Preparation

The materials needed to implement this investigation are listed in Table 20.1. This lab requires internet access to manipulate an online simulation. The *Bending Light* simulation was developed by PhET Interactive Simulations, University of Colorado (*http://phet.colorado.edu*), and is available at *https://phet.colorado.edu/en/simulation/bending-light.* It is free to use and can be run online using an internet browser on a school computer or a tablet. You should access the website and learn how the simulation works before beginning the lab investigation. In addition, it is important to check if students can access and use the simulation from a school computer, because some schools have set up firewalls and other restrictions on web browsing.

TABLE 20.1

Materials list for Lab 20

Item	Quantity
Computer (or tablet) with internet access	1 per group
Investigation Proposal C (optional)	1 per group
Whiteboard, 2' × 3'*	1 per group
Lab Handout	1 per student
Peer-review guide	1 per student
Checkout Questions	1 per student

* As an alternative, students can use computer and presentation software, such as Microsoft PowerPoint or Apple Keynote, to create their arguments.

Safety Precautions and Laboratory Waste Disposal

Remind students to follow all normal lab safety rules. There is no laboratory waste associated with this activity.

Topics for the Explicit and Reflective Discussion

Concepts That Can Be Used to Justify the Evidence

To provide an adequate justification of their evidence, students must explain why they included the evidence in their arguments and make the assumptions underlying their analysis and interpretation of the data explicit. In this investigation, students can use the following concepts to help justify their evidence:

- The nature and properties of light
- The nature and behavior of rays and waves
- The different physical properties of different materials

We recommend that you review these concepts during the explicit and reflective discussion to help students make this connection.

How to Design Better Investigations

It is important for students to reflect on the strengths and weaknesses of the investigation they designed during the explicit and reflective discussion. Students should therefore be encouraged to discuss ways to eliminate potential flaws, measurement errors, or sources of bias in their investigations. To help students be more reflective about the design of their investigation, you can ask the following questions:

1. What were some of the strengths of your investigation? What made it scientific?
2. What were some of the weaknesses of your investigation? What made it less scientific?
3. If you were to do this investigation again, what would you do to address the weaknesses in your investigation? What could you do to make it more scientific?

Crosscutting Concepts

This investigation is well aligned with two crosscutting concepts found in *A Framework for K–12 Science Education*, and you should review these concepts during the explicit and reflective discussion.

- *Patterns:* Scientists look for patterns in nature and attempt to understand the underlying cause of these patterns. Scientists, for example, often collect data and then look for patterns to identify a relationship between two variables, a trend over time, or a difference between groups. In this lab students will observe patterns in the behavior of light and rely on those patterns to make predictions.
- *Systems and system models:* Defining a system under study and making a model of it are tools for developing a better understanding of natural phenomena in science. Models can take many forms. In this lab students will work with a simulation, which is a model of the natural world.

The Nature of Science and the Nature of Scientific Inquiry

This investigation is well aligned with two important concepts related to the *nature of science* (NOS) and the *nature of scientific inquiry* (NOSI), and you should review these concepts during the explicit and reflective discussion.

- *Changes in scientific knowledge over time:* A person can have confidence in the validity of scientific knowledge but must also accept that scientific knowledge may be abandoned or modified in light of new evidence or because existing evidence has been reconceptualized by scientists. There are many examples in the history of science of both evolutionary changes (i.e., the slow or gradual refinement of ideas) and revolutionary changes (i.e., the rapid abandonment of a well-established idea) in scientific knowledge.
- *Science as a culture*: Scientists share a set of values, norms, and commitments that shape what counts as knowing, how to represent or communicate information, and how to interact with other scientists. The culture of science affects who gets to do science, what scientists choose to investigate, how investigations are conducted, how research findings are interpreted, and what people see as implications. People also view some research as being more important than others because of cultural values and current events.

Hints for Implementing the Lab

- Learn how to use the online simulation before the lab begins. It is important for you to know how to use the simulation so you can help students when they get stuck or confused.
- Lab Handout Figures L20.2 (p. 358) and L20.4 (p. 360) are most useful in color; full-color versions of these figures can be found at *www.nsta.org/adi-physicalscience.*
- A group of three students per computer or tablet tends to work well.
- Allowing students to design their own procedures for collecting data gives students an opportunity to try, to fail, and to learn from their mistakes. However, you can scaffold students as they develop their procedure by having them fill out an investigation proposal. These proposals provide a way for you to offer students hints and suggestions without telling them how to do it. You can also check the proposals quickly during a class period. For this lab we suggest using Investigation Proposal C.
- Allow the students to play with the simulation as part of the tool talk before they begin to design their investigation. This gives students a chance to see what they can and cannot do with simulation.
- Be sure that students record actual values (e.g., angle of incidence, angle of reflection, angle or refraction, light intensity, index of refraction) when they use the simulation, rather than just attempting to describe what they see on the computer screen (e.g., the angle was bigger).
- Encourage students to focus on changing only one factor at time (i.e., angle of incidence, material 1, material 2, or index of refraction) so it is easier to determine a relationship.
- This is a good lab for students to make mistakes during the data collection stage. Students will quickly figure out what they did wrong during the argumentation session, and it will only take them a short period of time to re-collect data. It will also create an opportunity for students to reflect on and identify ways to improve the way they design investigations (especially how they attempt to control variables as part of an experiment) during the explicit and reflective discussion.

Topic Connections

Table 20.2 (p. 356) provides an overview of the scientific practices, crosscutting concepts, disciplinary core ideas, and supporting ideas at the heart of this lab investigation. In addition, it lists the NOS and NOSI concepts for the explicit and reflective discussion. Finally, it lists literacy and mathematics skills (*CCSS ELA* and *CCSS Mathematics*) that are addressed during the investigation.

TABLE 20.2

Lab 20 alignment with standards

Scientific practices	• Asking questions and defining problems • Developing and using models • Planning and carrying out investigations • Analyzing and interpreting data • Using mathematics and computational thinking • Constructing explanations and designing solutions • Engaging in argument from evidence • Obtaining, evaluating, and communicating information
Crosscutting concepts	• Patterns • Systems and system models
Core ideas	• PS1.A: Structure and properties of matter • PS4.A: Wave properties • PS4.B: Electromagnetic radiation
Supporting ideas	• Nature and behavior of light • Reflection and refraction • The different physical properties of different materials
NOS and NOSI concepts	• Changes in scientific knowledge over time • Culture of science
Literacy connections (*CCSS ELA*)	• *Reading*: Key ideas and details, craft and structure, integration of knowledge and ideas • *Writing*: Text types and purposes, production and distribution of writing, research to build and present knowledge, range of writing • *Speaking and listening*: Comprehension and collaboration, presentation of knowledge and ideas
Mathematics connections (*CCSS Mathematics*)	• Reason abstractly and quantitatively • Construct viable arguments and critique the reasoning of others • Use appropriate tools strategically • Attend to precision

Lab Handout

Lab 20. Reflection and Refraction

How Can You Predict Where a Ray of Light Will Go When It Comes in Contact With Different Types of Transparent Materials?

Introduction

Our understanding of the nature of light and how it behaves has changed a great deal over the centuries. The first real explanations for the nature and behavior of light came from the ancient Greeks. Most of these early models describe the nature of light as a ray. A ray moves in a straight line from one point to another. Euclid and Ptolemy, for example, used ray diagrams to show how light bounces off a smooth surface or bends as it passes from one transparent medium to another. Other scholars took these ideas and refined them to explain the behavior of light when it strikes a mirror, a lens, or a prism. This field of study is now called geometrical optics. The most famous practitioner of geometrical optics was the 10th–11th century Arab scientist Ibn al-Haytham, who developed mathematical equations that describe how light bends as it travels through different media.

Scientists began to use different models to explain the nature of light in the 17th century. For example, Christiaan Huygens claimed that light is a wave that moves through an "invisible ether" that exists all around us. Isaac Newton, in contrast, claimed that light is composed of small particles, because it travels in a straight line and bounces off a mirror, much like a ball bounces off a wall. Most scientists continued to use a model that treated light as particle in their research until the early part of the 19th century. In 1801, however, Thomas Young showed that if light is made to travel through two slits in a card, it produces a series of light and dark bands on a screen. He argued that this observation would not be possible if light was composed of particles that travel in a straight line (see Figure L20.1[a]) but it would be possible if it traveled through space and time as a wave (see Figure L20.1[b]).

FIGURE L20.1

Appearance of light on screen if light is composed of particles (a) or waves (b)

(a)

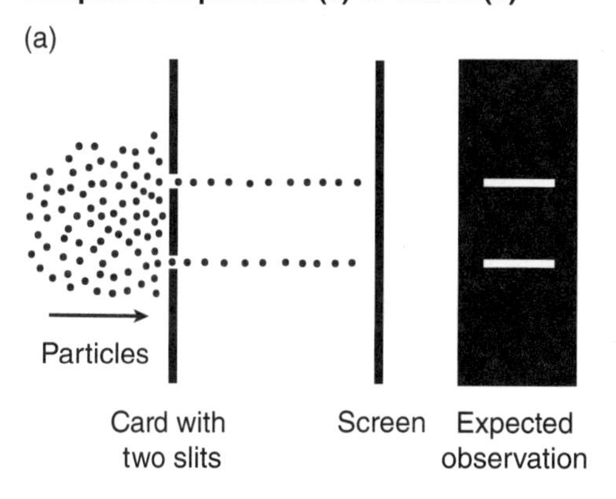

(b)

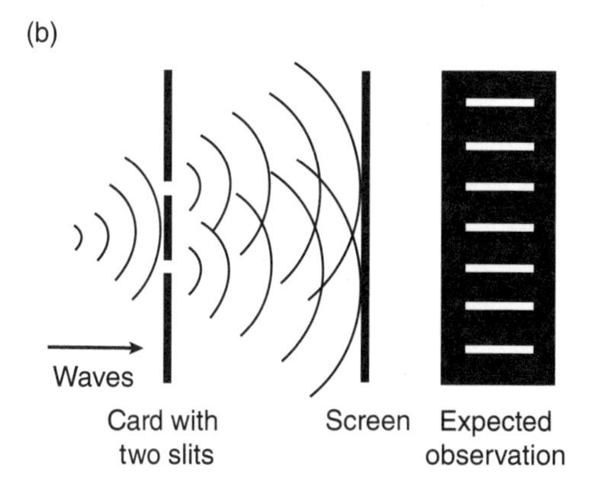

Then in the 1860s, James Maxwell created a new model that described the nature of light as electromagnetic radiation. Electromagnetic radiation does not need a medium to travel through like sound waves do, and when it is traveling in a vacuum (such as space), it moves at a speed of about 300,000 kilometers per second. According to this model, light waves come in many different sizes and these waves can be described in terms of wavelength and frequency (see Figure L20.2). The wavelengths of light that we can see are between 400 and 700 nanometers long, but all the different wavelengths in the electromagnetic spectrum range from 0.1 nanometer (gamma rays) to several meters (radio waves) in length. The frequency of a light wave is the number of waves that pass a point in space in a specific time interval. We measure frequency in hertz (cycles per second), abbreviated Hz. Red light has a frequency of 430 trillion Hz, and violet light has a frequency of 750 trillion Hz.

FIGURE L20.2

Wavelengths and frequencies of the different types of waves in the electromagnetic spectrum

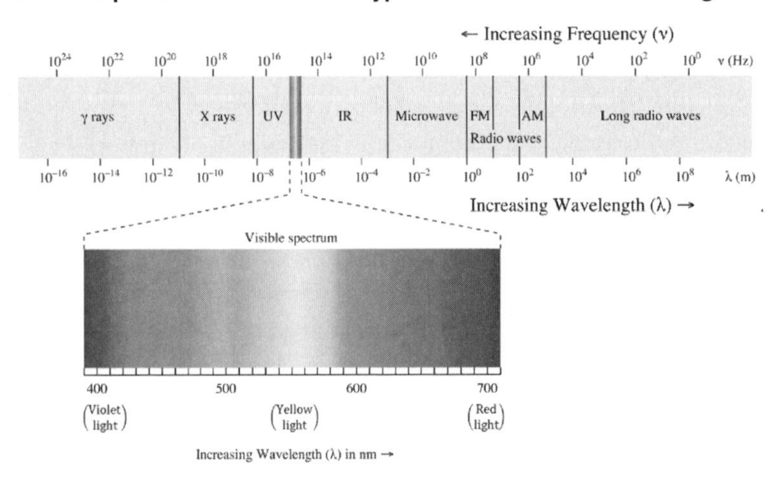

As it turns out, all of these models for the nature of light are both right and wrong at the same time, because they can only be used to explain or predict certain behaviors of light. Scientists now use a model that describes the nature of light as being a particle and a wave. In this investigation, however, you will use a ray model of light to investigate how light behaves when it comes in contact with different types of transparent materials. When a ray of light passes between two transparent materials (such as air, water, plastic, or glass), part of the ray is reflected and stays in the first material, while the rest of the ray is refracted as it passes into the second material. The ray of light refracts when it enters the second material because it changes speed (slows down or speeds up) as it begins to travel through the new materials.

Figure L20.3 shows a ray of light crossing the boundary between two transparent materials. In the field of optics, a line perpendicular to the boundary is used to measure the angles of the light rays. This line is called the surface normal. The angle the incoming ray makes with the surface normal is called the angle of incidence (θ_i). The angle the reflected ray makes with the normal is called the angle of reflection (θ_r), and the angle the refracted ray makes with the normal is called the angle of refraction (θ_R). Your goal in this investigation is to develop one or more rules that you can use to predict the behavior and path of the reflected and refracted rays, much like Ibn al-Haytham did when he created mathematical equations to describe the behavior of light when it strikes a mirror, a lens, or a prism.

The Task

Use what you know about light, uncovering patterns in nature, and the use of models in science to design and carry out an investigation using a simulation to determine how light behaves when it travels through one transparent material and then enters into a different one.

The guiding question of this investigation is, **How can you predict where a ray of light will go when it comes in contact with different types of transparent materials?**

FIGURE L20.3

A ray of light crossing the boundary between two transparent materials (air and plastic)

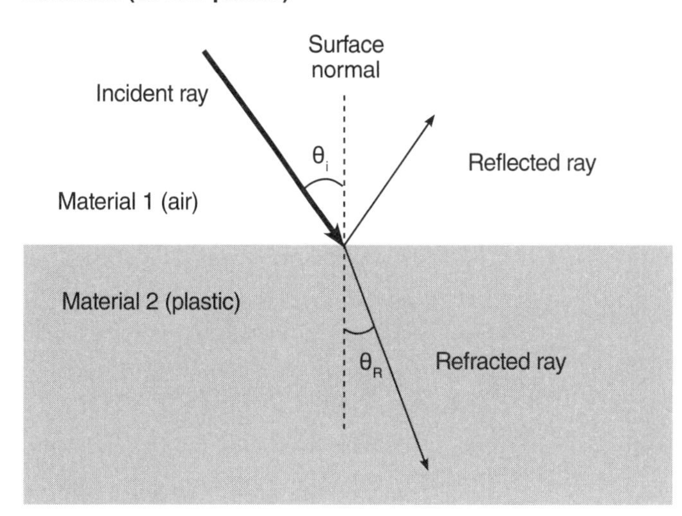

Materials

You will use an online simulation called *Bending Light* to conduct your investigation. You can access the simulation by going to the following website: *https://phet.colorado.edu/en/simulation/bending-light.*

Safety Precautions

Follow all normal lab safety rules.

Investigation Proposal Required? ☐ Yes ☐ No

Getting Started

The *Bending Light* simulation (see Figure L20.4, p. 360) enables you to change the angle of incidence of a light ray that crosses the boundary between two transparent materials and then measure the angle of reflection and refraction. You can also adjust the properties of the two materials and measure the light intensity of each light ray. To use this simulation, start by clicking on the "Intro" button. You will then see a laser pointer and a horizontal line that represents the boundary between two different materials. Click on the red button on the laser pointer to turn it on. This will allow you see a light ray and what happens to it as it crosses the boundary between the two transparent materials. You can change the angle of incidence of the light ray by clicking and dragging on the left end of the laser pointer. To measure the angle of incidence, the angle of reflection, and the angle of refraction, simply drag the protractor in the lower-left corner and drop it on the surface normal (which is represented by the dashed line). You can change the properties of the two transparent materials using the gray boxes on the right side of the screen. Finally, you can measure the light intensity of any ray by dragging and dropping the green light intensity meter where you need it. The green light intensity meter is located in the lower-left corner of the simulation.

FIGURE L20.4

A screen shot of the *Bending Light* simulation

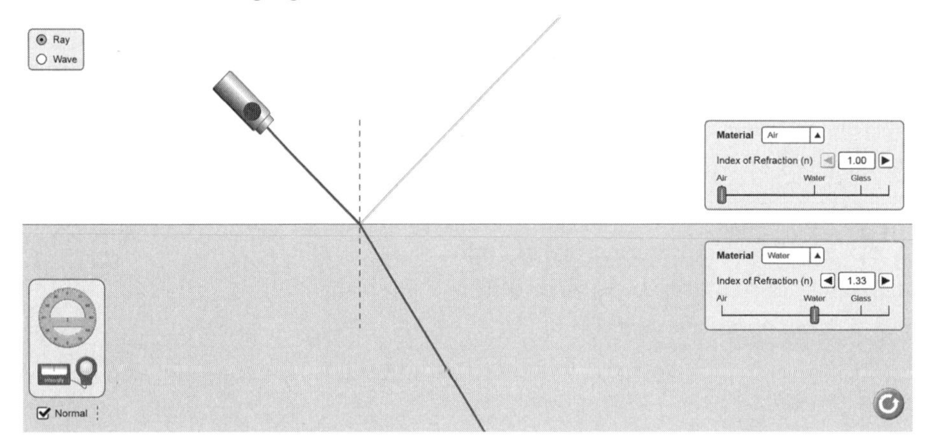

To answer the guiding question, you must determine what type of data you need to collect, how you will collect the data, and how you will analyze it. To determine *what type of data you need to collect,* think about the following questions:

- Which factors will you need to account for to be able to make accurate predictions?
- What type of measurements will you need to record?

To determine *how you will collect the data using the simulation,* think about the following questions:

- What will serve as your dependent variable or variables?
- What will serve as your independent variable or variables?
- How will you vary the independent variable?
- What will you do to hold the other variables constant during each experiment?
- What types of comparisons will you need to make using the simulation?
- How many comparisons will you need to make to determine a trend or a relationship?
- How will you keep track of the data you collect and how will you organize it?

To determine *how you will analyze the data,* think about the following questions:

- What type of calculations will you need to make?
- What type of graph could you create to help make sense of your data?

Once you have collected the data you need, your group will need to use your findings to develop an answer to the guiding question for this investigation. Your answer to the guiding question must explain how to predict the path of the ray as it crosses the boundary between two transparent materials. For your claim to be sufficient, your answer will need to include both the angle of reflection and the angle of refraction. You can then transform the data you collected using the simulation to support the validity of your overall explanation.

Connections to Crosscutting Concepts, the Nature of Science, and the Nature of Scientific Inquiry

As you work through your investigation, be sure to think about

- the importance of looking for and understanding patterns in data,
- the importance of using models to study natural phenomena in science,
- how scientific knowledge can change over time, and
- the culture of science and how it influences the work of scientists.

Initial Argument

Once your group has finished collecting and analyzing your data, your group will need to develop an initial argument. Your initial argument needs to include a *claim*, *evidence* to support your claim, and a *justification* of the evidence. The claim is your group's answer to the guiding question. The evidence is an analysis and interpretation of your data. Finally, the justification of the evidence is why your group thinks the evidence matters. The justification of the evidence is important because scientists can use different kinds of evidence to support their claims. Your group will create your initial argument on a whiteboard. Your whiteboard should include all the information shown in Figure L20.5.

FIGURE L20.5

Argument presentation on a whiteboard

The Guiding Question:	
Our Claim:	
Our Evidence:	Our Justification of the Evidence:

Argumentation Session

The argumentation session allows all of the groups to share their arguments. One member of each group will stay at the lab station to share that group's argument, while the other members of the group go to the other lab stations to listen to and critique the arguments developed by their classmates. This is similar to how scientists present their arguments to other scientists at conferences. If you are responsible for critiquing your classmates' arguments, your goal is to look for mistakes so these mistakes can be fixed and they can make their argument better. The argumentation session is also a good time to think about ways you can make your initial argument better. Scientists must share and critique arguments like this to develop new ideas.

To critique an argument, you might need more information than what is included on the whiteboard. You will therefore need to ask the presenter lots of questions. Here are some good questions to ask:

- How did you collect your data? Why did you use that method? Why did you collect those data?
- What did you do to make sure the data you collected are reliable? What did you do to decrease measurement error?
- How did your group analyze the data? Why did you decide to do it that way? Did you check your calculations?
- Is that the only way to interpret the results of your analysis? How do you know that your interpretation of your analysis is appropriate?
- Why did your group decide to present your evidence in that way?
- What other claims did your group discuss before you decided on that one? Why did your group abandon those alternative ideas?
- How confident are you that your claim is valid? What could you do to increase your confidence?

Once the argumentation session is complete, you will have a chance to meet with your group and revise your initial argument. Your group might need to gather more data or design a way to test one or more alternative claims as part of this process. Remember, your goal at this stage of the investigation is to develop the most acceptable and valid answer to the research question!

Report

Once you have completed your research, you will need to prepare an *investigation report* that consists of three sections. Each section should provide an answer to the following questions:

1. What question were you trying to answer and why?
2. What did you do to answer your question and why?
3. What is your argument?

Your report should answer these questions in two pages or less. This report must be typed, and any diagrams, figures, or tables should be embedded into the document. Be sure to write in a persuasive style; you are trying to convince others that your claim is acceptable and valid!

Checkout Questions

Lab 20. Reflection and Refraction

How Can You Predict Where a Ray of Light Will Go When It Comes in Contact With Different Types of Transparent Materials?

1. A student is conducting an investigation in which she wants to shine a laser pointer into a tank of water so that the beam of light hits the center of a target on the bottom of the tank. Which position for the laser pointer gives her the best chance of hitting the center of the target: position A, position B, or position C?

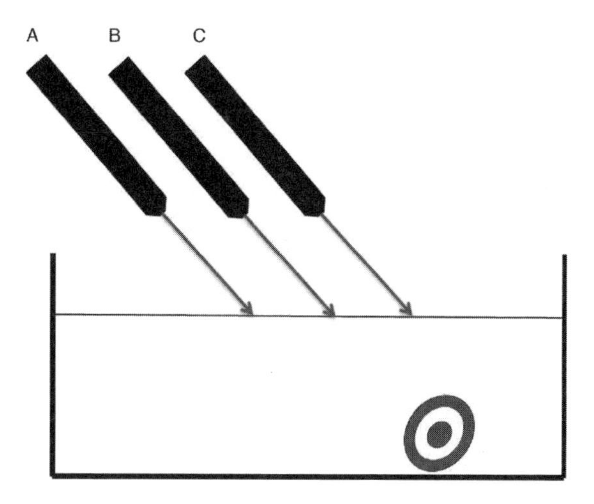

Explain why you chose that position as the best choice for the beam of light to hit the center of the target.

2. An engineer is designing a piece of equipment that will help change the path of a beam of light. The new piece of equipment needs to change the path of the light as much as possible. The table below has all of the materials, and the index of refraction for each one, that the engineer can choose from to help change the path of the light. Which material is the best choice?

Material	Index of refraction
Crown glass	1.52
Ice	1.31
Pyrex glass	1.47
Clear plastic	1.60
Liquid water	1.33

Explain why you chose that material as the best choice for the piece of equipment.

3. Once scientists learn about something new, their ideas do not change.

   a. I agree with this statement.
   b. I disagree with this statement.

   Explain your answer, using an example from your investigation about reflection and refraction.

4. Scientists are objective, so they are not influenced by the culture of society.

    a. I agree with this statement.
    b. I disagree with this statement.

    Explain your answer, using an example from your investigation about reflection and refraction.

5. Scientists study patterns in nature and patterns within the data that they collect. Explain why it is important for scientists to understand patterns, using an example from your investigation about reflection and refraction.

6. Models are useful tools that help scientists better understand what they are studying. These models can be conceptual, mathematical (such as equations or relationships), or physical (such as a drawing). Using an example from your investigation about reflection and refraction, explain why it is important for scientists to develop and use models in their work.

Application Labs

LAB 21

Teacher Notes

Lab 21. Light and Information Transfer

How Does the Type of Material Affect the Amount of Light That Is Lost When Light Waves Travel Down a Tube?

Purpose

The purpose of this lab is for students to *apply* what they know about light waves to their use in information transfer. This lab gives students an opportunity to track how energy moves through a system and the relationship between the structure of a cable and its function. Students will also learn about the culture of science and the importance of imagination and creativity in science.

The Content

Information theory is the study of information. This includes the study of what counts as information, ways to represent information, and ways to transmit information. The current definition of what counts as information is that information is either the content of a message or direct observation of something. This definition is agreed upon by most scientists, so current research focuses on how best to represent information and how best to transmit information from one place to another. These two lines of research have led to many of the technologies we take for granted, such as the internet, smartphones, and cable/satellite TV.

The question of representation of information involves how to use symbols to efficiently encode information. For example, in science and mathematics, the symbol π encodes quite a bit of information. It encodes information about its numerical value (3.1415…), but it also encodes information about the ratio of the circumference to the diameter of a circle. Thus, π is a very efficient symbolic representation, because it encodes a lot of information in only a single symbol.

There are also many symbolic representation systems. The most common symbolic representation system is an alphabet. The combination of letters from an alphabet form words, and combinations of words form sentences. These sentences encode information and allow meaning to transfer from one place to another. Alphabets are also efficient symbolic representation systems because a set of symbols (26 in the case of English) can be combined in an infinite number of ways. Computers use symbolic representations to encode information. Most computers use a binary system to encode information and programs that can read the binary code and present information on the computer screen.

The question of how best to transmit information is grounded in our need to communicate information with other people. Humans have been transmitting information for thousands of years. The Rosetta Stone, which was carved in 196 B.C. (though not discovered until 1799), is one of the earliest examples of encoding and transmitting information. Since that time, there have been many efforts made to improve communication technologies so we can transmit information more and more effectively and efficiently. Two notable and familiar inventions that changed communication were the electric telegraph in the first half of the 19th century, followed by the telephone in the second half of the century.

More recently, scientists in the field of information theory, pioneered in the 1940s, have been working on ways to improve our communication technologies so that we can transmit information more effectively and efficiently. This research led to several significant developments. First, information theorists recognized that nothing can travel faster than the speed of light. They further recognized that nothing other than light (and other electromagnetic waves) travels at the speed of light. This led information theorists to determine that light would be the quickest way to transmit a message.

The second significant development grew out of the first one. Specifically, scientists needed to transmit light signals in such a way that they (a) would not be blocked and (b) would not be interfered with by other light, such as sunlight. This led to the development of *fiber optic cables,* which are specifically designed to allow light signals to pass from one place to another without getting disturbed. Another advantage of fiber optic cables is that they can bend and still transmit the light effectively.

Fiber optic cables take advantage of a property of some materials called *total internal reflection.* To understand total internal reflection, first imagine a beam of light (from a laser pointer or flashlight) that you shine on a glass window. Some of the light will pass through the glass (called the refracted light), while some of it will be reflected (see Figure 21.1). If you shine the light at certain angles, however, none of the light will pass through the glass. Instead, all of the light will be reflected (see Figure 21.2). This is called total internal reflection.

Through a series of experiments, scientists were able to determine that for very

FIGURE 21.1

A beam of light shining on a glass window

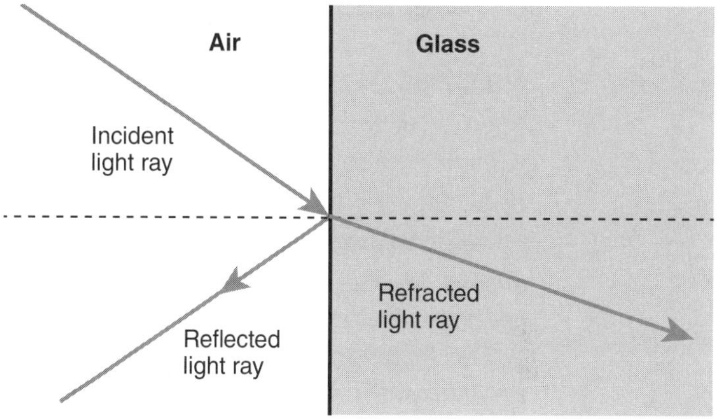

Air
Glass
Incident light ray
Refracted light ray
Reflected light ray

FIGURE 21.2

Total internal reflection of a beam of light

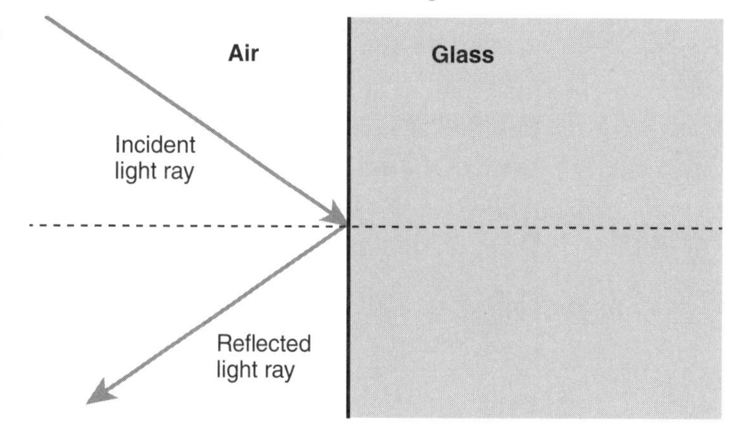

thin fibers of glass, total internal reflection would happen no matter at what angle light was shone into the end of the fiber. In Figure 21.3, light enters the end of the fiber and exhibits total internal reflection as it moves down the fiber. Only when it reaches the other end of the fiber does it exit the fiber. The fiber in this example is about the same diameter as a single human hair. When multiple fibers are bundled together, they are called fiber optic cables.

FIGURE 21.3

A single fiber optic strand with a beam of light that exhibits total internal reflection

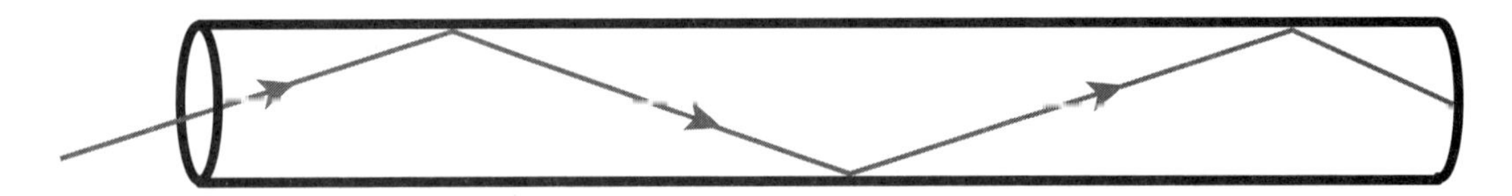

Timeline

The instructional time needed to complete this lab investigation is 200–280 minutes. Appendix 2 (p. 411) provides options for implementing this lab investigation over several class periods. Option E (280 minutes) should be used if students are unfamiliar with scientific writing, because this option provides extra instructional time for scaffolding the writing process. You can scaffold the writing process by modeling, providing examples, and providing hints as students write each section of the report. Option H (200 minutes) should be used if students are familiar with scientific writing and have developed the skills needed to write an investigation report on their own. In option H, students complete stage 6 (writing the investigation report) and stage 8 (revising the investigation report) as homework.

Materials and Preparation

The materials needed to implement this investigation are listed in Table 21.1. The equipment can be purchased from a science supply company such as Carolina, Flinn Scientific, or Ward's Science. We recommend that you use a set routine for distributing and collecting the materials during the lab investigation. For example, the equipment for each group can be set up at each group's lab station before class begins, or one member from each group can collect them from a table or a cart when needed during class.

TABLE 21.1

Materials list for Lab 21

Item	Quantity
Safety glasses or goggles	1 per student
Light source*	1 per group
Light sensor with interface	1 per group
Electrical tape (various colors)	1 roll per group
Fiber optic tubing	1 per group
Amber rubber tubing	1 per group
Red vacuum and pressure tubing	1 per group
Tygon laboratory tubing	1 per group
Protractor	1 per group
Investigation Proposal C (optional)	1 per group
Whiteboard, 2' × 3'†	1 per group
Lab Handout	1 per student
Peer-review guide	1 per student
Checkout Questions	1 per student

* Laser pointers work well, but some states do not allow use of laser pointers in schools. Check legal standards at the state level if you plan on using laser pointers.

† As an alternative, students can use computer and presentation software, such as Microsoft PowerPoint or Apple Keynote, to create their arguments.

Safety Precautions and Laboratory Waste Disposal

Remind students to follow all normal lab safety rules. In addition, tell students to take the following safety precautions:

1. Wear sanitized safety glasses or goggles during lab setup, hands-on activity, and takedown.
2. Use caution when working with the light source, because it can get hot and burn skin.
3. Use only GFCI-protected electrical receptacles for the lamp power source.
4. Do not shine a laser pointer at anyone's eyes or face.

5. Lightbulbs are made of glass. Be careful handling them. If they break, clean them up immediately and place in a broken glass box.
6. Wash hands with soap and water after completing the lab activity.
7. If students cut any of the tubing because they are controlling for length, the tubing can be disposed of in any trash receptacle.

Topics for the Explicit and Reflective Discussion

Concepts That Can Be Used to Justify the Evidence

To provide an adequate justification of their evidence, students must explain why they included the evidence in their arguments and make the assumptions underlying their analysis and interpretation of the data explicit. In this investigation, students can use the following concepts to help justify their evidence:

- Information transfer
- Information loss
- The speed of light
- Reflection, refraction, and absorption of light

We recommend that you review these concepts during the explicit and reflective discussion to help students make this connection.

How to Design Better Investigations

It is important for students to reflect on the strengths and weaknesses of the investigation they designed during the explicit and reflective discussion. Students should therefore be encouraged to discuss ways to eliminate potential flaws, measurement errors, or sources of bias in their investigations. To help students be more reflective about the design of their investigation, you can ask the following questions:

1. What were some of the strengths of your investigation? What made it scientific?
2. What were some of the weaknesses of your investigation? What made it less scientific?
3. If you were to do this investigation again, what would you do to address the weaknesses in your investigation? What could you do to make it more scientific?

Crosscutting Concepts

This investigation is well aligned with two crosscutting concepts found in *A Framework for K–12 Science Education,* and you should review these concepts during the explicit and reflective discussion.

- *Energy and matter: Flows, cycles, and conservation:* In science it is important to track how energy and matter move into, out of, and within systems. In this lab students will investigate how information is transferred within a system in the form of energy transmitted by electromagnetic waves.
- *Structure and function:* The way an object is shaped or structured determines many of its properties and functions. In this lab students will investigate how the properties of a substance influence its ability to transmit information in the form of electromagnetic waves.

The Nature of Science and the Nature of Scientific Inquiry

This investigation is well aligned with two important concepts related to the *nature of science* (NOS) and the *nature of scientific inquiry* (NOSI), and you should review these concepts during the explicit and reflective discussion.

- *Science as a culture*: Scientists share a set of values, norms, and commitments that shape what counts as knowing, how to represent or communicate information, and how to interact with other scientists. The culture of science affects who gets to do science, what scientists choose to investigate, how investigations are conducted, how research findings are interpreted, and what people see as implications. People also view some research as being more important than others because of cultural values and current events.
- *The importance of imagination and creativity in science:* Students should learn that developing explanations for or models of natural phenomena and then figuring out how they can be put to the test of reality is as creative as writing poetry, composing music, or designing video games. Scientists must also use their imagination and creativity to figure out new ways to test ideas and collect or analyze data.

Hints for Implementing the Lab

- Learn how to use the light sensor and interface before the lab begins. It is important for you to know how to use the equipment so you can help students when technical issues arise.
- Allowing students to design their own procedures for collecting data gives students an opportunity to try, to fail, and to learn from their mistakes. However, you can scaffold students as they develop their procedure by having them fill out an investigation proposal. These proposals provide a way for you to offer students hints and suggestions without telling them how to do it. You can also check the proposals quickly during a class period. For this lab we suggest using Investigation Proposal C.

- Allow the students to become familiar with the light sensor as part of the tool talk before they begin to design their investigation. This gives students a chance to see what they can and cannot do with the equipment.
- For the light source, laser pointers will generally work quite well, but some states do not allow use of laser pointers in schools. Check legal standards at state level if you plan on using laser pointers. If you are not allowed to use laser pointers or do not have access to them, another option is to use a lightbulb on a lamp without a lampshade.
- Ambient light may affect the results of the experiment, especially if you are not using a laser pointer. This will provide students with a good opportunity to think about how to isolate a system under study from the surrounding environment.
- The light sensor measures *illuminance*, which is the measure of visible light per unit area. This means that illuminance is calibrated to what the human eye can perceive. Thus, illuminance is a measure of how much light radiation hits a given area as opposed to how much total electromagnetic radiation hits the area. The unit of illuminance is lux (abbreviated as lx).
- The length of the tubes may affect the light loss during this lab. You can allow students to figure this out on their own and control for this by cutting tubes to equal lengths, or you can give students tubing of equal length to begin with.
- When purchasing tubing, we recommend using tubing that is of the same diameter as the fiber optic cable.

Topic Connections

Table 21.2 provides an overview of the scientific practices, crosscutting concepts, disciplinary core ideas, and supporting ideas at the heart of this lab investigation. In addition, it lists NOS and NOSI concepts for the explicit and reflective discussion. Finally, it lists literacy and mathematics skills (*CCSS ELA* and *CCSS Mathematics*) that are addressed during the investigation.

TABLE 21.2

Lab 21 alignment with standards

Scientific practices	• Asking questions and defining problems • Planning and carrying out investigations • Analyzing and interpreting data • Using mathematics and computational thinking • Constructing explanations and designing solutions • Engaging in argument from evidence • Obtaining, evaluating, and communicating information
Crosscutting concepts	• Energy and matter: Flows, cycles, and conservation • Structure and function
Core idea	• PS4.C. Information technologies and instrumentation
Supporting ideas	• Information transfer • Information loss • The speed of light • Reflection, refraction, and absorption of light
NOS and NOSI concepts	• Culture of science • Imagination and creativity in science
Literacy connections (*CCSS ELA*)	• *Reading*: Key ideas and details, craft and structure, integration of knowledge and ideas • *Writing*: Text types and purposes, production and distribution of writing, research to build and present knowledge, range of writing • *Speaking and listening*: Comprehension and collaboration, presentation of knowledge and ideas
Mathematics connections (*CCSS Mathematics*)	• Reason abstractly and quantitatively • Construct viable arguments and critique the reasoning of others • Model with mathematics • Attend to precision

Lab Handout

Lab 21. Light and Information Transfer

How Does the Type of Material Affect the Amount of Light That Is Lost When Light Waves Travel Down a Tube?

Introduction

Starting in the late 1940s, scientists and mathematicians began conducting experiments that led to a new field of study that today we call information theory. Scientists and mathematicians who conduct research in the field of information theory focus on answering a few important questions. The first of these questions is, how can we transfer information from one place to another? By information transfer, scientists and mathematicians mean how information is shared between people and things. For example, you might have watched a sporting event over the weekend and know who won the game, while your friend was unable to watch and does not know who won. If you tell your friend who won, that is transferring information from you to your friend. Although not a new question (humans have been transferring information for thousands of years), the formal study of this question is quite new. Telephones, fax machines, and even the internet grew out of this type of research. Furthermore, many of the cables that you connect to your TV or computer serve the purpose of transferring information from someplace else to your TV or computer.

A second question that scientist and mathematicians who study information theory ask is, what are the advantages and disadvantages to transferring information in different ways? For example, the oldest ways to transfer information from one person to another person is by talking to that person. The advantages of this type of information transfer are that (1) it happens very quickly and (2) you know who sent the message because you can see him or her in front of you. The disadvantage is that the message does not last very long. Another way to transfer information is by writing a letter. The advantage of the letter is that it lasts a long time. The disadvantage of the letter is that it also takes a long time to mail a letter to a friend.

Related to the question about the advantages and disadvantages of information transfer is the question, how can we get a message to another person in the least amount of time possible? Scientists who study physics (another branch of science) have determined that light moves faster than anything else in the universe. Information scientists used this finding to answer their question about transmitting a message as fast as possible. If light is the fastest thing in the universe, then maybe light can be used to transfer information.

Another question that information scientists ask about information transfer is, how can we limit the loss of information when transferring it from one place to another? Information scientists have determined that all messages lose some information between being sent and being received. Sometimes this is not a problem; for example, if you write a letter to a friend, the letter will not transfer information about whether you wrote it while sitting inside or wrote it while sitting outside (unless you say so in the letter). Other times, however, loss of information is a problem. If you have ever heard static while you talked to another person on the phone, this is an example of information loss.

Your Task

Use what you know about light, tracking energy and matter, and the relationship between structure and function to design and carry out an investigation that will allow you to determine how much light is lost when it shines down different types of tubing. This will allow you to make a recommendation about what type of materials we should use for transferring information with light. It is also important to recognize that many of the cables we use to transfer information are able to bend, so that we can get the cables to go in whatever direction we want. To complete this task, you will need to test how much light makes it from one end of a tube to the other end, when the tube has a 45° bend in the middle.

The guiding question of this investigation is, **How does the type of material affect the amount of light that is lost when light waves travel down a tube?**

Materials

You may use any of the following materials during your investigation:

- Light source
- Light sensor with interface
- Electrical tape
- Fiber optic tubing
- Amber rubber tubing
- Red vacuum and pressure tubing
- Tygon laboratory tubing
- Protractor
- Safety glasses or goggles

Safety Precautions

Follow all normal lab safety rules. In addition, take the following safety precautions:

1. Wear sanitized safety glasses or goggles during lab setup, hands-on activity, and takedown.
2. Use caution when working with the light source, because it can get hot and burn skin.
3. Use only GFCI-protected electrical receptacles for the lamp power source.
4. Do not shine a laser pointer at anyone's eyes or face.
5. Lightbulbs are made of glass. Be careful handling them. If they break, clean them up immediately and place in a broken glass box.
6. Wash hands with soap and water after completing the lab activity.

Investigation Proposal Required? ☐ Yes ☐ No

Getting Started

To answer the guiding question, you will need to design and conduct an investigation to measure the amount of light that is lost when it shines down a tube. To accomplish this task, you must determine what type of data you need to collect, how you will collect it, and how you will analyze it before you begin.

To determine *what type of data you need to collect,* think about the following questions:

- How will you determine how much light enters the tube?
- How will you determine how much light exits the tube?

To determine *how you will collect your data,* think about the following questions:

- What equipment will you need to collect the data you need?
- How will you make sure that your data are of high quality (i.e., how will you reduce error)?
- Are there different ways you can measure the amount of light transferred?
- How will you keep track of the data you collect?
- How will you organize your data?

To determine *how you will analyze your data,* think about the following questions:

- How will you determine the amount of light lost?
- What type of table or graph could you create to help make sense of your data?

Connections to Crosscutting Concepts, the Nature of Science, and the Nature of Scientific Inquiry

As you work through your investigation, be sure to think about

- the importance of tracking how energy and matter move into, out of, and within systems;
- the relationship between structure and function in nature;
- science as a culture and how it influences the work of scientists; and
- the importance of imagination and creativity in science.

Initial Argument

Once your group has finished collecting and analyzing your data, your group will need to develop an initial argument. Your initial argument needs to include a *claim, evidence* to support your claim, and a *justification* of the evidence. The claim is your group's answer to the guiding question. The evidence is an analysis and interpretation of your data. Finally, the justification of the evidence is why your group thinks the evidence matters. The justification of the evidence is important because scientists can use different kinds of evidence to support their claims. Your group will create your initial argument on a whiteboard. Your whiteboard should include all the information shown in Figure L21.1.

FIGURE L21.1

Argument presentation on a whiteboard

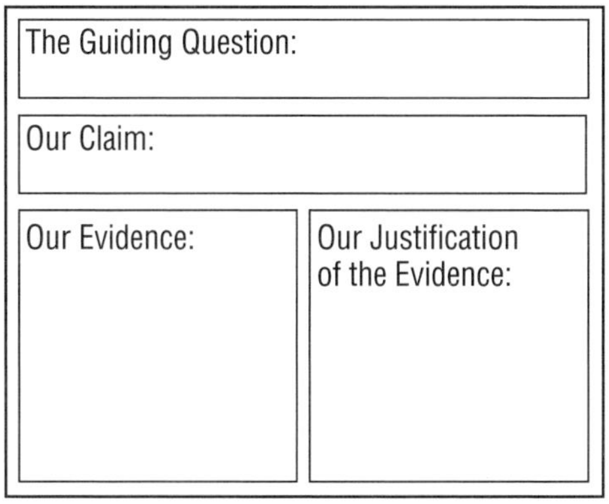

<table>
<tr><td colspan="2">The Guiding Question:</td></tr>
<tr><td colspan="2">Our Claim:</td></tr>
<tr><td>Our Evidence:</td><td>Our Justification of the Evidence:</td></tr>
</table>

Argumentation Session

The argumentation session allows all of the groups to share their arguments. One member of each group will stay at the lab station to share that group's argument, while the other members of the group go to the other lab stations to listen to and critique the arguments developed by their classmates. This is similar to how scientists present their arguments to other scientists at conferences. If you are responsible for critiquing your classmates' arguments, your goal is to look for mistakes so these mistakes can be fixed and they can make their argument better. The argumentation session is also a good time to think about ways you can make your initial argument better. Scientists must share and critique arguments like this to develop new ideas.

To critique an argument, you might need more information than what is included on the whiteboard. You will therefore need to ask the presenter lots of questions. Here are some good questions to ask:

- How did you collect your data? Why did you use that method? Why did you collect those data?
- What did you do to make sure the data you collected are reliable? What did you do to decrease measurement error?
- How did your group analyze the data? Why did you decide to do it that way? Did you check your calculations?
- Is that the only way to interpret the results of your analysis? How do you know that your interpretation of your analysis is appropriate?
- Why did your group decide to present your evidence in that way?
- What other claims did your group discuss before you decided on that one? Why did your group abandon those alternative ideas?
- How confident are you that your claim is valid? What could you do to increase your confidence?

Once the argumentation session is complete, you will have a chance to meet with your group and revise your initial argument. Your group might need to gather more data or design a way to test one or more alternative claims as part of this process. Remember, your goal at this stage of the investigation is to develop the most acceptable and valid answer to the research question!

Report

Once you have completed your research, you will need to prepare an *investigation report* that consists of three sections. Each section should provide an answer to the following questions:

1. What question were you trying to answer and why?
2. What did you do to answer your question and why?
3. What is your argument?

Your report should answer these questions in two pages or less. This report must be typed, and any diagrams, figures, or tables should be embedded into the document. Be sure to write in a persuasive style; you are trying to convince others that your claim is acceptable and valid!

Checkout Questions

Lab 21. Light and Information Transfer

How Does the Type of Material Affect the Amount of Light That Is Lost When Light Waves Travel Down a Tube?

1. A student shines a flashlight on a glass window from across the room. Using arrows, draw the different paths that the beam of light might take when they reach the window.

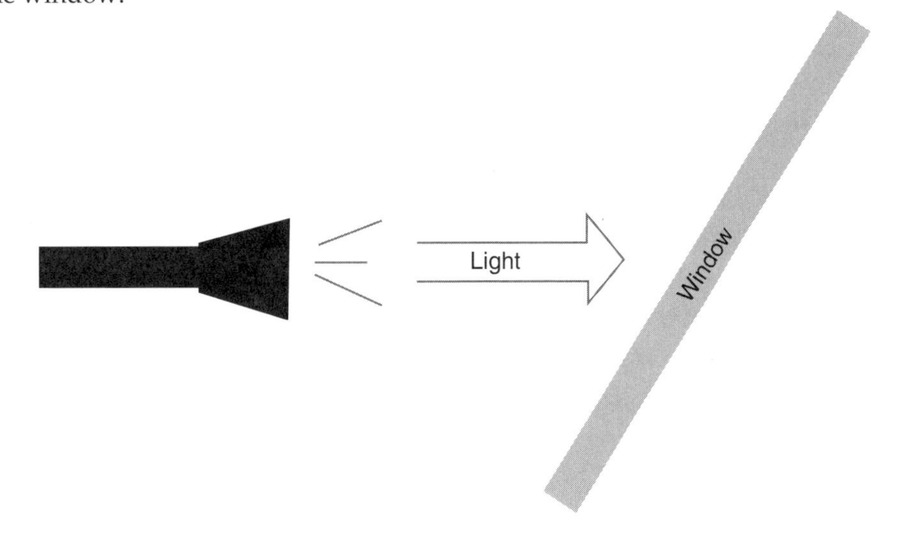

Explain why you drew the arrows in those directions.

2. Two engineers are trying to decide how to send information signals and messages between two different locations. One engineer wants to use electrical cables to send the signals and messages because that technique has been used for a very long time. The other engineer wants to use fiber optic cables and the newer technique of sending signals and messages with light. Compare and contrast the two approaches, and make a recommendation to the engineers on which approach they should take to send their signals and messages.

3. Current events influence the research that scientist do.

   a. I agree with this statement.
   b. I disagree with this statement.

   Explain your answer, using an example from your investigation on light and information transfer.

4. Imagination and creativity are traits that only artists need, not scientists.

   a. I agree with this statement.
   b. I disagree with this statement.

   Explain your answer, using an example from your investigation on light and information transfer.

5. Solving problems in science or engineering often means developing a new structure, tool, or piece of equipment. Each of these objects will have a specific function that helps solve the problem. Explain why it is important for scientists and engineers to understand the relationship between a structure and its function, using an example from your investigation about light and information transfer.

6. Two of the most important laws related to the natural world are the law of conservation of matter and the law of conservation of energy. Using an example from your investigation on light and information transfer, explain why it is important for scientists to understand and track the flow of matter and energy into, out of, and within systems.

Teacher Notes

Lab 22. Design Challenge

How Should Eyeglasses Be Shaped to Correct for Nearsightedness and Farsightedness?

Purpose

The purpose of this lab is for students to *apply* what they know about the behavior of light to a novel situation that allows them to solve a common problem related to vision. In addition, this lab gives students an opportunity to develop models of the eye and eyeglasses to explain how the structure and function of the human eye and eyeglasses system work together to effectively manipulate light to result in clear vision. Students will also learn about the role of creativity and imagination in science and engineering related to designing solutions and how the culture of science influences the problems that scientists and engineers investigate and attempt to solve.

The Content

This lab centers on the behavior of light rays when they interact with a medium such as a lens. Unimpeded light rays travel in a straight path unless they interact with something. When light rays interact with a medium, the rays can be reflected, transmitted, or absorbed. In many cases, these processes happen together. For example, if you shine a light out a window at night, it is possible for you to see objects outside, but there will also be light reflected back in your eyes. In this example, light rays are both reflected by the window and transmitted through the window.

When light rays are reflected off a surface, they behave in very predictable patterns; specifically, the angle of incidence is equal to the angle of reflection. In other words, the light will reflect off the surface at the same angle at which it hit the surface. When light rays are transmitted through a material, there are a variety of scenarios that can occur. Most often, transmitted light rays are refracted when they pass through a transparent material. When light rays are refracted, they leave the material at a different angle than they entered; to put it another way, the light rays are bent. One important property that influences how light rays are refracted is how the transparent material affects the speed of light as it travels through the material. The *index of refraction* for a transparent material is defined as the ratio of the speed of light in a vacuum to the speed of light in the material. The index of refraction for all transparent materials is greater than 1.0. The index of refraction of air is about 1.0003; the index of refraction of acrylic is 1.49.

When a ray of light passes between any two transparent materials, part of the ray is reflected and stays in the first material, while the rest of the ray is refracted and passes into the second material. Figure 22.1 shows a ray of light crossing the boundary between two transparent materials. The index of refraction for the first material is n_1 and for the second material is n_2. A line that runs perpendicular to the boundary is called the *surface normal,* and this line is used to measure the angles of the light rays. The angle the incident ray makes with the normal is called the *angle of incidence* (θ_i), the angle the reflected ray makes with the normal is called the *angle of reflection* (θ_r) and the angle the refracted ray makes with the normal is called the *angle of refraction* (θ_R).

FIGURE 22.1

A ray of light crossing the boundary between two transparent media when $n_1 < n_2$

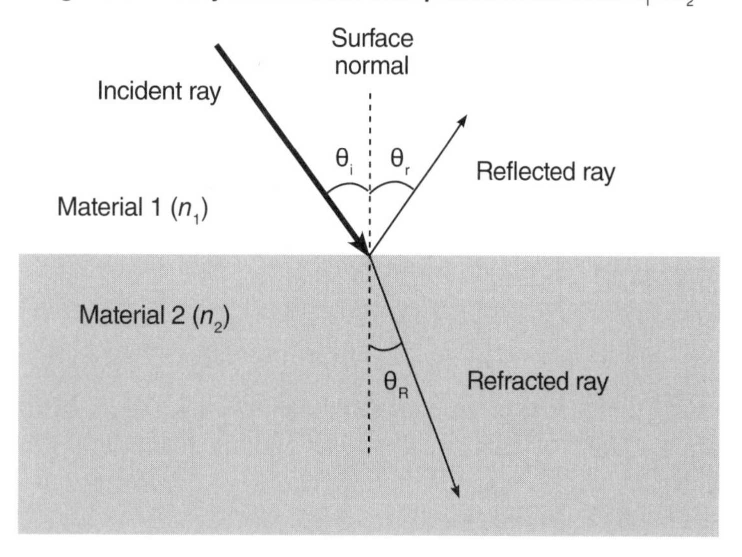

Scientists use the law of reflection and the law of refraction to describe and predict the path a ray of light will take when it reaches the boundary between two transparent materials. The *law of reflection* states that the angle of incidence (θ_i) equals the angle of reflection (θ_r), and the reflected ray will be on the opposite side of the normal from the incident ray. The *law of refraction* states that the sine of the angle of incidence (θ_i) and the sine of the angle of refraction (θ_R) are in a constant ratio to each other, and the refracted ray lies on the opposite side of the normal from the incident ray. The law of refraction is also known as Snell's law. The constant in Snell's law is the ratio of the indices of refraction for two materials, n_1 and n_2. The law of refraction is usually written as

$$\frac{\sin \theta_i}{\sin \theta_R} = \frac{n_2}{n_1}$$

When $n_1 < n_2$ (see Figure 22.1), a light ray passes from a medium or material with a low index of refraction, such as air, to a medium or material with a higher index of refraction, such as water, plastic, or glass. The reflected portion of the light remains in the first

FIGURE 22.2

Refraction of light rays when they pass through a convex lens

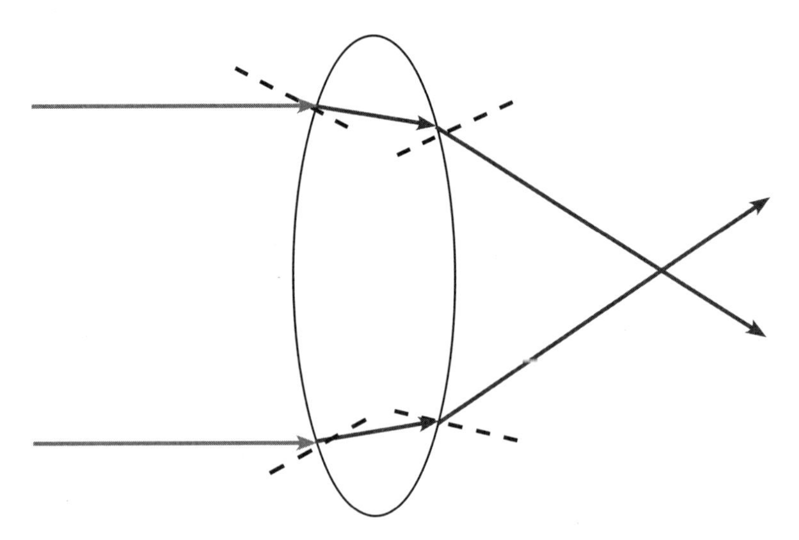

medium. This is called *external reflection.* Based on the law of reflection, the incident ray and the reflected ray both are at the same angle with respect to the normal ($\theta_i = \theta_r$). The refracted part of the light ray enters the material with the higher index of refraction at an angle that can be calculated using Snell's law. The angle of refraction is smaller than the incident angle ($\theta_R < \theta_i$) in this situation.

When $n_1 = n_2$, the angle of incidence will always equal the angle of refraction ($\theta_i = \theta_R$). This happens when light passes through a boundary between two materials with the same index of refraction or when it moves through a single material without a boundary. The path of the incident ray and the path of the refracted ray are therefore identical.

When $n_1 > n_2$, a light ray moves from a medium or material with a higher index of refraction, such as water or plastic, to a medium or material that has a lower index of refraction, such as air. The reflected part of the light will stay inside the medium or material with the higher index of refraction. This is called *internal reflection.* Given the law of reflection, the angle of incidence (θ_i) will equal the angle of reflection (θ_r). The refracted part of the incident ray will leave the medium or material with the higher index of refraction and enter the medium or material with the lower index of refraction at an angle (θ_R) that can be predicted using Snell's law.

When light rays pass through a lens, they make two transitions, first from air to glass and then from glass back to air. Each time the light ray makes one of these transitions, its path is bent according to Snell's law. Lenses, however, have curved surfaces, so even though all the light rays enter the lens parallel to each other and each is bent, the light rays do not emerge from the lens parallel to each other. When light rays pass through a convex

FIGURE 22.3

Refraction of light rays when they pass through a concave lens

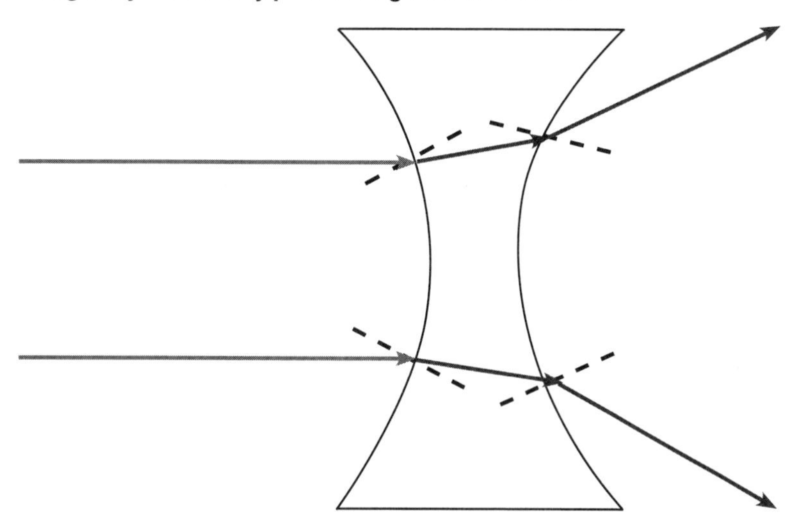

lens, the exiting light rays converge with each other. In contrast, when light rays pass through a concave lens, the exiting light rays diverge from each other.

Figure 22.2 shows how light rays get refracted twice as they pass through a convex lens. When the light rays first pass from the air into the glass lens, they are moving from a less dense medium to a more dense medium; glass has a higher index of refraction than air. When the light rays make this first transition, they are refracted toward the normal line. Then the light rays must pass from the glass lens back into the air, which is a transition into a less dense medium. When the light rays make this second transition, they are refracted away from the normal line. Because the lens is convex (i.e., curves outward), this refracting process causes the light rays to converge. If the lens had flat surfaces, the incoming rays would still be refracted but the exiting rays, like the incoming rays, would remain parallel.

Figure 22.3 shows how light rays get refracted twice as they pass through a concave lens. When the light rays first pass from the air into the glass lens, they are moving from a less dense medium to a more dense medium. Just like the convex lens above, this transition causes the light rays to be refracted toward the normal line. The second transition is when the light rays pass from the glass lens back into the air, which causes the rays to be refracted away from the normal line. Due to Snell's law and the relationship of the index of refraction of the lens material and air, the light rays are refracted in similar ways with respect to the normal line drawn from the surface of each lens. However, the concave lens curves inward, and that change in curvature compared with the convex lens results in diverging rays that exit the concave lens.

Scientists have long used combinations of convex and concave lenses to manipulate the path of light rays. For example, passing light through a concave lens prior to a convex lens can extend the focal length of the convex lens. The focal length is the distance between the lens and the point at which the light rays converge. Alternatively, coupling two convex lenses in a system can serve to shorten the focal length of one of the lenses. There are many ways that these types of lenses can be used together within a system, and their relative positions can have dramatic impacts on the path light rays take and the resulting images that are seen if looking through these systems of lenses.

Engineering Connection

This investigation engages students in a design challenge to develop a solution to the conditions of nearsighted vision and farsighted vision. Unlike a typical scientific investigation, in this lab students are not trying to explain a natural phenomenon; rather, they are determining the best solution to a problem. To determine the best solution, students will need to develop multiple solutions and test them using an iterative process of design, test, refine, and optimize. During this iterative cycle of design, students' solutions are tested and refined based on data related to how well the design helps solve the given problem. The optimization stage of the design cycle is where the final design is improved based on trading off more or less important features of the design given the constraints of the task. Design constraints may include the size of the device, cost of the materials, the ratio of cost to outcome, or acceptable margin of error for a specified outcome. We suggest that you and your students work collaboratively to determine what the constraints will be for this design challenge.

The outcome for this investigation will be a design solution that meets the constraints identified for the given problem. At the conclusion of the design cycle, student will still generate an argument that includes a claim, evidence, and justification; however, the arguments will be slightly modified with respect to a more typical argument-driven inquiry investigation. Students may make a general claim about the features of a successful design, or their claim may be the specific design developed by their group. The evidence portion of the argument will include data that have been analyzed and interpreted to support the success of their design. Finally, the justification of their evidence will include a connection to scientific ideas like a typical scientific argument, with the addition of how their design also addresses the constraints of the problem they are solving.

Timeline

The instructional time needed to complete this lab investigation is 200–230 minutes. Appendix 2 (p. 411) provides options for implementing this lab investigation over several class periods. Option C (230 minutes) should be used if students are unfamiliar with scientific writing, because this option provides extra instructional time for scaffolding the writing process. You can scaffold the writing process by modeling, providing examples, and providing hints as students write each section of the report. Option F (200 minutes) should be used if students are familiar with scientific writing and have developed the skills needed to write an investigation report on their own. In option F, students complete stage 6 (writing the investigation report) and stage 8 (revising the investigation report) as homework.

Materials and Preparation

The materials needed to implement this investigation are listed in Table 22.1. The consumables and equipment can be purchased from a science supply company such as Carolina, Flinn Scientific, or Ward's Science. We recommend that you use a set routine for distributing and collecting the materials during the lab investigation. For example, the consumables and equipment for each group can be set up at each group's lab station before class begins, or one member from each group can collect them from a table or a cart when needed during class.

TABLE 22.1

Materials list for Lab 22

Item	Quantity
Consumable	
Large paper	3 sheets per group
Equipment and other materials	
Safety glasses or goggles	1 per student
Light box with power supply	1 per group
Lens kit	1 per group
Protractor	1 per group
Investigation Proposal C (optional)	1 per group
Ruler	1 per group
Whiteboard, 2' × 3'*	1 per group
Lab Handout	1 per student
Peer-review guide	1 per student
Checkout Questions	1 per student

* As an alternative, students can use computer and presentation software, such as Microsoft PowerPoint or Apple Keynote, to create their arguments.

Safety Precautions and Laboratory Waste Disposal

Remind students to follow all normal lab safety rules. In addition, tell students to take the following safety precautions:

1. Wear sanitized safety glasses or goggles during lab setup, hands-on activity, and takedown.
2. Use caution when working with the light source. It can get hot and burn skin.
3. Do not look directly at the light coming from the light box.
4. Use only GFCI-protected electrical receptacles for the light box power supply.
5. Use caution when handling glass. It can have sharp edges, which can cut skin.
6. Lightbulbs are made of glass. Be careful handling them. If they break, clean them up immediately and place in a broken glass box.
7. Wash hands with soap and water after completing the lab activity.

There is no laboratory waste associated with this activity.

Topics for the Explicit and Reflective Discussion

Concepts That Can Be Used to Justify the Evidence

To provide an adequate justification of their evidence, students must explain why they included the evidence in their arguments and make the assumptions underlying their analysis and interpretation of the data explicit. In this investigation, students can use the following concepts to help justify their evidence:

- Refraction
- Focal length of lenses
- Nature and behavior of light rays (convergence and divergence)

We recommend that you review these concepts during the explicit and reflective discussion to help students make this connection.

How to Design Better Investigations

It is important for students to reflect on the strengths and weaknesses of the investigation they designed during the explicit and reflective discussion. Students should therefore be encouraged to discuss ways to eliminate potential flaws, measurement errors, or sources of bias in their investigations. To help students be more reflective about the design of their investigation, you can ask the following questions:

1. What were some of the strengths of your investigation? What made it scientific?
2. What were some of the weaknesses of your investigation? What made it less scientific?
3. If you were to do this investigation again, what would you do to address the weaknesses in your investigation? What could you do to make it more scientific?

4. Did you meet the goal of the design challenge?
5. Did you ensure that your solution is consistent with the design parameters?

Crosscutting Concepts

This investigation is well aligned with two crosscutting concepts found in *A Framework for K–12 Science Education,* and you should review these concepts during the explicit and reflective discussion.

- *Systems and system models:* Scientists often need to use models to understand complex phenomena. In this lab students will have the opportunity to model the human eye to better understand vision and how the eye manipulates light rays.
- *Structure and function:* The way an object is shaped or structured determines many of its properties and functions. In this lab students will investigate how the structure of the human eye can support vision or in some cases hinder it due to subtle differences in its structure.

The Nature of Science and the Nature of Scientific Inquiry

This investigation is well aligned with two important concepts related to the *nature of science* (NOS) and the *nature of scientific inquiry* (NOSI), and you should review these concepts during the explicit and reflective discussion.

- *Science as a culture:* Scientists share a set of values, norms, and commitments that shape what counts as knowing, how to represent or communicate information, and how to interact with other scientists. The culture of science affects who gets to do science, what scientists choose to investigate, how investigations are conducted, how research findings are interpreted, and what people see as implications. People also view some research as being more important than others because of cultural values and current events.
- *The importance of imagination and creativity in science:* Students should learn that developing explanations for or models of natural phenomena and then figuring out how they can be put to the test of reality is as creative as writing poetry, composing music, or designing video games. Scientists must also use their imagination and creativity to figure out new ways to test ideas and collect or analyze data.

Hints for Implementing the Lab

- Allowing students to design their own procedures for collecting data gives students an opportunity to try, to fail, and to learn from their mistakes. However, you can scaffold students as they develop their procedure by having them fill out an investigation proposal. These proposals provide a way for you to offer students hints and suggestions without telling them how to do it. You can also check the proposals quickly during a class period. For this lab we suggest using Investigation Proposal C.

- In this lab it is important for students to develop their own models for eyes that represent normal vision, nearsightedness, and farsightedness before developing a solution to correct the vision in the different models. It is important to allow your students to do this step rather than you drawing specific models for them before the investigation.
- Even though your students should be responsible for making their own models, it is important for you to be familiar with all the lenses available to the students and ensure that there are multiple combinations that will allow the students to generate a solution to the vision issues.
- Allow the students to become familiar with the light box and lens kit as part of the tool talk before they begin to design their investigation. This gives students a chance to see what they can and cannot do with the equipment. It also allows them to become familiar with concave and convex lenses if they have not used them previously. (A full-color version of Figure L22.2 [p. 395], which shows a convex lens, and Figure L22.3 [p. 396], which shows a concave lens, can be found at *www.nsta.org/adi-physicalscience*.)
- Most light boxes come with filters that create multiple beams of light from a single bulb. We recommend using a filter that creates three beams of light that shine through the lenses. It can be tempting to use more beams of light, but occasionally that can result in confounded data if not all the beams are passing through the different combinations of lenses.
- When using the light boxes, ensure that your students set them up in a manner such that the initial beams of light are indeed parallel to each other before entering any lens. If the beams of light are already converging or diverging prior to passing through a lens, your students' results may be affected.

Topic Connections

Table 22.2 provides an overview of the scientific practices, crosscutting concepts, disciplinary core ideas, and supporting ideas at the heart of this lab investigation. In addition, it lists NOS and NOSI concepts for the explicit and reflective discussion. Finally, it lists literacy skills (*CCSS ELA*) that are addressed during the investigation.

TABLE 22.2

Lab 22 alignment with standards

Scientific practices	• Asking questions and defining problems • Developing and using models • Planning and carrying out investigations • Analyzing and interpreting data • Using mathematics and computational thinking • Constructing explanations and designing solutions • Engaging in argument from evidence • Obtaining, evaluating, and communicating information
Crosscutting concepts	• Systems and system models • Structure and function
Core ideas	• PS4.A: Wave properties • PS4.B: Electromagnetic radiation • ETS1.B: Developing possible solutions
Supporting ideas	• Refraction • Focal length of lenses • Nature and behavior of light rays (convergence and divergence)
NOS and NOSI concepts	• Culture of science • Imagination and creativity in science
Literacy connections (*CCSS ELA*)	• *Reading*: Key ideas and details, craft and structure, integration of knowledge and ideas • *Writing*: Text types and purposes, production and distribution of writing, research to build and present knowledge, range of writing • *Speaking and listening*: Comprehension and collaboration, presentation of knowledge and ideas

LAB 22

Lab Handout

Lab 22. Design Challenge

How Should Eyeglasses Be Shaped to Correct for Nearsightedness and Farsightedness?

Introduction

The study of light, an area in physics known as optics, dates back to the times of the ancient Mesopotamians, Egyptians, Greeks, and Romans. It is believed that the first lenses were made as early as 750 B.C. (see the Nimrud lens at *www.britishmuseum.org/research/collection_online/collection_object_details.aspx?objectId=369215&partId=1&searchText=lens&page=1*). Early lenses were used to manipulate light and likely most often used to start fires by focusing light in a small area to generate enough heat to ignite flammable material. Over time, scientists have used their understanding of the properties of light to develop many useful instruments for their investigations and for society, including telescopes, microscopes, magnifying glasses, and eyeglasses. Each of these instruments uses at least one lens to change the path of light rays to be more beneficial to the person using the instrument.

Light rays behave in predictable ways. There are three general ways that light rays can behave when they interact with, or pass through, a lens: they can be reflected, transmitted, or absorbed. When light rays are reflected, that means they come into contact with a surface and bounce back in the direction they came from; this is called reflection. When light rays come into contact with a surface and continue on, passing through the surface, it is called transmission. The third behavior for light rays is that when they hit a surface they may not be reflected or transmitted but instead are absorbed. In many cases when light rays hit a surface, a combination of these behaviors happens. For example, some rays may get reflected while others are transmitted. Also, when light rays are reflected or transmitted, it is common for them to change direction. When light rays are transmitted through a substance but change direction on the other side, it is called refraction. Figure L22.1 shows examples of what happens when light rays are reflected, transmitted, or absorbed.

When light rays are transmitted through an object, such as a lens, the light is refracted in specific ways based on the shape of the object. Scientists have conducted many investigations to understand how light rays behave when they pass through a lens. Two major

FIGURE L22.1

Examples of how light behaves when it interacts with a medium: (a) reflection, (b) transmission, (c) transmission with refraction, and (d) absorption. The arrows represent light rays.

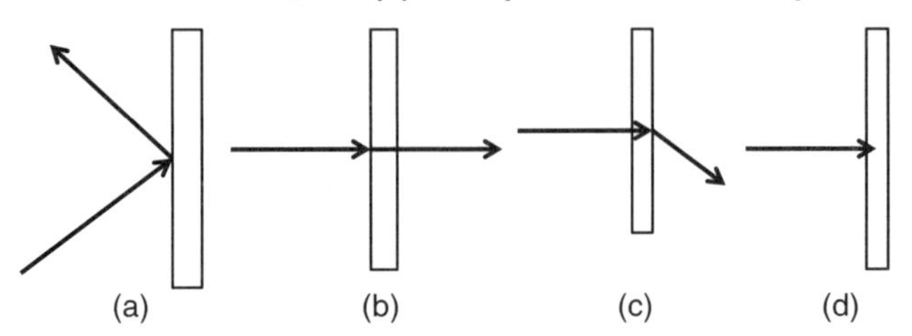

findings from these investigations are as follows: When light rays are transmitted through a convex lens, the light rays come together, or converge on the other side; when light rays are transmitted through a concave lens, the light rays spread out, or diverge, on the other side. Figure L22.2 shows how light rays behave when they pass through a convex or concave lens. Glass lenses used in instruments like the ones described earlier in this section are very common, and even our own eyes have lenses that collects light rays to help us see.

FIGURE L22.2

A convex (a) and a concave (b) lens refracting light

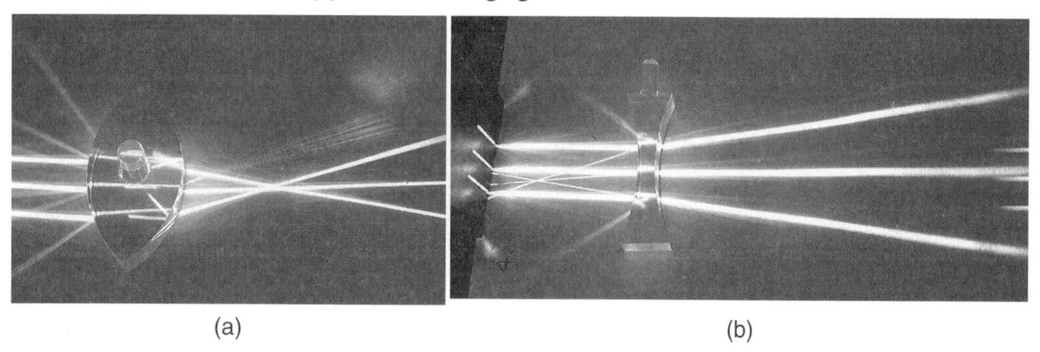

(a) (b)

The lens in a human eye focuses the incoming light rays on the retina, which is the back portion of the eyeball. However, there are times when a person's eye does not focus the light correctly, resulting in the person being nearsighted or farsighted. When a person is nearsighted, the lenses in the eyes focus the incoming light rays before they have a chance to reach the retina. When a person is farsighted, the lenses in the eyes do not focus the incoming light rays fast enough and they are still spread out when the light rays reach the retina.

Eyeglasses or contact lenses are used to correct the vision of people with nearsightedness or farsightedness. It is believed that eyeglasses were first invented in the 1200s and then gained popularity in the mid-1400s with the invention of the printing press and the rise in the number of people that had access to books and began learning to read (see "Timeline of Eyeglasses" at *www.museumofvision.org/exhibitions/?key=44&subkey=4&relkey=35*). The lenses of the eyeglasses (or modern contact lenses) work together with the lenses of the eye to change the path of the incoming light rays to ensure that they focus on the retina, resulting in clear vision. Figure L22.3 (p. 396) shows examples of eyes and incoming light rays that represent normal vision, nearsightedness, and farsightedness.

FIGURE L22.3

Eyeball models for (a) normal vision, (b) nearsightedness, and (c) farsightedness

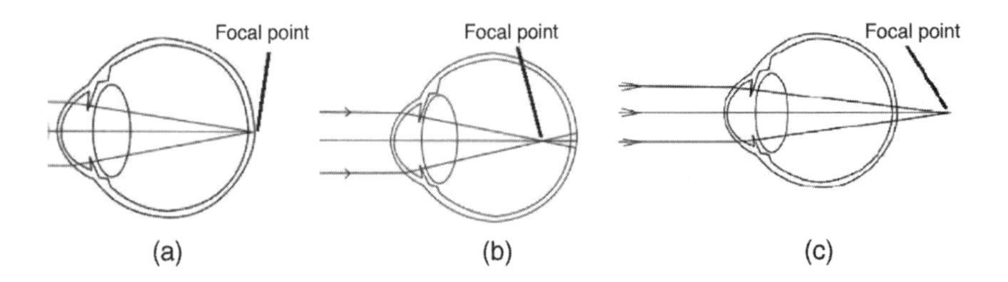

Your Task

Use what you know about the behavior of light and the relationship between the structure and function of a lens to develop a model that helps you explain how different shapes of eyeglasses will correct the vision of someone who is nearsighted and someone who is farsighted. Your model should demonstrate the two types of vision conditions as well as show how your solution corrects each of the vision conditions.

The guiding question of this investigation is, **How should eyeglasses be shaped to correct for nearsightedness and farsightedness?**

Materials

You may use any of the following materials during you investigation:

Consumable	**Equipment**
Large paper	• Light box with power supply • Lens kit • Protractor • Ruler • Safety glasses or goggles

Safety Precautions

Follow all normal lab safety rules. In addition, take the following safety precautions:

1. Wear sanitized safety glasses or goggles during lab setup, hands-on activity, and takedown.
2. Use caution when working with the light source. It can get hot and burn skin.
3. Do not look directly at the light coming from the light box.
4. Use only GFCI-protected electrical receptacles for the light box power supply.
5. Use caution when handling glass. It can have sharp edges, which can cut skin.

6. Lightbulbs are made of glass. Be careful handling them. If they break, clean them up immediately and place in a broken glass box.
7. Wash hands with soap and water after completing the lab activity.

Investigation Proposal Required? ☐ Yes ☐ No

Getting Started

The first step in developing your vision models is to use lenses from your kit to determine how to draw a model eyeball that represents normal vision. Then, draw a model eyeball that represents nearsightedness and a model eyeball that represents farsightedness. Remember that each model eyeball needs a lens that remains part of the model eyeball (this lens represents the lens portion of the human eye, and removing it would be the same as doing surgery on your model eyeball), and that lens cannot be used to represent the eyeglasses intended to correct the vision in that model. Work with each model separately; this allows you to use a lens from your kit more than once if necessary.

To determine *what type of data you need to collect,* think about the following questions:

- What information do you need to make your models?
- What measurements will you take during your investigation?
- How will you know if the vision has been corrected in your models?

To determine *how you will collect your data,* think about the following questions:

- What equipment will you need to collect the data you need?
- How will you make sure that your data are of high quality (i.e., how will you reduce error)?
- How will you keep track of the data you collect?
- How will you organize your data?

To determine *how you will analyze your data,* think about the following questions:

- How can you show that the vision or light rays in your model have changed?
- What type of diagrams or images could you create to help make sense of your data?

Connections to Crosscutting Concepts, the Nature of Science, and the Nature of Scientific Inquiry

As you work through your investigation, be sure to think about

- how scientists use models to understand complex systems;

- how the structure and function of an object are related;
- science as a culture and how it influences the work of scientists; and
- how scientists must use imagination and creativity when developing models and explanations.

Initial Argument

Once your group has finished collecting and analyzing your data, your group will need to develop an initial argument. Your initial argument needs to include a *claim*, *evidence* to support your claim, and a *justification* of the evidence. The claim is your group's answer to the guiding question. The evidence is an analysis and interpretation of your data. Finally, the justification of the evidence is why your group thinks the evidence matters. The justification of the evidence is important because scientists can use different kinds of evidence to support their claims. Your group will create your initial argument on a whiteboard. Your whiteboard should include all the information shown in Figure L22.4.

FIGURE L22.4

Argument presentation on a whiteboard

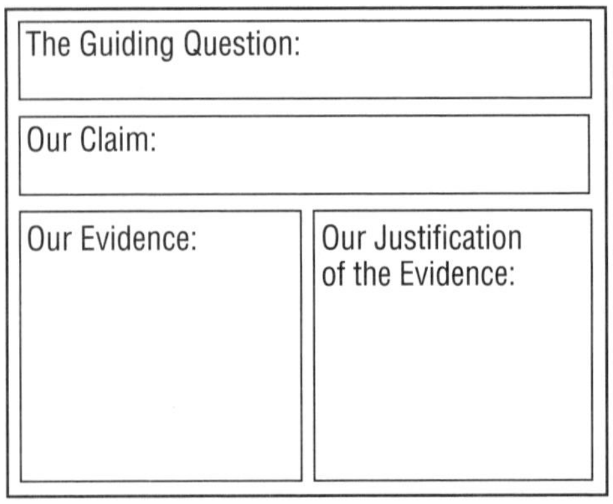

The Guiding Question:	
Our Claim:	
Our Evidence:	Our Justification of the Evidence:

Argumentation Session

The argumentation session allows all of the groups to share their arguments. One member of each group will stay at the lab station to share that group's argument, while the other members of the group go to the other lab stations to listen to and critique the arguments developed by their classmates. This is similar to how scientists present their arguments to other scientists at conferences. If you are responsible for critiquing your classmates' arguments, your goal is to look for mistakes so these mistakes can be fixed and they can make their argument better. The argumentation session is also a good time to think about ways you can make your initial argument better. Scientists must share and critique arguments like this to develop new ideas.

To critique an argument, you might need more information than what is included on the whiteboard. You will therefore need to ask the presenter lots of questions. Here are some good questions to ask:

- How did you collect your data? Why did you use that method? Why did you collect those data?
- What did you do to make sure the data you collected are reliable? What did you do to decrease measurement error?
- How did you group analyze the data? Why did you decide to do it that way? Did you check your calculations?

- Is that the only way to interpret the results of your analysis? How do you know that your interpretation of your analysis is appropriate?
- Why did your group decide to present your evidence in that way?
- What other claims did your group discuss before you decided on that one? Why did your group abandon those alternative ideas?
- How confident are you that your claim is valid? What could you do to increase your confidence?

Once the argumentation session is complete, you will have a chance to meet with your group and revise your initial argument. Your group might need to gather more data or design a way to test one or more alternative claims as part of this process. Remember, your goal at this stage of the investigation is to develop the most acceptable and valid answer to the research question!

Report

Once you have completed your research, you will need to prepare an *investigation report* that consists of three sections. Each section should provide an answer to the following questions:

1. What question were you trying to answer and why?
2. What did you do to answer your question and why?
3. What is your argument?

Your report should answer these questions in two pages or less. This report must be typed, and any diagrams, figures, or tables should be embedded into the document. Be sure to write in a persuasive style; you are trying to convince others that your claim is acceptable and valid!

Checkout Questions

Lab 22. Design Challenge

How Should Eyeglasses Be Shaped to Correct for Nearsightedness and Farsightedness?

1. Below are two different convex lenses. Show how the path of the light rays would change when they pass through each lens. The arrows represent the incoming light rays.

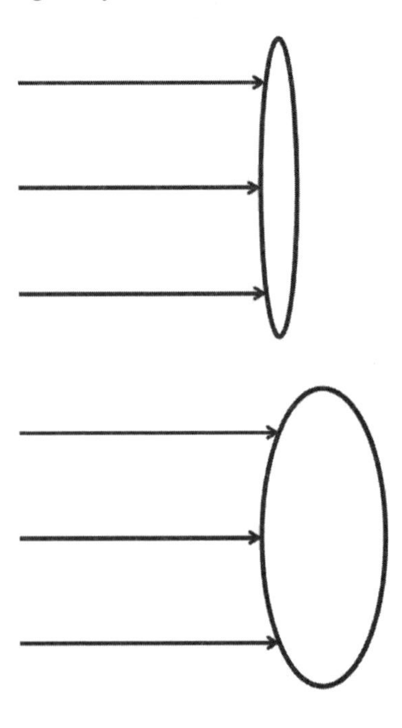

Explain why you chose those paths for the light rays for each lens.

2. Eyeglasses are worn close to your eye. Use what you know about concave and convex lenses and the behavior of light to describe how the curvature of lenses in eyeglasses would need to change to maintain clear vision if they were worn farther away from your eye.

3. Scientists only study topics that have direct application to society.

   a. I agree with this statement.
   b. I disagree with this statement.

   Explain your answer, using an example from your investigation about eyeglasses and the behavior of light.

4. Imagination and creativity are important characteristics for scientists.

   a. I agree with this statement.
   b. I disagree with this statement.

   Explain your answer, using an example from your investigation about eyeglasses and the behavior of light.

5. When the printing press became popular, the number of people reading books increased, and this is about the same time that eyeglasses started to gain popularity as a means to correct blurry vision. Eyeglasses have a specific structure and serve a distinct purpose. Explain why it is important for scientists and engineers to understand the relationship between a structure and its function, using an example from your investigation about eyeglasses and the behavior of light.

6. Scientists and engineers often make models to help them understand complex systems and phenomena. These models can be conceptual, mathematical, or physical—like a drawing. Using an example from your investigation about eyeglasses and the behavior of light, explain why it is important for scientists and engineers to develop models in their work.

SECTION 6
Appendixes

APPENDIX 1

Standards Alignment Matrixes

Alignment of the Argument-Driven Inquiry Lab Investigations With the Scientific Practices, Crosscutting Concepts, and Core Ideas in *A Framework for K–12 Science Education* (NRC 2012)

Aspect of the NRC *Framework*	Lab investigation																					
	Lab 1. Thermal Energy and Matter	Lab 2. Chemical and Physical Changes	Lab 3. Physical Properties of Matter	Lab 4. Conservation of Mass	Lab 5. Design Challenge: Koozie	Lab 6. Strength of Gravitational Force	Lab 7. Mass and Free Fall	Lab 8. Force and Motion	Lab 9. Mass and Motion	Lab 10. Magnetic Force	Lab 11. Design Challenge: Electromagnet	Lab 12. Unbalanced Forces	Lab 13. Kinetic Energy	Lab 14. Potential Energy	Lab 15. Thermal Energy and Specific Heat	Lab 16. Electrical Energy and Lightbulbs	Lab 17. Rate of Energy Transfer	Lab 18. Radiation and Energy Transfer	Lab 19. Wave Properties	Lab 20. Reflection and Refraction	Lab 21. Light and Information Transfer	Lab 22. Design Challenge: Eyeglasses
Scientific practices																						
Asking questions and defining problems	□	□	□	□	□	□	□	□	□	□	□	□	□	□	□	□	□	□	□	□	□	□
Developing and using models	■					■		□	□		■		■						■	■		■
Planning and carrying out investigations	■	■	■	■	■	■	■	■	■	■	■	■	■	■	■	■	■	■	■	■	■	■
Analyzing and interpreting data	■	■	■	■	■	■	■	■	■	■	■	■	■	■	■	■	■	■	■	■	■	■
Using mathematics and computational thinking		■	■	■	■	■	■	■	■	■	■	■	■	■	■	■	■	■	■	■	■	■
Constructing explanations and designing solutions	■	■	■	■	■	■	■	■	■	■	■	■	■	■	■	■	■	■	■	■	■	■
Engaging in argument from evidence	■	■	■	■	■	■	■	■	■	■	■	■	■	■	■	■	■	■	■	■	■	■
Obtaining, evaluating, and communicating information	■	■	■	■	■	■	■	■	■	■	■	■	■	■	■	■	■	■	■	■	■	■

Key: ■ = strong alignment; □ = moderate alignment.

Alignment of the Argument-Driven Inquiry Lab Investigations With the Scientific Practices, Crosscutting Concepts, and Core Ideas in *A Framework for K–12 Science Education* (NRC 2012) *(continued)*

Aspect of the NRC *Framework*	Lab investigation																					
	Lab 1. Thermal Energy and Matter	Lab 2. Chemical and Physical Changes	Lab 3. Physical Properties of Matter	Lab 4. Conservation of Mass	Lab 5. Design Challenge: Koozie	Lab 6. Strength of Gravitational Force	Lab 7. Mass and Free Fall	Lab 8. Force and Motion	Lab 9. Mass and Motion	Lab 10. Magnetic Force	Lab 11. Design Challenge: Electromagnet	Lab 12. Unbalanced Forces	Lab 13. Kinetic Energy	Lab 14. Potential Energy	Lab 15. Thermal Energy and Specific Heat	Lab 16. Electrical Energy and Lightbulbs	Lab 17. Rate of Energy Transfer	Lab 18. Radiation and Energy Transfer	Lab 19. Wave Properties	Lab 20. Reflection and Refraction	Lab 21. Light and Information Transfer	Lab 22. Design Challenge: Eyeglasses
Crosscutting concepts																						
Patterns		■	■			■	■		■			■								■		
Cause and effect: Mechanism and explanation											■		■					■	■			
Scale, proportion, and quantity	■		■			■	■															
Systems and system models	■			■				■			■			■	■					■		■
Energy and matter: Flows, cycles, and conservation	■			■	■					■			■	■	■	■	■	■	■		■	
Structure and function					■					■						■	■				■	■
Stability and change		■						■	■			■										
Core ideas																						
PS1: Matter and its interactions	■	■	■	■	■										■		■			■		
PS2: Motion and stability: Forces and interactions						■	■	■	■	■	■	■	■									
PS3: Energy	■				■								■	■	■		■	■				
PS4: Waves and their applications in technologies for information transfer																		■	■	■	■	■
ETS1: Engineering design					■						■											□

Key: ■ = strong alignment; □ = moderate alignment.

Alignment of the Argument-Driven Inquiry Lab Investigations With the *Common Core State Standards*, in English Language Arts and Mathematics (NGAC and CCSSO 2010)

Common Core State Standards	Lab investigation																					
	Lab 1. Thermal Energy and Matter	Lab 2. Chemical and Physical Changes	Lab 3. Physical Properties of Matter	Lab 4. Conservation of Mass	Lab 5. Design Challenge: Koozie	Lab 6. Strength of Gravitational Force	Lab 7. Mass and Free Fall	Lab 8. Force and Motion	Lab 9. Mass and Motion	Lab 10. Magnetic Force	Lab 11. Design Challenge: Electromagnet	Lab 12. Unbalanced Forces	Lab 13. Kinetic Energy	Lab 14. Potential Energy	Lab 15. Thermal Energy and Specific Heat	Lab 16. Electrical Energy and Lightbulbs	Lab 17. Rate of Energy Transfer	Lab 18. Radiation and Energy Transfer	Lab 19. Wave Properties	Lab. 20. Reflection and Refraction	Lab 21. Light and Information Transfer	Lab 22. Design Challenge: Eyeglasses
Reading																						
Key ideas and details	■	■	■	■	■	■	■	■	■	■	■	■	■	■	■	■	■	■	■	■	■	■
Craft and structure	■	■	■	■	■	■	■	■	■	■	■	■	■	■	■	■	■	■	■	■	■	■
Integration of knowledge and ideas	■	■	■	■	■	■	■	■	■	■	■	■	■	■	■	■	■	■	■	■	■	■
Writing																						
Text types and purposes	■	■	■	■	■	■	■	■	■	■	■	■	■	■	■	■	■	■	■	■	■	■
Production and distribution of writing	■	■	■	■	■	■	■	■	■	■	■	■	■	■	■	■	■	■	■	■	■	■
Research to build and present knowledge	■	■	■	■	■	■	■	■	■	■	■	■	■	■	■	■	■	■	■	■	■	■
Range of writing	■	■	■	■	■	■	■	■	■	■	■	■	■	■	■	■	■	■	■	■	■	■
Speaking and listening																						
Comprehension and collaboration	■	■	■	■	■	■	■	■	■	■	■	■	■	■	■	■	■	■	■	■	■	■
Presentation of knowledge and ideas	■	■	■	■	■	■	■	■	■	■	■	■	■	■	■	■	■	■	■	■	■	■

Key: ■ = strong alignment; □ = moderate alignment.

Alignment of the Argument-Driven Inquiry Lab Investigations With the *Common Core State Standards*, in English Language Arts and Mathematics (NGAC and CCSSO 2010) *(continued)*

Common Core State Standards	Lab investigation																					
	Lab 1. Thermal Energy and Matter	Lab 2. Chemical and Physical Changes	Lab 3. Physical Properties of Matter	Lab 4. Conservation of Mass	Lab 5. Design Challenge: Koozie	Lab 6. Strength of Gravitational Force	Lab 7. Mass and Free Fall	Lab 8. Force and Motion	Lab 9. Mass and Motion	Lab 10. Magnetic Force	Lab 11. Design Challenge: Electromagnet	Lab 12. Unbalanced Forces	Lab 13. Kinetic Energy	Lab 14. Potential Energy	Lab 15. Thermal Energy and Specific Heat	Lab 16. Electrical Energy and Lightbulbs	Lab 17. Rate of Energy Transfer	Lab 18. Radiation and Energy Transfer	Lab 19. Wave Properties	Lab. 20. Reflection and Refraction	Lab 21. Light and Information Transfer	Lab 22. Design Challenge: Eyeglasses
Mathematics																						
Make sense of problems and persevere in solving them																						
Reason abstractly and quantitatively	■	■	■	■	■	■	■	■	■	■	■	■	■	■	■	■	■	■	■	■	■	
Construct viable arguments and critique the reasoning of others	■	■	■	■	■	■	■	■	■	■	■	■	■	■	■	■	■	■	■	■	■	
Model with mathematics											■		■				■				□	
Use appropriate tools strategically	■	■	■	■		■	■						■	■	■	■				■		
Attend to precision	■	■	■	■		■	■	■	■					■	■				■	■	■	
Look for and make use of structure								■	■									■				
Look for and express regularity in repeated reasoning																		■				

Key: ■ = strong alignment; □ = moderate alignment.

Alignment of the Argument-Driven Inquiry Lab Investigations With the Nature of Science (NOS) and the Nature of Scientific Inquiry (NOSI) Concepts*

NOS or NOSI concept	Lab investigation																					
	Lab 1. Thermal Energy and Matter	Lab 2. Chemical and Physical Changes	Lab 3. Physical Properties of Matter	Lab 4. Conservation of Mass	Lab 5. Design Challenge: Koozie	Lab 6. Strength of Gravitational Force	Lab 7. Mass and Free Fall	Lab 8. Force and Motion	Lab 9. Mass and Motion	Lab 10. Magnetic Force	Lab 11. Design Challenge: Electromagnet	Lab 12. Unbalanced Forces	Lab 13. Kinetic Energy	Lab 14. Potential Energy	Lab 15. Thermal Energy and Specific Heat	Lab 16. Electrical Energy and Lightbulbs	Lab 17. Rate of Energy Transfer	Lab 18. Radiation and Energy Transfer	Lab 19. Wave Properties	Lab 20. Reflection and Refraction	Lab 21. Light and Information Transfer	Lab 22. Design Challenge: Eyeglasses
Observations and inferences		■							■						■		■					
Changes in scientific knowledge over time						■														■		
Scientific laws and theories	■						■			■			■	■				■				
Culture of science						■														■	■	■
Difference between data and evidence			■	■			■		■	■		■				■			■			
Methods used in scientific investigations			■		■			■			■	■		■	■		■		■			
Imagination and creativity in science	■	■		■	■						■		■								■	■
Nature and role of experiments								■								■		■				

Key: ■ = strong alignment; □ = moderate alignment.

*The NOS/NOSI concepts listed in this matrix are based on the work of Abd-El-Khalick and Lederman 2000; Akerson, Abd-El-Khalick, and Lederman 2000; Lederman et al. 2002, 2014; and Schwartz, Lederman, and Crawford 2004.

References

Abd-El-Khalick, F., and N. G. Lederman. 2000. Improving science teachers' conceptions of nature of science: A critical review of the literature. *International Journal of Science Education* 22: 665–701.

Akerson, V., F. Abd-El-Khalick, and N. Lederman. 2000. Influence of a reflective explicit activity-based approach on elementary teachers' conception of nature of science. *Journal of Research in Science Teaching* 37 (4): 295–317.

Lederman, N. G., F. Abd-El-Khalick, R. L. Bell, and R. S. Schwartz. 2002. Views of nature of science questionnaire: Toward a valid and meaningful assessment of learners' conceptions of nature of science. *Journal of Research in Science Teaching* 39 (6): 497–521.

Lederman, J., N. Lederman, S. Bartos, S. Bartels, A. Meyer, and R. Schwartz. 2014. Meaningful assessment of learners' understanding about scientific inquiry: The Views About Scientific Inquiry (VASI) questionnaire. *Journal of Research in Science Teaching* 51 (1): 65–83.

National Governors Association Center for Best Practices and Council of Chief State School Officers (NGAC and CCSSO). 2010. *Common core state standards.* Washington, DC: NGAC and CCSSO.

National Research Council (NRC). 2012. *A framework for K–12 science education: Practices, crosscutting concepts, and core ideas.* Washington, DC: National Academies Press.

Schwartz, R. S., N. Lederman, and B. Crawford. 2004. Developing views of nature of science in an authentic context: An explicit approach to bridging the gap between nature of science and scientific inquiry. *Science Education* 88: 610–645.

APPENDIX 2

Options for Implementing ADI Lab Investigations

Option A: 6 days (280 minutes), no homework

Day	Stage	Time
1	1: Introduce the task and the guiding question	20 minutes
	2: Design a method	30 minutes
2	2: Collect data	50 minutes
3	3: Develop an initial argument	20 minutes
	4: Argumentation session (and revise initial argument)	30 minutes
4	5: Explicit and reflective discussion	20 minutes
	6: Write investigation report (draft)	30 minutes
5	7: Double-blind peer review	50 minutes
6	8: Revise and submit the investigation report	30 minutes

Option B: 5 days (230 minutes), writing done as homework

Day	Stage	Time
1	1: Introduce the task and the guiding question	20 minutes
	2: Design a method	30 minutes
2	2: Collect data	50 minutes
3	3: Develop an initial argument	20 minutes
	4: Argumentation session (and revise initial argument)	30 minutes
4	5: Explicit and reflective discussion	20 minutes
	6: Write investigation report (draft)	Homework
5	7: Double-blind peer review	50 minutes
	8: Revise and submit the investigation report	Homework

Option C: 5 days (230 minutes), no homework

Day	Stage	Time
1	1: Introduce the task and the guiding question	20 minutes
	2: Design a method and collect data	30 minutes
2	3: Develop an initial argument	20 minutes
	4: Argumentation session (and revise initial argument)	30 minutes
3	5: Explicit and reflective discussion	20 minutes
	6: Write investigation report (draft)	30 minutes
4	7: Double-blind peer review	50 minutes
5	8: Revise and submit the investigation report	30 minutes

Option D: 4 days (170 minutes), writing done as homework

Day	Stage	Time
1	1: Introduce the task and the guiding question	20 minutes
	2: Design a method and collect data	30 minutes
2	3: Develop an initial argument	20 minutes
	4: Argumentation session (and revise initial argument)	30 minutes
3	5: Explicit and reflective discussion	20 minutes
	6: Write investigation report (draft)	Homework
4	7: Double-blind peer review	50 minutes
	8: Revise and submit the investigation report	Homework

Option E: 6 days (280 minutes), no homework

Day	Stage	Time
1	1: Introduce the task and the guiding question	20 minutes
	2: Design a method	30 minutes
2	2: Collect data	30 minutes
	3: Develop an initial argument	20 minutes
3	4: Argumentation session (and revise initial argument)	30 minutes
	5: Explicit and reflective discussion	20 minutes
4	6: Write investigation report (draft)	50 minutes
5	7: Double-blind peer review	50 minutes
6	8: Revise and submit the investigation report	30 minutes

Option F: 4 days (200 minutes), writing done as homework

Day	Stage	Time
1	1: Introduce the task and the guiding question	20 minutes
	2: Design a method	30 minutes
2	2: Collect data	30 minutes
	3: Develop an initial argument	20 minutes
3	4: Argumentation session (and revise initial argument)	30 minutes
	5: Explicit and reflective discussion	20 minutes
	6: Write investigation report (draft)	Homework
4	7: Double-blind peer review	50 minutes
	8: Revise and submit the investigation report	Homework

Option G: 6 days (280 minutes), no homework

Day	Stage	Time
1	1: Introduce the task and the guiding question	20 minutes
	2: Design a method	30 minutes
2	2: Collect data	50 minutes
3	3: Develop an initial argument	20 minutes
	4: Argumentation session (and revise initial argument)	30 minutes
4	5: Explicit and reflective discussion	20 minutes
	6: Write investigation report (draft)	30 minutes
5	7: Double-blind peer review	50 minutes
6	8: Revise and submit the investigation report	30 minutes

Option H: 5 days (200 minutes), writing done as homework

Day	Stage	Time
1	1: Introduce the task and the guiding question	20 minutes
	2: Design a method	30 minutes
2	2: Collect data	30 minutes
3	3: Develop an initial argument	20 minutes
	4: Argumentation session (and revise initial argument)	30 minutes
4	5: Explicit and reflective discussion	20 minutes
	6: Write investigation report (draft)	Homework
5	7: Double-blind peer review	50 minutes
	8: Revise and submit the investigation report	Homework

APPENDIX 3

Investigation Proposal Options

This appendix presents three investigation proposals that may be used in most labs. The development of these proposals was supported by the Institute of Education Sciences, U.S. Department of Education, through grant R305A100909 to Florida State University.

The format of investigation proposals A and B is modeled after a hypothetical deductive-reasoning guide described in *Exploring the Living World* (Lawson 1995) and modified from an investigation guide described in an article by Maguire, Myerowitz, and Sampson (2010).

References

Lawson, A. E. 1995. *Exploring the living world: A laboratory manual for biology.* McGraw-Hill College.

Maguire, L., L. Myerowitz, and V. Sampson. 2010. Diffusion and osmosis in cells: A guided inquiry activity. *The Science Teacher* 77 (8): 55–60.

Investigation Proposal A

The Guiding Question …

Hypothesis 1

IF …

Hypothesis 2

IF …

The Test

AND …

Procedure

What data will you collect?

How will you analyze the data?

What safety precautions will you follow?

Predicted Result if hypothesis 1 is valid

THEN …

Predicted Result if hypothesis 2 is valid

THEN …

The Actual Results

AND …

I approve of this investigation. ______________________ ____________

Instructor's Signature Date

The development of this investigation proposal was supported by the Institute of Education Sciences, U.S. Department of Education, through Grant R305A100909 to the Florida State University. The format of the proposal is modeled after a hypothetical deductive-reasoning guide described in *Exploring the Living World* (Lawson 1995) and modified from an investigation guide described in Macquire, Myerowitz, and Sampson (2010).

Investigation Proposal B

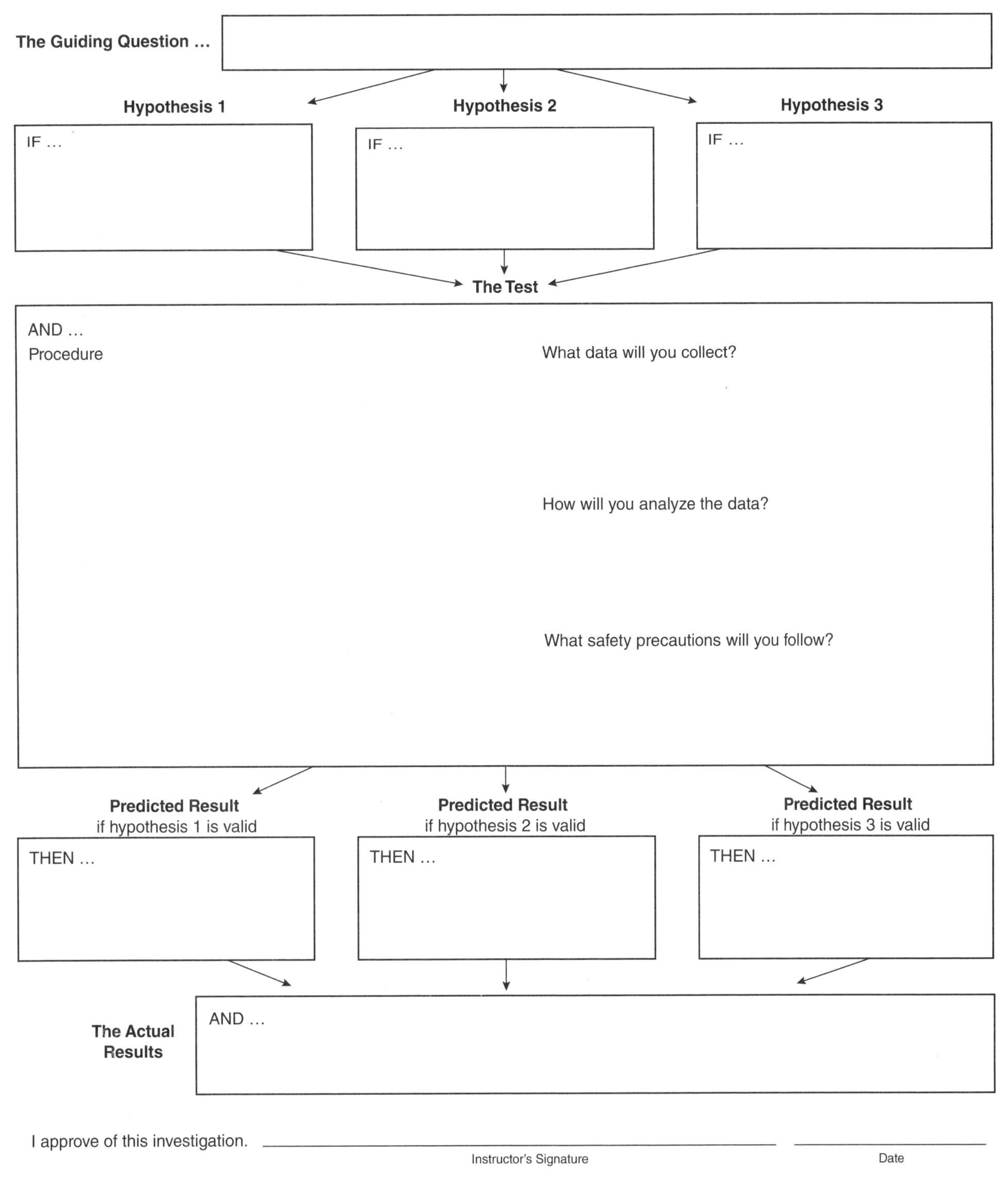

I approve of this investigation. ______________________________ ____________

Instructor's Signature Date

The development of this investigation proposal was supported by the Institute of Education Sciences, U.S. Department of Education, through Grant R305A100909 to the Florida State University. The format of the proposal is modeled after a hypothetical deductive-reasoning guide described in *Exploring the Living World* (Lawson 1995) and modified from an investigation guide described in Macquire, Myerowitz, and Sampson (2010).

Investigation Proposal C

The Guiding Question ...	
What data will you collect?	
How will you collect your data?	Your Procedure / What safety precautions will you follow?
How will you analyze your data?	
Your actual data	

I approve of this investigation. ______________________ ____________

Instructor's Signature Date

The development of this investigation proposal was supported by the Institute of Education Sciences, U.S. Department of Education, through Grant R305A100909 to the Florida State University.

APPENDIX 4

Investigation Report Peer-Review Guide: Middle School Version

Report By: ______________________ (ID Number)

Author: Did the reviewers do a good job? 1 2 3 4 5 (Rate the overall quality of the peer review)

Reviewed By: ______________ (ID Number) ______________ (ID Number) ______________ (ID Number) ______________ (ID Number)

Section 1: Introduction and Guiding Question	Reviewer Rating			Instructor Score
1. Did the author provide enough ***background information***?	☐ No	☐ Partially	☐ Yes	0 1 2
2. Is the background information ***correct***?	☐ No	☐ Partially	☐ Yes	0 1 2
3. Did the author make the ***goal of the investigation*** clear?	☐ No	☐ Partially	☐ Yes	0 1 2
4. Did the author make the ***guiding question*** clear?	☐ No	☐ Partially	☐ Yes	0 1 2
Reviewers: If your group made any "No" or "Partially" marks in this section, please ***explain how the author could improve*** this part of his or her report.	**Author:** What revisions did you make in your report? Is there anything you decided to keep the same even though the reviewers suggested otherwise? Be sure to explain why.			

Section 2: Method	Reviewer Rating			Instructor Score
1. Did the author provide a clear description of what he or she did during the investigation to ***collect data*** (the method)?	☐ No	☐ Partially	☐ Yes	0 1 2
2. Did the author describe ***how*** he or she ***analyzed*** the data?	☐ No	☐ Partially	☐ Yes	0 1 2
3. Did the author use the ***correct term*** to describe his or her investigation (e.g., experiment, observations, interpretation of a data set)?	☐ No	☐ Partially	☐ Yes	0 1 2

Section 2: Method *(continued)*	
Reviewers: If your group made any "No" or "Partially" marks in this section, please ***explain how the author could improve*** this part of his or her report.	**Author:** What revisions did you make in your report? Is there anything you decided to keep the same even though the reviewers suggested otherwise? Be sure to explain why.

Section 3: The Argument	Reviewer Rating			Instructor Score
1. Did the author provide a ***clear and complete claim*** that answers the guiding question?	☐ No	☐ Partially	☐ Yes	0 1 2
2. Did the author use ***evidence*** to support his or her claim? Evidence is an analysis of data and an explanation of what the analysis means.	☐ No	☐ Partially	☐ Yes	0 1 2
3. Did the author ***present the evidence*** in an appropriate manner by				
• including a correctly formatted and labeled graph (or table);	☐ No	☐ Partially	☐ Yes	0 1 2
• using correct metric units (e.g., m/s, g, ml); and	☐ No	☐ Partially	☐ Yes	0 1 2
• referencing the graph or table in the body of the text?	☐ No	☐ Partially	☐ Yes	0 1 2
4. Is the ***evidence support the author's claim***?	☐ No	☐ Partially	☐ Yes	0 1 2
5. Did the author use a scientific concept to ***justify the evidence***? The justification of the evidence explains why the evidence matters.	☐ No	☐ Partially	☐ Yes	0 1 2
6. Is the ***justification of the evidence*** acceptable?	☐ No	☐ Partially	☐ Yes	0 1 2
7. Did the author ***use scientific terms correctly*** (e.g., *hypothesis* vs. *prediction*, *data* vs. *evidence*) and r***eference the evidence in an appropriate manner*** (e.g., *supports* or *suggests* vs. *proves*)?	☐ No	☐ Partially	☐ Yes	0 1 2

Section 3: The Argument *(continued)*	
Reviewers: If your group made any "No" or "Partially" marks in this section, please ***explain how the author could improve*** this part of his or her report.	**Author:** What revisions did you make in your report? Is there anything you decided to keep the same even though the reviewers suggested otherwise? Be sure to explain why.

Mechanics	Reviewer Rating			Instructor Score
1. ***Organization:*** Is each section easy to follow? Do paragraphs include multiple sentences? Do paragraphs begin with a topic sentence?	☐ No	☐ Partially	☐ Yes	0 1 2
2. ***Grammar:*** Are the sentences complete? Is there proper subject-verb agreement in each sentence? Are there run-on sentences?	☐ No	☐ Partially	☐ Yes	0 1 2
3. ***Conventions:*** Did the author use appropriate spelling, punctuation, and capitalization?	☐ No	☐ Partially	☐ Yes	0 1 2
4. ***Word Choice:*** Did the author use the appropriate word (e.g., *there* vs. *their*, *to* vs. *too*, *than* vs. *then*)?	☐ No	☐ Partially	☐ Yes	0 1 2

Instructor Comments:

Total: __________/40

IMAGE CREDITS

Chapter 1

Figure 5: Authors

Figure 6: Authors

Figure 7: Authors

Figure 8: Authors

Lab 1

Figure 1.1: User:Yelod, Wikimedia Commons, CC BY-SA 3.0. *https://commons.wikimedia.org/wiki/File:States_of_matter_En.svg*

Figure 1.2: User:F l a n k e r, User:penubag, Wikimedia Commons, Public domain. *https://commons.wikimedia.org/wiki/File:Phase_change_-_en.svg*

Figure 1.3: Authors

Figure L1.1: Authors

Figure L1.2: Authors

Figure L1.3: Authors

Checkout Questions figures: Authors

Lab 2

Figure L2.1: Chris Dlugosz, Wikimedia Commons, CC BY 2.0. *https://commons.wikimedia.org/wiki/File:Red_crayon_melting.jpg*

Figure L2.2: Authors

Lab 3

Figure L3.1: Authors

Figure L3.2: Authors

Lab 4

Figure L4.1: Authors

Figure L4.2: Authors

Checkout Questions figures: Authors

Lab 5

Figure 5.1: Authors

Figure L5.1: Authors

Figure L5.2: Authors

Checkout Questions figures: Vacuum-style container: User:Dhscommtech, Wikimedia Commons, CC BY-SA 3.0, GFDL 1.2. *https://commons.wikimedia.org/wiki/File:Thermos.JPG*; Cross-section of a vacuum-style container: Authors.

Lab 6

Figure 6.1: Authors

Figure L6.1: Adapted from *http://manashsubhaditya.blogspot.com/2012/06/gravity-and-upthurst-two-opposite.html*.

Figure L6.2: Wikimedia Commons, Public domain. *https://commons.wikimedia.org/wiki/File:Nikolaus_Kopernikus.jpg*

Figure L6.3: PhET Interactive Simulations, University of Colorado Boulder *http://phet.colorado.edu*; *http://phet.colorado.edu/en/simulation/gravity-force-lab*

Figure L6.4: Authors

Checkout Questions figures: Authors

Lab 7

Figure 7.1: User: DVIDSHUB, Wikimedia Commons, CC BY 2.0. *https://commons.wikimedia.org/wiki/File:Flickr_-_DVIDSHUB_-_Master_Chief_Skydives_With_Golden_Knights.jpg*

Figure L7.2: Authors

Lab 8

Figure 8.1: Authors

Figure 8.2: Authors

Figure L8.1: User:Cdang, Wikimedia Commons, Public domain. *https://commons.wikimedia.org/wiki/File:Tir_a_la_corde_equilibre.svg*

Figure L8.2: Authors

Figure L8.3: Authors

Checkout Questions figure: Authors

Lab 9

Figure 9.1: Authors

Figure 9.2: Authors

Figure 9.3: Authors

Figure L9.1: GOGO Visual, Wikimedia Commons, CC BY 2.0. *https://commons.wikimedia.org/wiki/File:Toni_El%C3%ADas.jpg*

Figure L9.2: Authors

Figure L9.3: Authors

Lab 10

Figure 10.1: User:pumbaa, Wikimedia Commons, CC BY-SA 2.0 UK. *https://commons.wikimedia.org/wiki/File:Electron_shell_026_Iron_-_no_label.svg*

Figure 10.2: a: Jens Böning, Wikimedia Commons, Public domain. *https://commons.wikimedia.org/wiki/File:Paramagnetic_probe_without_magnetic_field.svg*; b: Jens Böning, Wikimedia Commons, Public domain. *https://commons.wikimedia.org/wiki/File:Paramagnetic_probe_with_strong_magnetic_field.svg*

Figure 10.3: a: Adapted from Victor Blacus, Wikimedia Commons, CC BY-SA 3.0. *https://commons.wikimedia.org/wiki/File:Magnetic_field_of_a_steady_current.svg*; b: Adapted from *https://commons.wikimedia.org/wiki/File:Solenoid-1_%28vertical%29.png*

Figure L10.1: Adapted from Victor Blacus, Wikimedia Commons, CC BY-SA 3.0. *https://commons.wikimedia.org/wiki/File:Magnetic_field_of_a_steady_current.svg*

Figure L10.2: Adapted from Zuriks, Wikimedia Commons, Public domain. *https://commons.wikimedia.org/wiki/File:Solenoid-1_%28vertical%29.png*

Figure L10.3: Authors

Figure L10.4: Authors

Checkout Questions figure: Authors

Lab 11

Figure 11.1: User:pumbaa, Wikimedia Commons, CC BY-SA 2.0 UK. *https://commons.wikimedia.org/wiki/File:Electron_shell_026_Iron_-_no_label.svg*

Figure 11.2: a: Jens Böning, Wikimedia Commons, Public domain. *https://commons.wikimedia.org/wiki/File:Paramagnetic_probe_without_magnetic_field.svg*; b: Jens Böning, Wikimedia Commons, Public domain. *https://commons.wikimedia.org/wiki/File:Paramagnetic_probe_with_strong_magnetic_field.svg*

Figure 11.3: a: Adapted from Victor Blacus, Wikimedia Commons, CC BY-SA 3.0. *https://commons.wikimedia.org/wiki/File:Magnetic_field_of_a_steady_current.svg*; b: Adapted from *https://commons.wikimedia.org/wiki/File:Solenoid-1_%28vertical%29.png*

Figure L11.1: Authors

Figure L11.2: Authors

Checkout Questions figures: Authors

Lab 12

Figure 12.1: Authors

Figure 12.2: Authors

Figure 12.3: Authors

Figure L12.1: Authors

Figure L12.2: Authors

Lab 13

Figure L13.1: Rob Young, Wikimedia Commons, CC BY 2.0. *https://commons.wikimedia.org/wiki/File:Car_Crash_@_Shinjuku_(9407085955).jpg*

Figure L13.2: Authors

Checkout Questions figure: Authors

Lab 14

Figure L14.1: User:The light and the dark, Wikimedia Commons, CC BY-SA 3.0, GFDL 1.2. *https://commons.wikimedia.org/wiki/File:Korean_teeterboard.JPG*

Figure L14.2: Authors

Checkout Questions figure: Authors

Lab 15

Figure 15.1: Authors

Figure L15.1: Authors

Figure L15.2: Authors

Figure L15.3: Authors

Checkout Questions figure: Authors

Lab 16

Figure 16.1: Authors

Figure 16.2: Authors

Figure L16.1: Authors

Figure L16.2: Authors

Checkout Questions figure: Authors

Lab 17

Figure L17.1: Authors

Figure L17.2: Authors

Checkout Questions figures: Authors

Lab 18

Figure 18.1: Jonathan S Urie, Wikimedia Commons, CC BY-SA 3.0. *https://commons.wikimedia.org/wiki/Category:Electromagnetic_spectrum#/media/File:BW_EM_spectrum.png*

Figure 18.2: Miquel Perelló Nieto, Wikimedia Commons, CC BY 4.0. *https://commons.wikimedia.org/wiki/File:Visible_color_wavelengths.svg*

Figure L18.1: same as Figure 18.1

Figure L18.2: Authors

Figure L18.3: Authors

Checkout Questions figure: Authors

Lab 19

Figure 19.1: a: Will Thomas; b: Authors

Figure 19.2: Will Thomas

Figure L19.1: Will Thomas

Figure L19.2: PhET Interactive Simulations, University of Colorado Boulder *http://phet.colorado.edu*; *https://phet.colorado.edu/sims/html/wave-on-a-string/latest/wave-on-a-string_en.html*

Figure L19.3: Authors

Checkout Questions figures: Authors

Lab 20

Figure 20.1: Authors

Figure 20.2: User:Josell7, Wikimedia Commons, CC BY-SA 3.0. *https://commons.wikimedia.org/wiki/File:RefractionReflextion.svg*

Figure L20.1: Authors

Figure L20.2: Modified from Philip Ronan, Wikimedia Commons, GFDL 1.2. *https://en.wikipedia.org/wiki/Frequency#/media/File:EM_spectrum.svg*

Figure L20.3: Authors

Figure L20.4: PhET Interactive Simulations, University of Colorado Boulder *http://phet.colorado.edu*; *https://phet.colorado.edu/en/simulation/bending-light*

Figure L20.5: Authors

Checkout Questions figure: Authors

Lab 21

Figure 21.1: Authors

Figure 21.2: Authors

Figure 21.3: Authors

Figure L21.1: Authors

Checkout Questions figure: Authors

Lab 22

Figure 22.1: Authors

Figure 22.2: Authors

Figure 22.3: Authors

Figure L22.1: Authors

Figure L22.2: a: User:Leridant, Wikimedia Commons, CC BY-SA 3.0, GFDL 1.2. *https://commons.wikimedia.org/wiki/File:Large_convex_lens.jpg*; b: User:Leridant, Wikimedia Commons, CC BY-SA 3.0, GFDL 1.2. *https://en.wikipedia.org/wiki/File:Concave_lens.jpg*

Figure L22.3: a: Adapted from CryptWizard, Wikimedia Commons, CC SA 1.0. *http://en.wikipedia.org/wiki/File:Hypermetropia.svg*; b: Adapted from A. Baris Toprak, Wikimedia Commons, CC SA 1.0. *http://en.wikipedia.org/wiki/File:Myopia.svg*; c: Adapted from CryptWizard, Wikimedia Commons, CC SA 1.0. *http://en.wikipedia.org/wiki/File:Hypermetropia.svg*

Figure L22.4: Authors

Checkout Questions figure: Authors

INDEX

Page numbers printed in **boldface** *type refer to figures or tables.*